化工过程强化关键技术丛书
CHEMICAL PROCESS INTENSIFICATION SERIES

中国化工学会组织编写

HiGee Chemical Separation Engineering

超重力分离工程

刘有智（Youzhi Liu） 等著

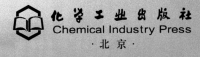

化学工业出版社
Chemical Industry Press
·北京·

ELSEVIER

内容简介

HiGee Chemical Separation Engineering is one of the books in Chemical Industry Press's Chemical Process Intensification series. The book introduces the basic concepts and technical terms of high-gravity (HiGee) separation technology and in a systematic way analyzes and expounds the differences between centrifugal separation technology and high-gravity separation technology. This book takes the "problem elicitation–theory–principle–key technology–application case" as the main theme and details the operation and technical contents of high-gravity chemical separation, such as absorption, desorption, distillation, extraction, and adsorption. Significantly, the book also highlights academic innovation and illustrative examples that are closely combined with practical production.

图书在版编目（CIP）数据

超重力分离工程 = HiGee Chemical Separation Engineering：英文 / 中国化工学会组织编写；刘有智等著． -- 北京：化学工业出版社，2024. 11. -- ISBN 978-7-122-46170-4

Ⅰ．TQ028

中国国家版本馆CIP数据核字第2024KP1322号

本书由化学工业出版社与爱思唯尔（Elsevier）出版公司合作出版。版权由化学工业出版社所有。本版本仅限在中华人民共和国境内（不包括中国台湾地区和中国香港、澳门特别行政区）销售。

责任编辑：任睿婷　杜进祥　吴　刚　　　文字编辑：曹　敏
责任校对：王鹏飞　　　　　　　　　　　　装帧设计：张　辉

出版发行：化学工业出版社
　　　　　（北京市东城区青年湖南街13号　邮政编码100011）
印　　装：北京建宏印刷有限公司
710mm×1000mm　1/16　印张20¾　字数440千字
2025年1月北京第1版第1次印刷

购书咨询：010-64518888　　　　　　　售后服务：010-64518899
网　　址：http://www.cip.com.cn
凡购买本书，如有缺损质量问题，本社销售中心负责调换。

定　　价：498.00元　　　　　　　　　　版权所有　违者必究

About the author

Professor Youzhi Liu is the former president of North University of China, and the chief academic leader of the first-level discipline doctoral degree authorization point of North University of China. He is the first batch fellow of the Chemical Industry and Engineering Society of China, and enjoys the State Council special allowance. Professor Liu is the standing vice minister of the Chemical Process Intensification Committee of CIESC, executive director of China Ordnance Society, deputy director of the National Key Laboratory of Coal and Coalbed Methane Co-production, and associate editor-in-chief of "Chemical Industry and Engineering Progress". He is also the vice chairman of the National Administry Committee on Teaching Chemical Engineering to Majors in Higher Education under the Ministry of Education, and the undergraduate teaching assessment expert.

Professor Liu has long been engaged in theoretical and applied research on chemical process intensification and high-gravity technologies to solve the key problems of chemical industry and chemical plants, especially in gas purification and high-gravity equipment development. He has hosted 46 scientific research projects, including the National Key Research and Development Program of China and the National Natural Science Foundation of China. In order to extend the application fields of high-gravity technology, professor Liu has made significant breakthroughs in the innovation of a new mechanism for the intensification of high-gravity chemical processes. He has developed impinging stream-rotating packed bed (IS-RPB) to extend the application of high-gravity technology from gas-liquid system to liquid-liquid system. He has also developed counter airflow shear-rotating packed bed (CAS-RPB) to enhance gas-film-controlled mass transfer processes, which expands and enriches the theory of intensified gas-film-controlled transfer processes. Professor Liu led his team in making breakthroughs in the industrial applications of high-gravity technology, and formulated the "High-gravity device" industry standard. He has solved a series of major problems in high-gravity device industrial applications, such as stability, safety, and air tightness. He led his team in forming a complete set of high-gravity separation engineering equipment and technology, and successfully applied these technologies in chemical, energy, metallurgy, environmental protection and other fields, which has achieved remarkable economic and environmental benefits, and greatly promoted the scientific and technological progress of the chemical industry. He has owned 98 patents and has published more than 400 papers and 4 books.

He is the winner of one Second National Prize for Progress in Science and

Technology, ten provincial and ministerial science and technology awards, the Science and Technology Innovation Award by Ho Leung Ho Lee Foundation, the Hou Debang Chemical Science and Technology Achievement Award, the Guanghua Engineering Science and Technology Prize, and the Third National Award for Excellence in Innovation.

Preface

Separation is of paramount importance in process industries such as chemical, oil refining, pharmaceutical, food, energy, metallurgic, and material industries. However, it is currently an energy-intensive process, and the investment and operating costs account for a significant proportion of the total cost. A high separation efficiency is essential to reduce energy consumption and waste production, undesired components should be sufficiently separated from the mixture to improve the product quality. Mass transfer separation is the main separation method in chemical industry because of its various advantages such as high separation efficiency, capacity and stability, but currently, available equipment for separation is not as good as expected because of the low mass transfer rate, large size, and high investment and operating cost caused by low turbulence intensity, low flow velocity, and small specific surface area. A common solution to these problems is to change the flow behavior of fluid and increase interphase mass transfer rate. To this end, it is important to understand how the flow pattern, state, scale, surface area, and renewing affect mass transfer rate.

In high-gravity separation, liquid is dispersed into micro/nanoscale films, filaments, and droplets with a large specific surface area as it flows through high-speed rotating packings. These liquid elements collide with each other and form larger droplets to be dispersed again. The repeated dispersion-coagulation process of the liquid in packings leads to a larger interfacial area and more rapid renewal of the surface for mass transfer. Therefore the mass transfer rate is greatly improved. In recent years, high-gravity separation has received considerable interest from both academics and practitioners, and a number of innovative technologies have been developed for distillation, absorption, desorption, adsorption, capture of fine particulate matter, and other chemical separation processes. Some processes have been successfully scaled up from laboratory to field applications. Based on our experience over the past years, we believe that high-gravity separation is a promising technology that can meet the demand for low-carbon development, energy saving and pollution reduction, and sustainable development because of its numerous advantages such as high separation efficiency and rate, small equipment size, low cost, high safety, and low energy consumption. Until now, there has been no book published on high-gravity separation. This book may provide readers with an exhaustive overview of high-gravity separation technologies and their applications.

In this book, we have reviewed recent progresses in high-gravity separation technologies, especially the differences between centrifugal separation and high-gravity separation. Each chapter is organized following the order of theory, principle, key technology, and application. We first highlight the importance of high-gravity intensification of separation and then introduce the theory, principle, and characteristics of key technologies and related equipment. Finally, several typical application examples are presented to demonstrate the technological, economic, and environmental advantages of high-gravity separation. This book may have some academic and practical contributions to chemical separation processes.

This book is supported by the National Natural Science Foundation of China, the Ministry of Science and Technology of the People's Republic of China, the Ministry of Education of the People's Republic of China, Shanxi Provincial Science and Technology Department, Shanxi Provincial Education Department, Shanxi Development and Reform Commission, Shanxi Provincial Finance Department, and various enterprises, research institutes, and designing institutes, to which we are very grateful. We are solely responsible for any errors in this book, and any comments and suggestions from readers of this book would be very much appreciated.

Youzhi Liu

Contents

1. Introduction   1
    1.1 Overview   1
    1.2 Operating principles and unit operations of high-gravity separation   7
    1.3 Equipment for high-gravity separation   12
    References   22

2. Absorption   23
    2.1 Overview   23
    2.2 Principles of high-gravity absorption   24
    2.3 Key techniques and challenges   31
    2.4 Application examples   33
    2.5 Future perspectives   72
    References   72

3. Desorption   75
    3.1 Intensification of heat and mass transfer in thermal desorption by high gravity   76
    3.2 Key technologies   80
    3.3 Application examples   82
    References   99

4. Distillation   101
    4.1 Overview   101
    4.2 Principles of high-gravity distillation   103
    4.3 Key technologies   104
    4.4 Characteristics of high-gravity distillation   111
    4.5 Application examples   146
    4.6 Prospects   153
    References   154

5. Liquid-liquid extraction — 159
- 5.1 Overview — 159
- 5.2 Mechanism of process intensification in the impinging stream-rotating packed bed — 165
- 5.3 Extraction operation in the impinging stream-rotating packed bed — 180
- 5.4 Application examples — 186
- 5.5 Prospects — 204
- References — 204

6. Liquid membrane separation — 207
- 6.1 Overview — 207
- 6.2 Mechanism of intensification of liquid membrane preparation and separation by impinging stream-rotating packed bed — 209
- 6.3 Key technologies of emulsion liquid membrane separation in the impinging stream-rotating packed bed — 216
- 6.4 Application examples — 221
- 6.5 Prospects — 238
- References — 239

7. Adsorption — 243
- 7.1 Overview — 243
- 7.2 Adsorption and separation technologies — 245
- 7.3 High-gravity adsorption — 259
- 7.4 Application examples — 264
- References — 280

8. Gas-solid separation — 283
- 8.1 Overview — 283
- 8.2 Key technologies and principles of high-gravity gas-solid separation — 289
- 8.3 Performance of high-gravity gas-solid separation — 291
- 8.4 Application examples — 308
- 8.5 Prospects — 311
- References — 312

Index — 315

CHAPTER 1

Introduction

Contents

1.1 Overview   1
    1.1.1 High gravity and high-gravity separation   2
    1.1.2 High-gravity factor   5
    1.1.3 Classification and characteristics of high-gravity separation   6
1.2 Operating principles and unit operations of high-gravity separation   7
    1.2.1 Operating principles of high-gravity separation   7
    1.2.2 Unit operations of high-gravity separation   8
1.3 Equipment for high-gravity separation   12
    1.3.1 Structures and types of rotating packing bed for intensification of gas-liquid mass transfer   12
    1.3.2 Structures and types of rotating packing bed for intensification of liquid-liquid mass transfer   20
    1.3.3 Structures and types of rotating packing bed for intensification of gas-solid mass transfer   21
References   22

1.1 Overview

The chemical industry is one of the most important pillars of China's economy and it makes a significant contribution to the economic and social development of the country. However, this industry has come under increasing scrutiny in recent years because of the pollution caused by the generation of "three wastes" (waste gas, waste water, and waste residue), underutilization of resources, and high energy consumption. This is not surprising when one considers that China was once an economically backward country and lacked the necessary technology and equipment. From a strategic perspective, it becomes increasingly important to reduce energy consumption and generation of "three wastes" and to achieve low-carbon development for the chemical industry.

    The reactor is the central part of the setup for separation. Separation is a complex process involving the supply of raw materials for chemical reaction, separation and purification of products, and treatment of waste generated. Technically speaking, separation is the key to producing high-quality chemical products, making the best use of resources, and controlling pollution. It is also economically important as far as investment and operating costs are concerned because it requires a large number of equipment and energy that account for a major portion of the total cost. Overall, separation plays a decisive role in both the technical and economic aspects of chemical processing.

The rapid development of the chemical industry in China has resulted in numerous new applications, as well as new technical challenges, for separation. Examples include high-purity substance production, biomedical separation and purification, coal gas purification, chemical product processing, and implementation of new environmental regulations. Currently, much effort is made to develop better separation techniques and reactors that are more efficient, energy-saving, environmentally friendly, and highly integrated. It is against this background that high-gravity separation has emerged as a novel and efficient separation technique with a wide range of potential applications.

Separation processes fall into either mechanical separation or mass transfer separation (including separation with chemical reaction). Mechanical separation is applicable to heterogeneous mixtures, while mass transfer separation is applicable to homogeneous mixtures. Note that this book will focus primarily on the mass transfer separation of homogeneous mixtures. Most mass transfer separation processes (e.g., distillation, absorption, and extraction) used in the industry are equilibrium separation processes based on differences in the distribution of the components of a mixture between two phases that are insoluble in one another at equilibrium. In contrast, rate separation processes (e.g., membrane separation, thermal diffusion, and gas diffusion) are based on differences in the mass transfer rate of the components of a mixture through a given medium (e.g., semipermeable membrane) driven by pressure, concentration, or potential gradient.

The high gravity can improve the separation process because of its potential to significantly enhance the mass transfer in multiphase flows. As a special kind of equilibrium separation, high-gravity separation also requires the contact and mass transfer between two phases in order to separate the components of a homogeneous mixture in a high-gravity reactor. For this purpose, it is necessary to select an appropriate separation medium. Typical separation media include energy (e.g., distillation), solvents or solid adsorbents (e.g., absorption, extraction, and adsorption), and both energy and solvents (e.g., extractive and azeotropic distillation). All separation processes involve contact and mass transfer of two- or multi-phase flows. The distribution of the components to be separated in the two phases at equilibrium is controlled by thermodynamic properties, and the process to reach equilibrium is controlled by the interphase mass transfer rate.

1.1.1 High gravity and high-gravity separation

The high-gravity is a force in nature and physics, and in mathematics, it can be represented as a vector that has both direction and magnitude. When one thinks of high-gravity, the first thing that comes to mind is probably gravity. As the Earth rotates about its axis, an object everywhere on the sphere except at the north and south poles undergoes approximately uniform circular motion about the axis of the Earth. This is attributed to the centripetal force directed perpendicularly to the axis of the Earth and only provided by the attractive force of the Earth on the object. This force can be decomposed into two components. One component (F_r) is directed perpendicularly to the axis of the Earth and its magnitude is equal to the centripetal force required for the approximately uniform circular motion of the object ($F_r = Mr\omega^2$, where ω is the angular velocity of rotation of the Earth, and r is the rotation radius of the object). The other component (F_g) is the gravity acting on the object. The magnitude of the centripetal force (F_r) is zero at both north and south poles and it increases as the latitude decreases and reaches a maximum at the equator. Given that the centripetal force is negligibly small, the magnitude of gravity can be assumed to be equal to that of universal gravitation. That is, the effect of Earth's rotation can be neglected in most circumstances. The gravity component of the attractive force provides the gravitational acceleration, while the centripetal force component provides the centripetal acceleration.

Thus, gravity can be defined more precisely as the attractive force of the Earth on the object rotates with the Earth, which means that:
1. Gravity is derived from the attractive force of the Earth on the object.
2. Gravity is an apparent concept representing the attractive force of the Earth on the object that rotates with the Earth.
3. Gravity is equal to the vector difference between the attractive force and the centripetal force needed for the object to rotate about the axis of the Earth.
4. Gravity is always directed vertically (not perpendicularly) downward.
5. Gravity is caused by the attractive force of the Earth, but it is not correct to say that gravity is the attractive force.

The magnitude of gravity on an object at the same location is proportional to the mass of the object (m). Thus, for an object with a given mass (m), the gravitational acceleration (g) is proportional to the magnitude of gravity ($F_g = mg$, where the gravitational acceleration is taken to be 9.8 m/s^2 on the Earth's surface).

A high-gravity field can be induced by the high-speed rotation of the rotor in a rotating packing bed (RPB). The rotor is the rotating part of RPB, and the packing is placed in the rotor and rotates synchronously with the rotor. Let the angular velocity be ω and the rotation radius is r. The centripetal force for the circular motion F_r provides the acceleration for the rotating object. Circular motion is the variable accelerated motion, and an object is said to be moving with a variable acceleration if the acceleration at different points along the moving path differs in magnitude, direction, or both. In uniform circular motion where an object travels a circular path at a constant speed, the linear velocity is constantly changing because the direction is always changing, but the angular velocity is kept constant. For any mass point $M(r, \varphi, z)$ of a rotating object with a mass of m in the cylindrical coordinate system, the resultant force is the vector difference between the centripetal force F_r and the gravity F_g.

Note that what we are talking about here is the rotating reference system (or non-inertial reference system). The inertial force (centrifugal force) should be introduced if we take into consideration the inertial reference system. The centrifugal force is a fictitious force that can make the object move away from the center of rotation. In order for Newton's laws to be applicable in a rotating frame of reference (or non-inertial reference system), the centrifugal force that is equal and opposite to the centripetal force must be included in motion equations.

In the cylindrical coordinate system, any mass point M (r, φ, z) of a rotating object is subjected to gravity and centrifugal force and the resultant force is the vector sum of these two forces. For vertically mounted RPB, the gravity is directed vertically downwards (opposite to the z axis), and the centrifugal force is directed along the r axis. In this case, gravity and centrifugal force are always perpendicular to each other. As long as rotation continues ($\omega \neq 0$), the force acting on a given mass point M ($r \neq 0$, φ, z) is always greater than gravity, and its direction is similar to the direction of the resultant force. For horizontally mounted RPB, the cylindrical coordinate system is rotated clockwise, where the gravity is always directed vertically downwards and the centrifugal force is directed toward the r axis. The resultant force is greater than the centrifugal force in the lower semicircle ($\varphi \in [0, \pi]$) but lower than the centrifugal force in the upper semicircle ($\varphi \in [\pi, 2\pi]$). The resultant force is along the direction of gravity and reaches a maximum at $\varphi = \frac{\pi}{2}$, but it is opposite to the direction of gravity and reaches a minimum at $\varphi = \frac{3\pi}{2}$. In order for the resultant force acting on any mass point M (r, φ, z) of a rotating object to be greater than gravity, the minimum force obtained at $\varphi = \frac{3\pi}{2}$ should be greater than gravity. More precisely, the centrifugal force should be at least two times greater than gravity, which can be achieved by increasing the rotation speed of the rotor.

It is concluded that in both vertically and horizontally mounted RPB, the force acting on any mass point M (r, φ, z) is the resultant force of gravity and centrifugal force, and the direction is similar to the direction of the resultant force.

The rotating system is a non-inertial system in which a mass point is subjected to an inertial force (centrifugal force). If we consider the inertial force field as the overall distribution of inertial force acting on each mass point of a rotating object at any moment, then it is a simulated force field. In this case, the acceleration is the centrifugal acceleration G ($G = r\omega^2$, where ω is the angular velocity of the rotor, rad/s, and r is the radius of the rotor, m). For any mass point M (r, φ, z), if the ratio of centrifugal force to gravity $\frac{F_r}{F_g} \gg 1$ (or $\frac{F_g}{F_r} \approx 0$), that is, if the centrifugal acceleration is much greater than the gravitational acceleration, then the effect of gravity is assumed to be negligible. In this case, the inertial force field is called the high-gravity field, and the force acting on a mass point is called the high gravity. In view of this, the high-gravity field generated by high-speed rotation of the rotor is a virtual force field in which the centrifugal acceleration is much greater than the gravitational acceleration and the effect of gravity is negligible.

The similarity between high-gravity separation and centrifugal separation is the centrifugal force caused by the high-speed rotation of the rotor. However, high-gravity separation is a mass transfer separation process that depends on the contact and mixing of two-phase flows, and the better the contact and mixing are, the better the separation efficiency will be; whereas centrifugal separation is a mechanical separation process that depends on the density difference of the components of a mixture, and the larger the density difference is, the better the separation efficiency will be. This is the fundamental difference between high-gravity separation and centrifugal separation.

1.1.2 High-gravity factor

In order to compare RPB of different sizes and rotation speeds, the intensity of high-gravity field is characterized by a dimensionless parameter called the high-gravity factor β. It is defined as the ratio of the inertial acceleration (centrifugal acceleration) G to the gravitational acceleration g (9.81 m/s^2):

$$\beta = \frac{G}{g} = \frac{r\omega^2}{g} \tag{1.1}$$

It can be simplified into:

$$\beta = \frac{n^2 r}{900} \tag{1.2}$$

where ω is the angular velocity of the rotor, rad/s;

r is the radius of the rotor, m;

n is the rotation speed of the rotor, r/min.

At a given rotation speed, the high-gravity factor varies linearly with the radius of the rotor. Thus, the high-gravity factor increases linearly in the radial direction (Fig. 1.1).

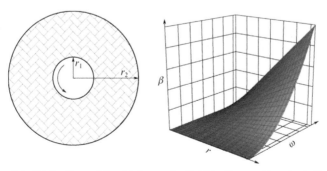

Figure 1.1 The radial distribution of the high-gravity field in the packing (r_1 is the inner diameter of the packing, and r_2 is the outer diameter of the packing).

Because the intensity of the high-gravity field varies in radial direction, it is more convenient to use the average high-gravity factor to represent the intensity of overall the high-gravity field. Although the high-gravity field is three-dimensional, it can be simplified into a two-dimensional plane field when the packing is uniformly distributed in the axial direction of the rotor. Then, the average area can be used to represent the average intensity of the high-gravity field:

$$\bar{\beta} = \frac{\int_{r_1}^{r_2} \beta \cdot 2\pi r dr}{\int_{r_1}^{r_2} 2\pi r dr} = \frac{2\omega^2(r_1^2 + r_1 r_2 + r_2^2)}{3(r_1 + r_2)g} \tag{1.3}$$

The high-gravity factor indicates that the acceleration of the high-gravity field is β times the gravitational acceleration, and this dimensionless parameter is useful for the comparison of RPB of different sizes and rotation speeds. The high-gravity factor can also be understood as the ratio of high-gravity (mG) to gravity (mg) acting on the same mass point.

Thus, the high-gravity field is the centrifugal force field induced by high-speed rotation and a mass point in the high-gravity field is subjected to high-gravity. It is noted that the force acting on the fluid in the high-gravity field is much greater than that in the gravity field, and because of this, more drastic change in fluids is expected. The mass point appears and then disappears quickly, and it becomes smaller in size and the surface is renewed more rapidly.

1.1.3 Classification and characteristics of high-gravity separation

Many separation processes, including absorption, desorption, distillation, extraction, gas-solid separation, and adsorption, can be performed in a high-gravity field. High-gravity separation processes can be classified into gas separation, liquid–liquid separation, gas-solid separation, and gas-liquid-solid separation according to the phases involved, and they can also be classified into countercurrent-flow, cross-flow, and concurrent-flow separation according to the contact of multiphase flows in RPB. It is argued that adequate mixing is a necessary prerequisite for the success of high-gravity separation, and the separation efficiency is highly dependent on the mixing efficiency of the phases. A major advantage of high-gravity separation is its potential to increase the mass transfer rate and consequently the mixing efficiency.

In RPB, the gas-liquid mass transfer takes place in the high-gravity field. RPB has the following advantages over traditional separation equipment such as packed columns, bubble columns, and sieve-plate columns [1-3]:

1. High mass transfer rate. The mass transfer coefficient is expected to be 1.3 orders of magnitude higher.
2. Low gas-phase pressure drop and energy consumption.

3. Low liquid holdup. Less online material stock is needed and it is intrinsically safer, which makes RPB particularly useful for treatment of expensive, toxic, flammable, and combustible materials.
4. Short residence time. RPB is suitable for applications that require fast mixing and reaction and makes it easier to control the reaction selectivity.
5. Short time is needed to reach equilibrium. Thus, it is easier to start/stop operations and replacement of materials.
6. Small size, low cost, low space requirement, and ease of installation and maintenance.
7. Ease of miniaturization and industrial scaling-up.
8. Low probability of scale formation and blockage because of the high self-cleaning capacity of the packing.
9. Wide applicability, good generalization, and high operating flexibility.

1.2 Operating principles and unit operations of high-gravity separation

1.2.1 Operating principles of high-gravity separation

This section describes the operating principles of high-gravity separation taking countercurrent-flow RPB as an example (Fig. 1.2). The operation of RPB mainly involves gas–liquid contact and reaction. The rotor filled with the packing rotates at a given speed driven by a motor. The liquid phase is introduced into RPB through the pipe in the central cavity enclosed by the inner edge of the packing and then sprayed onto the inner edge through the liquid distributor. Within the packing, liquid is driven radially outward by centrifugal force from the inner edge to the outer edge of

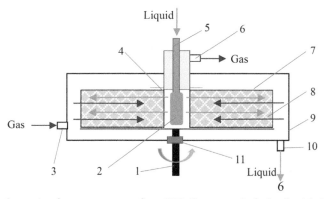

Figure 1.2 The schematic of countercurrent-flow RPB (1—rotor shaft; 2—liquid distributor; 3—gas inlet; 4—airtight seal; 5—liquid inlet; 6—gas outlet; 7—rotor; 8—packing; 9—shell; 10—liquid outlet; 11—shaft seal). *RPB*, rotating packing bed.

the packing. After that, liquid is splashed onto the interior wall of the shell and leaves the bed through the liquid outlet at the bottom of the bed. The gas phase is introduced into RPB through the gas inlet on the shell, and then it is driven radially inward by centrifugal force from the outer edge to the inner edge of the packing. Finally, gas is accumulated in the central cavity and leaves the bed through the gas outlet at the top of the bed. It is seen that gas and liquid pass through the packing in opposite directions and come into contact with the rotating packing.

Note that the required high-gravity factor can be easily achieved by adjusting the rotation speed of the rotor, and the mixing and separation efficiency can be maximized by properly adjusting the gas-liquid ratio and flow rate.

RPB is similar to a packed column in which gas flows countercurrently to liquid and thus comes into intimate contact with the liquid. However, they differ in two important aspects. First, the packing rotates at high speed in RPB, but it is kept stationary in packed column. Second, liquid flows downward and gas flows upward in packed column, so countercurrent contact takes place between liquid and gas in the axial direction, but gas and liquid flow radially in RPB.

The unit operations of high-gravity separation, such as absorption, desorption, rectification, and extraction, involve contact and mass transfer of two- or multi-phase flows. In order to be consistent throughout this book, the flows containing the components to be separated are called feed flows, and other flows that participate in high-gravity separation are called media flows. High-gravity is capable of improving mass transfer between feed and media flows, and as a result the components to be separated in the feed flow can be transferred to the media flow more effectively and rapidly. Thus, the interphase mixing and contact are critical for separation efficiency, and as expected, the better the mixing and the more intimate the contact, the higher the separation efficiency.

It should be noted that high-gravity separation can be coupled with other separation methods to form new separation methods. A typical example is the emulsion liquid membrane separation technique, in which emulsion is formed in a high-gravity field and then dispersed in the external phase in the form of droplets to obtain the emulsion liquid membrane system.

1.2.2 Unit operations of high-gravity separation
1.2.2.1 High-gravity absorption
High-gravity absorption can be used to separate unwanted components from a gas mixture (feed flow) based on differences in solubility in the liquid solvent (media flow). Unlike conventional absorption processes, high-gravity absorption takes place in a high-gravity field that can substantially increase the absorption rate. High-gravity absorption is applicable to the separation of gas mixtures, gas purification, and manufacturing of liquid products. High-gravity absorption processes can

be divided into physical absorption and chemical absorption depending on whether there is a chemical reaction; single-component absorption, and multi-component absorption depending on the number of components in the gas mixture; and isothermal absorption and non-isothermal absorption depending on whether there is a thermal effect.

1.2.2.2 High-gravity desorption

High-gravity desorption is a mass transfer separation process by which some solutes of a liquid mixture (feed flow) are transferred to the gas phase (media flow) in contact with the liquid phase. High-gravity desorption can be viewed as the reverse process of high-gravity absorption, and it is also referred to as gas stripping. In industrial applications, an absorption process usually includes a desorption step to regenerate adsorbent so that it can be recycled back into the absorption process. However, desorption can also be used alone to remove the gas dissolved in a liquid. High-gravity desorption processes can be divided into physical desorption and chemical desorption according to whether there is a chemical reaction.

1.2.2.3 High-gravity distillation

High-gravity distillation is a physical separation process by which a liquid mixture or a liquid-solid mixture is heated in a high-gravity reactor to force its components with different boiling points into a gaseous state, which is subsequently condensed to the liquid state. High-gravity distillation involves evaporation and condensation and characterized by mass transfer between the gas phase (feed flow) formed in the evaporation process and the liquid phase (media flow) formed in the condensation process. Simple distillation involves only one vaporization-condensation cycle, in which a liquid mixture is heated in a high-gravity reactor to vaporize preferred components and subsequently the vapors are condensed. It can be used to prepare distilled liquor from wine or other fermented fruit juice. In contrast, fractional distillation involves repeated distillations and condensations, leading to better separation than simple distillation because vapors can condense, then re-evaporate, and then re-condense. Fractional distillation is often used for the separation of crude oil into different fractions, such as gasoline, diesel, kerosene, and mazout. High-gravity distillation processes can be divided into continuous distillation and batch distillation according to the mode of operation, two-component distillation and multi-component distillation according to the number of components in the mixture, ordinary distillation and special distillation (e.g., extractive distillation, azeotropic distillation and salt distillation) according to whether there are additives in the mixture that can affect gas-liquid equilibrium. Reactive distillation is the simultaneous implementation of reaction and distillation within a single unit of column.

1.2.2.4 High-gravity emulsion liquid membrane separation

The high-gravity emulsion liquid membrane process consists of preparation of emulsion liquid membrane in an impinging stream-rotating packed bed (IS-RPB) and membrane separation. The internal aqueous phase and the membrane phase containing surfactant and stabilizer are mixed in IS-RPB to form an emulsion, and then the resultant emulsion is dispersed in the external phase in the form of droplets. The components to be separated are transported from the external phase to the membrane phase and then to the internal phase. After extraction, the loaded emulsion is separated from the feed solution and demulsified to yield the membrane phase for reuse. A small-sized, uniform, and stable emulsion could be obtained in IS-RPB by properly adjusting the intensity of the high-gravity field. The emulsion liquid membrane process involves simultaneous extraction and stripping in one step and it has many unique advantages, such as high mass transfer rate and low consumption of extractant. Therefore, high-gravity emulsion liquid membrane process has potential applications in hydrometallurgy, petrochemical engineering, environmental protection, and gas separation.

1.2.2.5 High-gravity extraction

High-gravity extraction is also known as high-gravity solvent extraction and liquid–liquid extraction. A solute can be transferred from one solvent to another based on differences in the solubility (or partition coefficient) between the two solvents that are immiscible or partially miscible with each other. All high-gravity extraction processes involve rapid and vigorous mixing of feed solution with extractant in a high-gravity reactor, which significantly increases the mass transfer from the feed solution to the extractant. Equilibrium is reached between the two phases. After settlement, two liquid phases are formed. One contains the components which are soluble in the solvent, and the other holds primarily the components which are not soluble in the solvent. It should be pointed out that for high-gravity extraction in IS-RPB, the mixing efficiency is approximately 40 times that of traditional continuous stirred-tank reactor (CSTR), and the micromixing characteristic time is about 10 μs. As a result, the liquid–liquid mass transfer is significantly improved, and equilibrium can be reached at a single stage. This is one of the most important advantages of IS-RPB in extraction.

1.2.2.6 High-gravity adsorption

In high-gravity adsorption, porous solid adsorbents are used as the packing of RPB for selective adsorption of one or more components (adsorbates) in the fluid as it flows through the packing. It differs from adsorption in conventional devices, such as stirred tank, fixed bed, fluidized bed, and moving bed, in that the mass transfer is greatly enhanced as the adsorbent surface is renewed and the liquid in the pores is replaced at a much faster rate. Note that the intensity of the high-gravity field can be controlled by adjusting the rotation speed, which makes the adsorption process more controllable and effective.

1.2.2.7 High-gravity gas-solid separation

High-gravity gas-solid separation has emerged as an efficient wet dedusting technique. The gas containing particulate matter such as dust and liquid droplets is brought into intimate contact with liquid (generally water) in a high-gravity device, and subsequently, particulate matter are removed from the gas. This technique is capable of removing solid and liquid particulate suspensions, as well as gaseous pollutants, from gas, and reducing the gas temperature. As these fine dust particles can be contacted, wetted, and captured by liquid droplets through different mechanisms, including inertial impact, interception, diffusion and condensation, they are agglomerated into larger particles that can be removed more easily.

High-gravity has the potential to disintegrate the liquid into micro elements (i.e., droplet, thread, film, and mist) with sizes similar to dust particles, and these elements can be well dispersed in gas in the narrow channels of the packing. As dust particles are vigorously impacted by the rotating packing, they are wetted and subsequently condensate into larger particles. Under the action of high-gravity, the liquid with dust particles can be broken into micro elements again for capture of free dust particles in the narrow channels of the packing. Such a repeated process leads to effective removal of dust particles in RPB.

High-gravity gas-solid separation has the following features: (1) It turns the traditional "washing" process into the "capturing" process as liquid is broken into microelements for capturing fine particles. As expected, low water consumption is demanded and the liquid-gas ratio is 0.3-1.3 L/m^3, resulting in a lower operating cost. (2) It is a state-of-the-art technology for deep purification of particulate matter with a cut diameter of 0.08 μm. (3) The removal rate is over 95% in engineering practice. High-gravity gas-solid separation brings novel ideas for removal of fine particles in gas by means of mechanical impact, capturing with micro-liquid elements, and condensation. The traditional "washing" dedusting technique requires large quantities of water, while high-gravity gas-solid separation leads to low water consumption and improved removal efficiency.

1.2.2.8 Other separation processes

The high-gravity separation technique can also be used for the treatment of organic pollutants in wastewater because of the potential of high-gravity to improve advanced oxidation processes, such as ozonation, electrocatalysis, and photocatalysis. In RPB, the rotating packing can shear liquid into micro/nano elements, leading to continuous renewal of liquid surface and strong turbulence of the liquid. Because of the intimate gas-liquid and liquid–liquid contact, organic pollutants are more likely to contact liquid and subsequently removed from wastewater. In the high-gravity field, the use of ozonation for the removal of organic pollutants in wastewater can significantly increase the mass transfer of ozone from the gas phase to the liquid phase, leading to an increase in the contact between the liquid phase ozone and the catalyst and subsequently formation of more ·OH for mineralization of organic pollutants. The coupling of high-gravity technique with electrochemistry leads to improved gas-liquid-solid separation by increasing the slip velocity of gas bubbles with solid electrodes and wastewater and then accelerating the escape of gas bubbles from electrode surface and wastewater. In photocatalyzed removal of organic pollutants from wastewater, high-gravity can increase the effective contact between nanometer photocatalysts and organic pollutants and more ·OH are generated, which can improve the utilization rate of photons and consequently the photocatalytic efficiency.

1.3 Equipment for high-gravity separation

The high-gravity equipment is initially designed for intensification of gas-liquid mass transfer, and gas flows countercurrently to liquid and comes into contact with liquid in a countercurrent-flow RPB (Fig. 1.2). Later, crosscurrent-flow RPB and concurrent-flow RPB are designed and the high-gravity technique is also used for intensification of liquid–liquid and gas-solid mass transfer. This section introduces the structures, classifications, and operating principles of RPB for intensification of gas-liquid, liquid-liquid, and gas-solid mass transfer.

1.3.1 Structures and types of rotating packing bed for intensification of gas-liquid mass transfer

According to the rotor structure, the high-gravity equipment for intensification of gas-liquid mass transfer falls into two categories: packed type and plate type. The packed type is further divided into single and split types or single-stage and multi-stage types; while the plate type is divided into single-stage and multi-stage types. According to the contact mode, it is divided into countercurrent-flow, cross-flow, and concurrent-flow types. According to the installation mode, it is divided into vertical type (rotor shaft is perpendicular to the horizon) and horizontal type (rotor shaft is parallel to the horizon).

1.3.1.1 Rotating packing bed

An RPB mainly consists of rotor, packing, liquid distributor, rotor shaft (motor), airtight seal, and shell on which gas inlet, gas outlet, liquid inlet, and liquid outlet are located (Fig. 1.2) [1]. The performance of RPB depends critically on the structure and property of the packing. The rotor is filled with packing and acts as a support to provide the required mechanical strength for rotation. RPB uses either single-block packing that is continuous and uniform or split packing that is split into annular rings with small gaps between two adjacent rings.

Two arrangement modes are discussed in the following section.

1.3.1.1.1 Rotating packing bed with single-block packing

In RPB with single-block packing, the packing is continuously arranged in the rotor. The packing includes structured packing, random packing, and specifically designed packing [2]. The next section will introduce the structures and operating principles of countercurrent-flow RPB, cross-flow RPB and concurrent-flow RPB with single-block packing and split packing for intensification of gas–liquid mass transfer.

Countercurrent-flow RPB: The structure of countercurrent-flow RPB is schematically shown in Fig. 1.2. The packing is placed in a rotor that is fixed to the shaft and rotates at high speed driven by the motor. The rotation speed of rotor is controlled by adjusting the rotation speed of the motor. Liquid is uniformly sprayed on the inner edge of the packing via a liquid distributor, and as it flows radially outward from the inner edge to the outer edge of the packing, it is sheared into microdroplets, threads, or films by the centrifugal force. Liquid is splashed onto the interior wall of the shell and leaves the bed via the liquid outlet at the bottom of the bed. Gas is introduced into RPB via the gas inlet, and as it flows radially inward from the outer edge to the inner edge of the packing driven by pressure, it comes into countercurrent contact with liquid. Finally, gas leaves the bed via the gas outlet at the center of the packing. In countercurrent-flow RPB, the improvement in mass transfer is due to the large interfacial areas of microdroplets, threads and films, and rapid surface renewal [3]. However, it should be noticed that the internal diameter of the packing is large because the gas outlet is arranged at the center of the packing and the gas pipe is large, thus leaving less room for the packing. Caution should also be taken to prevent possible leakage of the airtight seal that can cause a short circuit of gas and subsequently reduce the contact between gas and liquid.

Cross-flow RPB: In cross-flow RPB (Fig. 1.3), gas and liquid are brought into contact in a cross-flow configuration within the rotating packing. The packing is placed in a rotor that is fixed to the shaft and rotates at high speed driven by the motor [4]. The rotation speed of the rotor is controlled by adjusting the rotation speed of the motor. Liquid is uniformly sprayed on the inner edge of the packing via a liquid distributor, and as it flows radially outward from the inner edge to the outer edge of the packing, it is sheared into microdroplets and threads or films by the centrifugal force. After that, liquid is driven to the interior wall of the shell and leaves the bed via the liquid outlet at the bottom of the bed. Gas is introduced into RPB via the gas inlet, and as it flows upward from the bottom to the top of the packing driven by pressure, it comes into cross-contact with the liquid. Finally, the gas leaves the bed via

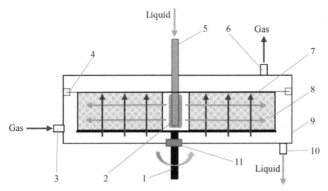

Figure 1.3 The schematic of cross-flow RPB (1—rotor shaft; 2—liquid distributor; 3—gas inlet; 4—airtight seal; 5—liquid inlet; 6—gas outlet; 7—rotor; 8—packing; 9—shell; 10—liquid outlet; 11—shaft seal). *RPB*, rotating packing bed.

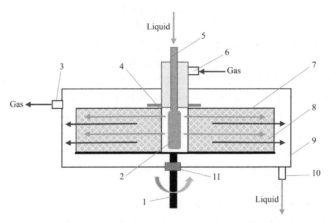

Figure 1.4 The schematic of concurrent-flow RPB (1—rotor shaft; 2—liquid distributor; 3—gas outlet; 4—airtight seal; 5—liquid inlet; 6—gas inlet; 7—rotor; 8—packing; 9—shell; 10—liquid out-

the gas outlet at the top of the shell. In cross-flow RPB, the improvement in mass transfer is also attributed to the large interfacial areas of microdroplets, threads and films, and rapid surface renewal.

Concurrent-flow RPB: The concurrent-flow RPB (Fig. 1.4) is structurally identical to countercurrent-flow RPB except that the positions of the gas inlet and outlet are reversed. The packing is arranged in a rotor that is fixed to the shaft and rotates at high speed driven by the motor. The rotation speed of the rotor is controlled by adjusting the rotation speed of the motor [5]. Liquid is uniformly sprayed on the inner edge of the packing via a liquid distributor, and as it flows radially outward from the inner edge to the outer edge of the packing, it is sheared into microdroplets, threads, or films by the centrifugal force. Finally, liquid is driven to the interior wall of the shell and leaves the bed via the liquid outlet at the bottom of the bed. Gas is introduced into the central cavity enclosed by the inner edge of the packing via the gas inlet, and as it flows radially outward from the inner edge to the outer edge of the packing driven by pressure, it comes into concurrent contact with the liquid. Finally, gas leaves the bed via the gas outlet at the shell.

It is concluded that gas and liquid flow countercurrently in a countercurrent RPB but concurrently in a concurrent RPB, and this seemingly trivial difference has important consequences for gas-liquid mass transfer and gas pressure drop. It should be emphasized that both liquid and gas are affected by high-gravity as they flow in the rotating packing. Liquid always flows radially outward from the inner edge to the outer edge of the packing, but the gas flow direction is not fixed. It may flow radially inward from the outer edge to the inner edge of the packing, forming countercurrent contact with liquid, or axially outward from the inner edge to the outer edge of the packing, forming concurrent contact with liquid, or upward from the bottom to the top of the packing, forming cross-contact with liquid. For countercurrent contact, gas-liquid mass transfer is improved because of high gas-liquid relative velocity and rapid surface renewal. A considerable pressure drop is observed for gas because of the existence of centrifugal force, and the dumping of liquid is moderately inhibited by high-gravity. For cross-current contact, gas turbulence takes place because of the shear force of the rotating packing, leading to high gas-liquid mass transfer and low-pressure drop, but liquid film entrainment occurs occasionally. Given the same size, cross-current RPB has a larger rotor diameter, and higher treatment capacity, but lower equipment investment.

1.3.1.1.2 Rotating packing bed with split packing

It is expected that the overall mass transfer coefficient is significantly improved in RPB. This is not entirely true as the liquid-side mass transfer coefficient is indeed significantly improved, but there is no notable improvement in the gas-side mass transfer coefficient. This is understandable because the drag force exerted by the packing with a high specific surface area on gas is low in RPB, and the slip velocity between gas and packing is small and gas rotates almost synchronously with the packing. Similar to the case in a packed column, gas flows through the packing in an integral or strand manner with low turbulence. Thus, the gas-liquid interface could not be renewed quickly, and the mass transfer in the gas film is limited. In order to solve this problem, Liu et al. [6] proposed a novel RPB called the counter airflow shear-rotating packed bed (CAS-RPB) based on the gas film-controlled mass transfer. As the name suggests, CAS-RPB imposes high shear stresses on the counter airflow in order to increase the gas film-controlled mass transfer process and the gas-side mass transfer coefficient. In RPB, shear stress is provided by the centrifugal force and the high specific surface area of the packing, and it is controlled by adjusting the rotation speed of the rotor. Liquid can be sheared into microdroplets, threads, and films, but gas is less affected and it flows through the packing in an integral or strand manner. The structure of CAS-RPB is shown in Fig. 1.5.

The rotor of CAS-RPB consists of two concentric disks (upper disk and lower disk) that are connected to different shafts and thus rotate independently of each other. There are two groups of concentric circular packing rings of increasing radius that are attached alternately to the upper and lower disk with a gap between adjacent rings and between the free end of each ring and the disk. Note that there are some requirements for the axial height and radial thickness of the packing. The two disks rotate in opposite directions most of the time, but they can rotate in the same direction at different velocities or synchronously if necessary. The packing is arranged inside the support that has a large number of densely spaced holes for the passage of fluid. Baffles are arranged on the edge of packing rings, which can be rectangular, round, fan blade, and wing in shape. The rotation of adjacent packing rings in opposite directions causes high shear stresses on gas and the formation of a continuous vortex and strong turbulence. Thus, gas no longer flows in a strand manner as that in RPB. CAS-RPB has the potential to improve gas-phase mass transfer by increasing the slip velocity between gas and packing, gas turbulence, and the surface renewal rate of the gas film and reducing mass transfer resistance. In particular, baffles can increase the shear stress on gas and its deformation rate, facilitating the separation of the gas boundary layer, which leads to the formation of gas vortex and gas turbulence and accelerates the renewal of gas film surface.

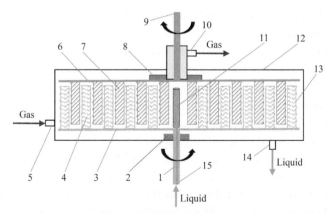

Figure 1.5 The schematic of CAS-RPB [1—rotor shaft for the lower disk, 2—shaft seal, 3—lower disk, 4—lower disk packing (support), 5—gas inlet, 6—upper disk, 7—upper disk packing (support), 8—airtight seal, 9—rotor shaft for the upper disk, 10—gas outlet, 11—liquid distributor, 12—shell, 13—baffle, 14—liquid outlet, 15—liquid inlet]. *CAS-RPB*, counter airflow shear-rotating packed bed.

In another configuration of CAS-RPB, circular rings are used instead of the packing and support and baffles are arranged around the circular rings. These circular rings are attached alternately to two separate disks with a gap between adjacent rings and between the free end of each ring and the disk. The adjacent circular rings rotate in opposite directions as the two disks rotate in opposite directions. Baffles can be installed on the surface of the circular rings in a concavo-convex or undulatory shape or around the circular rings in a flake (i.e., wing and paddle) or concavo-convex shape. As adjacent circular rings rotate in opposite directions, baffles can increase the shear stress on gas and its deformation rate and facilitate the separation of the gas boundary layer, which leads to the formation of gas vortex and turbulence and accelerates the renewal of the gas film surface. As a consequence, the gas-film-controlled mass transfer process is significantly improved. It is evident that the shape, size, position, and rotation speed of baffles have great impacts on the gas-phase mass transfer process.

In short, the split packing, baffles, and two separate disks that rotate in opposite directions in CAS-RPB contribute to breaking the gas into smaller fragments, and the gas no longer flows through the packing in an integral or strand manner. Therefore, CAS-RPB can improve gas-film-controlled mass transfer and gas mass transfer rate.

1.3.1.1.3 Rotating packing bed with two-stage or multi-stage packing

A typical structure of RPB with two-stage packing is shown in Figs. 1.6 and 1.7. RPB has two coaxial rotors with independent components (liquid distributor, airtight seal, etc.),

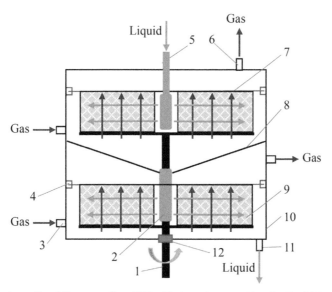

Figure 1.6 The schematic of the cross-flow RPB with two-stage packing for liquid treatment (1—rotor shaft, 2—liquid distributor, 3—gas inlet, 4—airtight seal, 5—liquid inlet, 6—gas outlet, 7—rotor, 8—separation board, 9—packing, 10—shell, 11—liquid outlet, 12—shaft seal). *RPB*, rotating packing bed.

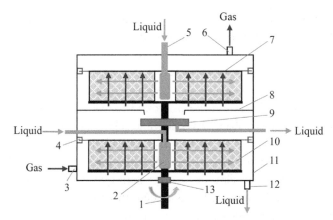

Figure 1.7 The schematic of the cross-flow RPB with two-stage packing for gas treatment (1—rotor shaft, 2—liquid distributor, 3—gas inlet, 4—airtight seal, 5—liquid inlet, 6—gas outlet, 7—rotor, 8—separation board, 9—water storage tank, 10—packing, 11—shell, 12—liquid outlet, 13—shaft seal). *RPB*, rotating packing bed.

and the structure differs slightly for gas and liquid treatment. The cross-flow RPB with two-stage packing for liquid treatment (e.g., desorption and devolatilization) is shown in Fig. 1.6. The liquid is introduced into the top rotor via the top liquid distributor, and it flows outward from the inner edge to the outer edge of the packing and comes into contact with the gas, which is introduced into the top rotor via the upper gas inlet and flows upward from the bottom to the top of the packing. After that, the gas leaves the top rotor via the upper gas outlet, and the liquid flows along the separation board to the second rotor. The liquid is introduced into the bottom rotor via the bottom liquid distributor, and it flows outward from the inner edge to the outer edge of the packing and comes into contact with gas, which is introduced into the bottom rotor via the lower gas inlet and flows upward from the bottom to the top of the packing. In this way, the liquid is treated twice in RPB and the gas–liquid mass transfer efficiency is significantly improved.

It is to be noticed that the two-stage packing can be readily extended to three or more stage packing with some minor modifications. This configuration makes the setup more compact and dramatically reduces the equipment size, leading to simple operation process, low equipment investment, and high automatic control.

The cross-flow RPB with two-stage packing for gas treatment (e.g., adsorption) is shown in Fig. 1.7. The gas is introduced into the bottom rotor and contacts in a cross-flow mode with fresh liquid flowing radially outward from the inner edge to the outer edge of the packing. The used liquid exits the bottom rotor via the bottom liquid outlet, and the treated gas flows upward into the top rotor through the annular passage between the separation board and the temporary water storage tank. Again, gas contacts in a cross-flow mode with fresh liquid flowing radially outward from the inner edge to the outer edge of the packing. In this way, gas contacts twice with fresh liquid in the cross-flow RPB. The used liquid flows into the temporary water storage tank and then exits the top rotor via the top liquid outlet.

In another configuration, the liquid used in the top rotor instead of fresh liquid is introduced into the bottom rotor. Gas first contacts with used liquid from the top rotor and then with fresh liquid introduced into the top rotor, while liquid first contacts with treated gas from the bottom rotor and then with untreated gas introduced into the bottom rotor. In this case, the average driving force is relatively large, which is favorable for operation. It is noted that the liquid process is flexible and can be adjusted if needed in order to better satisfy industrial requirements.

One important lesson learned from practical applications is that stability and economic feasibility should be considered in the design of cross-flow RPB with multistage packing.

1.3.1.2 Baffle rotating bed

The structure of baffle rotating bed is schematically shown in Fig. 1.8. It consists of a rotatable disk (lower disk) to which a set of rotational baffles (concentric circular sheets) are fixed and a stationary disk (upper disk) to which a set of stationary baffles (concentric circular sheets) are fixed. These rotational and stationary baffles are arranged alternatively with a gap between adjacent baffles and between baffles and disks. Both gas and liquid flow in a zigzag channel between rotational and stationary baffles [7]. The liquid undergoes repeated dispersion and aggregation in the zigzag channel and gas flows in a countercurrent direction, leading to an increase in contact time. Remarkably, the separation capacity and efficiency would increase exponentially for multi-stage baffle rotating bed. For this reason, it is widely used in distillation, absorption, and chemical reaction [8]. The baffle rotating bed is developed from RPB

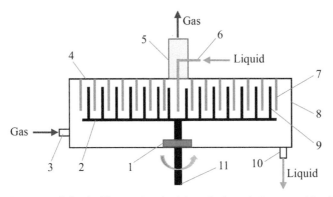

Figure 1.8 The schematic of the baffle rotating bed (1—shaft seal, 2—rotatable disk, 3—gas inlet, 4—stationary disk, 5—gas outlet, 6—liquid inlet, 7—stationary baffle, 8—shell, 9—rotational baffle, 10—liquid outlet, 11—rotor shaft).

with the use of baffles instead of packing. The increase of the gas-liquid contact area due to the rotation of baffles can improve the mass transfer rate and efficiency. The operation principle of the multi-stage baffle rotating bed is the same as that of the single-stage baffle rotating bed, and the structure is schematically shown in Fig. 4.14.

1.3.2 Structures and types of rotating packing bed for intensification of liquid-liquid mass transfer

In recent years, high-gravity technology has also been applied to improve liquid-liquid mixing and mass transfer. Based on the studies of Tamir et al. [9] and Wu et al. [10], we devised a novel reactor called the impinging stream-rotating packed bed (IS-RPB) to improve liquid-liquid mixing and contact processes (Fig. 1.9) [11]. The high-gravity technology provides innovative insights into liquid-liquid reaction, emulsification, mixing, extraction, liquid membrane separation, and other separation processes. We have conducted a series of experiments by using high-gravity technology for chemical process intensification (e.g., liquid-liquid mixing, extraction, liquid membrane separation) and reaction process intensification (e.g., synthesis of nanoparticles from liquid-liquid reaction) [12-15]. So far, our high-gravity equipment has been put into industrial operation.

In IS-RPB, an impinging stream unit is installed in the central cavity of the rotor with two sets of coaxial and concentric jet nozzles that are opposite to each other, and they are concentric or parallel to the rotation axis of the rotor. Axially, these nozzles should be symmetrical about the center line of the packing. The operating principle of

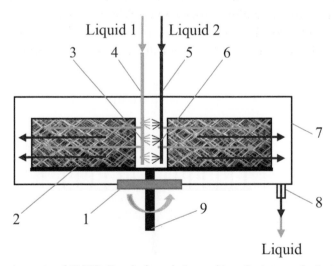

Figure 1.9 The schematic of IS-RPB (1—shaft seal, 2—packing, 3—jet nozzle 1, 4—liquid 1 pipe, 5—liquid 2 pipe, 6—jet nozzle 2, 7—shell, 8—liquid outlet, 9—rotor shaft). *RPB*, rotating packing bed.

IS-RPB is as follows. Two pressurized fluids are injected at high speed via nozzles in opposite directions and come into collision in the central cavity of the packing. An impinging zone develops as a result of such collision that allows for excellent micromixing and mass transfer between the two liquids. As the mixture flows radially outward from the inner edge to the outer edge of the packing, it is repeatedly sheared and dispersed by the rotating packing to improve the mixing efficiency. After that, the liquid is splashed onto the interior wall of the shell by centrifugal force and leaves the bed through the liquid outlet at the bottom of the bed under the effect of gravity. It is suggested that the match between the inner diameter (central cavity) of the packing and the impingement zone is the key to eliminating the edge effect of impinging stream so that the mixture at the edge of the impingement zone can also be well-mixed in the rotating packing.

1.3.3 Structures and types of rotating packing bed for intensification of gas-solid mass transfer

The RPB for intensification of gas-solid adsorption or desorption processes is schematically shown in Fig. 1.10. In a typical RPB, adsorbent is placed in the rotor as the packing and gas flows axially through the packing in a high-gravity field, leading to more intimate gas-solid contact and consequently higher gas-solid adsorption (desorption) rate and efficiency. The high-gravity adsorption mechanism is closely related to the high-gravity. In physical absorption processes, high-gravity can increase the van der Waals force and hydrogen bonding force by increasing the diffusion between the two phases and reducing ineffective adsorption layers. In chemical absorption processes, high-gravity can effectively increase the chemical reaction rate between adsorbate and adsorbent. The flow pattern of gas is altered in RPB and the residence time is reduced. The external diffusion between adsorbent and adsorbate is increased because of the high rotating force and the ineffective adsorption area is reduced. Adsorbate can be adsorbed on adsorbent surface in an instant and diffuse into

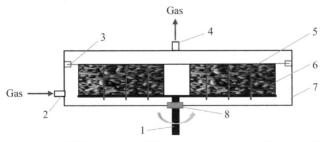

Figure 1.10 The schematic of RPB for gas-solid mass transfer intensification (1—rotor shaft, 2—gas inlet, 3—airtight seal, 4—gas outlet, 5—rotor, 6—adsorbent, 7—shell, 8—shaft seal). *RPB*, rotating packing bed.

the pores of the adsorbent, which increases the effective adsorption sites of adsorbent. In desorption processes, the excellent mass and heat transfer efficiency in RPB can effectively increase the regeneration of saturated adsorbent and extend its service life. In conclusion, RPB has wide applications, high flexibility, excellent regeneration capacity that allows replacement of adsorbent at a lower frequency, and high processing capacity [16].

References

[1] Chen JF. High-gravity Technology and Application. Beijing: Chemical Industry Press, 2002.
[2] Ouyang CB, Liu YZ, Qi GS. A new type reaction facility: rotating packed bed technology and its application. Science & Technology in Chemical Industry, 2002,10(4):50-53.
[3] Yuan ZG, Song W, Jin GL, et al. Comparison of pressure-drop and mass-transfer characteristics of two kinds of packing in countercurrent rotating packed bed. Chemical Engineering (China), 2015,43(7):7-11.
[4] Jiao WZ, Liu YZ, Diao JX. Mass-transfer characteristics of cross-flow rotating bed with protruded ripple plate packing. Chemical Industry and Engineering Progress, 2006,25(2):209-212.
[5] Yuan ZG, Liu YZ, Song W. Removal of sulfur dioxide from flue gas by sodium phosphate buffer solution in a co-current rotating packed bed. Chemical Industry and Engineering Progress, 2014,33(5):1327-1331.
[6] Liu Y, Gu D, Xu C. Mass transfer characteristics in a rotating packed bed with split packing. Chinese Journal of Chemical Engineering, 2015,23(5):868-872.
[7] Ji JB, Xu ZC, Bao TH. Rotating zigzag high-gravity bed. CN 2523482Y, 2002.
[8] Xu ZC, Yu YL, Ji JB. Rotating zigzag high-gravity bed and its application in distillation. Petrochemical Technology, 2005,34(8):778-781.
[9] Tamir A. Impinging-stream Reactors: Fundamentals and Applications. Beijing: Chemical Industrial Press, 1996.
[10] Wu Y. Impinging Stream: Principles, Properties and Applications. Beijing: Chemical Industrial Press, 2006.
[11] Liu YZ, Jiao WZ, Qi GS. Flow structures in impinging streams and impinging stream-rotating packed bed. CN 104226202A, 2014.
[12] Ouyang CB, Liu YZ, Qi GS. Function study on micro mixing in impinging stream-rotating packed bed reactor. Applied Chemical Industry, 2002,31(5):22-24.
[13] Liu YZ, Qi GS, Yang LR. Study on the mass transfer characteristics in impinging stream-rotating packed bed extractor. Chemical Industry and Engineering Progress, 2003,22(10):1108-1111.
[14] Liu YZ. Research progress of high-gravity technology of IS-RPB to intensify liquid-liquid contact. Chemical Industry and Engineering Progress, 2009,28(7):1101-1108.
[15] Fan HL, Zhou SF, Jiao WZ. Removal of heavy metal ions by magnetic chitosan nanoparticles prepared continuously via high-gravity reactive precipitation method. Carbohydrate Polymers, 2017,174:1192-1200.
[16] Guo Q, Liu Y, Qi G. Adsorption and desorption behaviour of toluene on activated carbon in a high-gravity rotating bed. Chemical Engineering Research and Design, 2019,143(3):47-55.

CHAPTER 2

Absorption

Contents

2.1 Overview	23
2.2 Principles of high-gravity absorption	24
2.2.1 Gas-liquid mass transfer theory	24
2.2.2 Principle of high-gravity technology for intensification of mass transfer	25
2.3 Key techniques and challenges	31
2.3.1 Absorbent and absorption technique	31
2.3.2 Engineering technology of rotating packing bed	32
2.4 Application examples	33
2.4.1 Removal of H_2S	33
2.4.2 Removal of SO_2	40
2.4.3 Removal of CO_2 from gas streams	49
2.4.4 Removal of NO_x from flue gas	54
2.4.5 Removal of ammonia in the production of nitrophosphate fertilizers	60
2.4.6 Purification of volatile organic compounds	65
2.5 Future perspectives	72
References	72

2.1 Overview

Absorption is a mass transfer process in which one or more components of a gas mixture are transferred to a liquid because of their differences in solubility or chemical reactivity in the liquid. Here, the liquid (solvent) is the absorbent, the gas component to be absorbed is the solute, and the resultant solution containing the solute is the absorbing solution. The absorption process may be a physical process with no chemical reaction between solute and absorbent, or a chemical process with a chemical reaction between solute and absorbent, or a physicochemical process that involves both physical and chemical absorption processes.

Absorption can be used for (1) removal of unwanted component(s) from a gas mixture and/or recovery of target component(s); (2) purification of feed gas for chemical synthesis. For instance, toxic or unnecessary components are often removed from the catalyst by absorption in the industry; (3) manufacturing of finished or semi-finished liquid products; (4) treatment of toxic and harmful gases. In most circumstances, absorption is intended to serve multiple purposes rather than a single purpose. It is also important to note that both absorption and desorption should be considered in the separation of a gas mixture.

The physical absorption process is the dissolution of a gas in a liquid with no chemical reaction occurring between them (e.g., absorption of ethanol and acetone vapor in water, absorption of gaseous hydrocarbons in liquid hydrocarbons, etc.), and it is mainly affected by operating pressure, temperature, and the solubility of the solute in the solvent. The absorption rate is predominantly dependent on the concentration gradient of the solute at the gas–liquid interface and its diffusion rate from the gas phase to the liquid phase.

The chemical absorption process involves a chemical reaction between the solute dissolved in a liquid and the liquid (or its active components). This process is irreversible if the reaction products could not be readily decomposed to release the absorbate (e.g., oxidative desulphurization, hydrated lime sorbents for SO_2 removal, etc.). While it is reversible if the reaction products could be readily decomposed and the sorbent could be recycled after sorption (e.g., removal of CO_2 by activated hot potassium carbonate solution, removal of CO by copper acetate monohydrate, etc.). The main factors affecting chemical absorption include pressure, temperature, gas–liquid equilibrium, and reaction equilibrium.

The physicochemical absorption process is achieved with the use of a mixture of physical and chemical solvents (e.g., desulphurization and decarburization by sulfolane process, etc.). The process encompasses both chemical and physical absorption processes and greatly improves the absorption capacity of the solvent.

The absorption rate is determined by the ratio of the driving force to the gas–liquid mass transfer resistance. The higher the ratio is, the higher the absorption rate will be. Increasing the driving force can substantially increase the absorption rate at the expense of high energy consumption. An economically more viable option is to reduce the gas–liquid mass transfer resistance. In traditional absorption processes, the absorbent moves from the top to the bottom of the column under the influence of gravity and comes into contact with the gas mixture in the form of a macroliquid film. In this case, the mass transfer rate is low and the column used is bulky with low space utilization and production capacity. The high-gravity technology allows liquid to be sheared into micro/nano elements because of the high-speed rotating packing, which leads to lower gas–liquid mass transfer resistance and larger gas–liquid interfacial area. The resulting volumetric mass transfer coefficient is 1–3 orders of magnitude higher than that in traditional columns [1]. The high-gravity absorption is also characterized by small size, low space requirement, high efficiency, and low operating cost.

2.2 Principles of high-gravity absorption

2.2.1 Gas-liquid mass transfer theory

The theoretical investigation of how high-gravity intensifies the gas–liquid mass transfer process can be traced back to the classical surface renewal theory developed by Danckwerts. The basic principle of this theory is that the liquid at the gas–liquid interface, where absorption takes place, is continuously renewed by fresh liquid elements from the bulk liquid, or in other words, the fresh liquid is continuously exposed to the gas to be absorbed. It is clear that the rapid and continuous surface renewal contributes greatly to gas-liquid mass transfer. It does not entail the slow diffusion of solute into the liquid and instead, the solute is brought into continuous contact with fresh liquid. The surface renewal frequency S is introduced to indicate the mean rate of production of the fresh surface. Then, the convective mass transfer coefficient is described as follows:

$$k_\mathrm{L} = \sqrt{DS} \qquad (2.1)$$

where k_L—convective mass transfer coefficient;
D—diffusion coefficient;
S—surface renewal frequency.

2.2.2 Principle of high-gravity technology for intensification of mass transfer

In a previous study on the absorption of CO_2 in NaOH solution in a cross-flow rotating packing bed (RPB) with stainless steel porous corrugated packing or plastic porous packing [2], the effective specific surface area is 490-1530 m^2/m^3 for the stainless steel porous corrugated packing and 290-975 m^2/m^3 for the plastic porous packing at a rotation speed of 500-1600 r/min, respectively. In a countercurrent-flow RPB with wire-gauze packing [3], the effective specific surface area is 700-1100 m^2/m^3 at a rotation speed of 0-1550 r/min and a liquid flow rate of 0-0.139 kg/s. In a helical rotating absorber [4], the effective specific surface area is 560-1100 m^2/m^3. In a B-25 rotating stream tray with an elevation angle of 25 degrees [5], the effective specific surface area is 29-50 m^2/m^3, which is approximately 1/25 of that of the stainless steel porous corrugated packing. In a countercurrent-flow RPB using plastic beads as the packing [6], the effective specific surface area is 372-1140 m^2/m^3 under conditions of rotation speed = 150-2100 r/min, liquid flow rate = 258-822 mL/min, and gas flow rate = 9.2-18.8 mL/min. The stainless steel porous corrugated packing has an excel-lent atomization performance in a high-gravity field and the effective specific surface area is increased by about 2.7 times, leading to a dramatic increase in the gas-liquid contact area and consequently the gas-liquid mass transfer efficiency. The reason why the effective specific surface area is significantly increased in RPB is illustrated in Fig. 2.1.

Fig. 2.1 clearly shows that the total specific interfacial area is $a_\mathrm{e} = a + a_1 + a_2 + a_3$, where a is the specific surface area of the packing, a_1 is the effective specific surface area for atomization from the liquid distributor to the inner edge of the packing, a_2 is the effective specific surface area for atomization between the packing, and a_3 is the

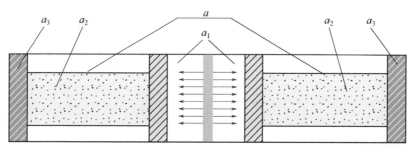

Figure 2.1 The distribution of the effective specific area in RPB. *RPB*, Rotating packing bed.

effective specific surface area for atomization from the outer edge of the packing to the casing wall. As a_1, a_2, and a_3 represent the specific surface areas for mass transfer between atomized liquid droplets, filaments, films and the gas, they are closely related to liquid dispersion, liquid flow rate, high-gravity factor, gas flow rate, packing structure, and liquid surface tension. In a normal gravity field, the effective surface area of the packing accounts for a small proportion of the total surface area. In a high-gravity field, the exceptionally high centrifugal force can not only allow the liquid to be distributed more uniformly in RPB and to contact with the gas in a highly dispersed state, but it can also lead to a significant reduction of liquid holdup. As the liquid flows from the inner edge to the outer edge of the rotating packing, a fraction of the liquid may leave the packing surface to form liquid droplets and mists that flow in a zigzag pattern in packing voids. As the liquid enters the cavity between the outer edge of the packing and the casing wall, it comes into contact with the gas that flows countercurrent to the liquid. Considering the liquid-gas contact areas represented by a_1, a_2, and a_3, the total interfacial area is undoubtedly higher than the specific surface area of the packing.

The gas–liquid mass transfer process in RPB is greatly enhanced because of the rapid surface renewal of the liquid. In the rotating packing, the liquid is subjected to repeated dispersion-coagulation-redispersion-recoagulation processes, resulting in high dispersion, high turbulence, and rapid surface renewal. The convective mass transfer coefficient is expected to increase by several orders of magnitude. Because of the large effective specific surface area and rapid surface renewal, the liquid-side volumetric mass transfer coefficient in the cross-flow RPB with the stainless steel porous corrugated packing is 1.4–2.1 times and 0.95–1.47 times that reported in the study [7,8], respectively, but it is slightly lower than that in a countercurrent-flow RPB [3]. A more impressive finding is that the mass transfer coefficient of RPB is 1–2 orders of magnitude higher than that in traditional mass transfer devices [9]. Table 2.1 summarizes the mass transfer coefficients of different absorption systems in RPB.

Table 2.1 The mass transfer coefficients of different absorption systems in RPB.

RPB	Absorption systems and operating conditions	Expressions of mass transfer coefficient	Mass transfer coefficient	Operating conditions of traditional equipment	Increasing extent
Cross-flow RPB	Absorption system: Water-NO_x [10] Liquid film-controlled mass transfer: Gas flow rate = 2 m^3/h Liquid-gas ratio = 20 L/m^3 High-gravity factor = 90 Inlet NO_x concentration = 18000 mg/m^3	$K_y a_e = 629.4650 \frac{pD_{ab}a_t}{RTD} Re_G^{0.2566} Re_L^{0.4283} \beta^{0.4593}$ where $K_y a_e$ = gas-phase volumetric mass transfer coefficient; D_{ab} = diffusion coefficient; D = equivalent diameter of the packing; a_t = specific surface area of the packing; Re_G = gas Reynolds number; Re_L = liquid Reynolds number; β = high-gravity factor; p = gas pressure, Pa; R = molar gas constant, $R = 8.314$ J/(mol·K); T = operating temperature, K	0.518 mol/(m^3·s)	Countercurrent sieve-plate column: Number of plate = 10 Column diameter = 1.4 m Column height = 14 m	480 times
	Absorption system: Fe II EDTA-NO: [11] Liquid film-controlled mass transfer: Gas flow rate = 3 m^3/h Gas-liquid ratio = 55 m^3/m^3 High-gravity factor = 82 Absorbing solution concentration = 0.02 mol/L Inlet NO concentration = 632 mg/m^3	$K_G a = \frac{2.625 c_B^{0.2798} \beta^{0.2798} G^{0.9144}}{c_A^{0.03201} L^{0.1427}}$ where $K_G a$ = total volumetric mass transfer coefficient; c_A = inlet NO concentration; c_B = absorbing solution concentration; β = high-gravity factor; G = gas flow rate; L = liquid flow rate	3.064 s^{-1}	Bubble column reactor: Height = 320 mm; Inner diameter = 45 mm; Gas distributor with micrometer aperture at the bottom	—

(Continued)

Table 2.1 (Continued)

RPB	Absorption systems and operating conditions	Expressions of mass transfer coefficient	Mass transfer coefficient	Operating conditions of traditional equipment	Increasing extent
	Absorption system: dilute nitric acid–ammonia [12] Gas film-controlled mass transfer: Gas flow rate = 15 m^3/h Gas–liquid ratio = 1000 m^3/m^3 High-gravity factor = 90	$k_y a_e = 6604 \dfrac{p D a_t}{R T R'} Re_G^{0.5} Re_L^{0.2118} \beta^{0.03354}$ where $k_y a_e$ = gas-phase volumetric mass transfer coefficient; D = equivalent diameter of the packing; a_t = specific surface area of the packing; Re_G = gas Reynolds number; Re_L = liquid Reynolds number; β = high-gravity factor; p = pressure; R' = geometric mean of inner and outer radius of the rotor	700 mol/(m$^3\cdot$s)	Scrubber: Radius = 2.6 m Height = 20.1 m Air flow: 5 × 10^4 m^3/h	850 times
Countercurrent-flow RPB	Absorption system: sodium phosphate–SO$_2$ [13] Liquid film-controlled mass transfer: High-gravity factor = 80 Inlet SO$_2$ volume concentration = 0.2% Phosphoric acid concentration = 1 mol/L Initial pH = 5.67 Empty bed gas flow rate = 0.73 m/s	$K_y a = 0.0744 \beta^{0.5046} u^{0.9150} q^{0.6801}$ where β = high-gravity factor; u = average empty bed gas flow rate, m/s; q = liquid spraying density, m^3/(m$^2\cdot$h); $K_y a$ = total volumetric mass transfer coefficient	0.749 kmol/(m$^3\cdot$s)	Packed column: Radius = 50 mm; Height = 2 m; Packing: ϕ6 mm × 6 mm × 0.1 mm stainless steel screen θ ring; Packing height = 1.1 m	9.86 times

(Continued)

Table 2.1 (Continued)

RPB	Absorption systems and operating conditions	Expressions of mass transfer coefficient	Mass transfer coefficient	Operating conditions of traditional equipment	Increasing extent
CAS-RPB	Absorption system: seawater–SO_2 [14] Liquid film-controlled mass transfer: Liquid–gas ratio 54 L/m^3 Gas flow rate = 40 m^3/h High-gravity factor = 66.3	The two rotors rotate in the same direction: $K_G a = 1.649 \beta^{0.13} R^{0.21} c_{in}^{0.082} U^{-3.158} G^{0.813}$ where $K_G a$ = total volumetric mass transfer coefficient; β = high-gravity factor; R = liquid flow rate; c_{in} = SO_2 initial concentration; U = absorbing solution concentration; G = gas volume The two rotors rotate in opposite directions: $K_G a = 1.492 \beta^{0.165} R^{0.307} c_{in}^{0.18} U^{-3.631} G^{0.759}$	9.5 s^{-1}	Spraying column: Radius = 26 mm Sprayed segment height = 80 cm	4.74 times
	Absorption system: water–NH_3 system [15] Liquid film-controlled mass transfer: Gas flow rate = 10–30 m^3/h Liquid–gas ratio = 5–30 L/m^3 High-gravity factor = 10.8–97.4	The two rotors rotate in the same direction: No form-drag baffle: $k_y a_e = 0.019587 Re_G^{0.902826} We_L^{0.146287} Ga^{0.123898}$ where $k_y a_e$ = gas-phase volumetric mass transfer coefficient; Re_G = gas Reynolds number; We_L = liquid Weber number; Ga = Galileo number	34.38–128.74 mol/(m^3·s) 36.58–138.36 mol/(m^3·s) 39.32–142.33 mol/(m^3·s)	—	—

(*Continued*)

Table 2.1 (Continued)

RPB	Absorption systems and operating conditions	Expressions of mass transfer coefficient	Mass transfer coefficient	Operating conditions of traditional equipment	Increasing extent
		Wire gauze form-drag baffle: $k_y a_e = 0.013098 Re_G^{0.946270} We_L^{0.112629} Ga^{0.115224}$ Plate form-drag baffle: $k_y a_e = 0.021614 Re_G^{0.874595} We_L^{0.136341} Ga^{0.131125}$ The two rotors rotate in opposite directions: No form-drag baffle: $k_y a_e = 0.016286 Re_G^{0.915434} We_L^{0.169247} Ga^{0.152470}$ Wire gauze form-drag baffle: $k_y a_e = 0.017324 Re_G^{0.915864} We_L^{0.172684} Ga^{0.153095}$ Plate form-drag baffle: $k_y a_e = 0.020621 Re_G^{0.891338} We_L^{0.175707} Ga^{0.157786}$	44.12–182.25 mol/(m³·s) 46.38–191.69 mol/(m³·s) 49.12–199.85 mol/(m³·s)		

2.3 Key techniques and challenges

The high-gravity technology utilizes RPB to improve the absorption process. Here, the absorbent, absorption technique, and device should be taken into account.

2.3.1 Absorbent and absorption technique

The high-gravity technology has the potential to increase the gas-liquid mass transfer rate by several orders of magnitude compared with conventional absorption processes and reduce the gas-liquid contact time to several seconds. In high-gravity adsorption processes, a major technical challenge is the selection of an appropriate absorbent to exploit the full potential of the high-gravity technology. A good absorbent should not only meet all requirements that are relevant to the absorbent itself, but it should also be well-matched with the high-gravity technology.

2.3.1.1 Solubility or absorption capacity

The physical absorption process requires the solute in the gas phase to have high solubility in the solvent, which is related to the operating pressure and temperature. The chemical absorption process involves the dissolution of a gaseous component in a solvent followed by a chemical reaction between them. This process may be reversible and the adsorbent can be regenerated for reuse. The chemical absorption process is mainly affected by pressure, temperature, gas-liquid equilibrium and chemical equilibrium. The high-speed rotating packing can shear the liquid into micro/nano droplets, filaments and films, resulting in a substantial increase in the specific surface area and consequently a higher transfer rate of the solute from the gas phase to the liquid phase. Less solvent is required in RPB due to the higher mass transfer rate compared with that in traditional columns, which yields a much lower liquid-gas ratio. Thus, the solvent should have high solubility (or absorption capacity) for the component to be separated.

2.3.1.2 Selectivity and cost

The solvent should be highly selective for the component to be separated so that other components are almost insoluble in the solvent. The solvent should also be cost-effective and readily available and can be regenerated for reuse.

2.3.1.3 Volatility

In RPB, the specific surface area is greatly increased as the liquid is sheared into micro/nano droplets, filaments, and films by the packing that rotates at a high speed, and the absorbent is highly susceptible to volatilization. It is recommended that the vapor pressure of the solvent should be as low as possible in order to avoid the volatility of solvent in the absorption process.

2.3.2 Engineering technology of rotating packing bed

RPB is an emerging absorption equipment and it may have a long way to go before it can work continuously, efficiently, and stably in laboratory or industrial settings. Of vital importance is the engineering technique, including structural optimization of RPB and packing.

Structural optimization of RPB: RPB falls into three distinct categories according to the flow directions of gas and liquid, including countercurrent-flow RPB, concurrent-flow RPB, and cross-flow RPB. Countercurrent-flow RPB is structurally identical to concurrent-flow RPB and both of them have a high moment of inertia that may adversely affect their long-term stable and efficient operation. In contrast, cross-flow RPB is simpler in structure and the moment of inertia is about 30% lower, making it more amenable for industrial scale-up, stable operation, and easy repair and maintenance in industrial applications. RPB can be installed either in a vertical position or a horizontal position. Nevertheless, the rotor would be less affected by gravity in the former case, so that more stable operation can be achieved. The accessories of RPB, such as liquid distributor and gas/liquid seal, should be compatible with the RPB in structural optimization.

Packing: It is a core component of RPB and plays a key role in the dynamic equilibrium, fluid mechanics, mass transfer efficiency, and service life of the RPB. For instance, the specific surface area of the packing determines the mass transfer rate, the porosity is related to the gas pressure drop, the geometric symmetry, density, filling load and mode have impacts on the dynamic equilibrium and power consumption, and the material properties have impacts on the anti-corrosion performance. Thus, much effort is being made to prepare packing materials with high uniformity, light weight, centrosymmetric structure, and high specific surface area.

The trade-off between weight and mechanical strength should be considered in the design of RPB rotor. In order for the rotor to have sufficiently high mechanical strength and operational stability, it is necessary to increase the rotor weight. However, as the rotor weight increases, a higher mechanical strength of the rotor is required, which leads to a further increase in rotor weight. At present, three possible options are available to deal with this problem. The first option is to increase the mechanical strength of the rotor and reduce the weight as much as possible by means of structural optimization. The second option is to use high-strength low-density materials, such as high-strength light metals, composite materials and engineering plastics, as a substitute for steel. The third option is to reduce the packing weight by optimizing the structure and using more appropriate packing materials. Every possible effort should be made to make RPB more lightweight by means of structural optimization and the development of new materials in order to minimize the moment of inertia, energy consumption, and production cost of the RPB.

2.4 Application examples

The chemical industry has always been a major source of waste gas emissions and there is always a need for more efficient and cost-effective methods for treating industrial gas emissions. Many traditional techniques for this purpose are not environmentally friendly, so we have sought to use high-gravity technology for waste gas purification. In 2011, we won the Second Prize of the National Prize for Progress in Science and Technology. Over the past decades, the high-gravity technology has found numerous applications in the removal of hydrogen sulfide (H_2S), sulfur dioxide (SO_2), carbon dioxide (CO_2), and nitric oxide (NO_x), as shown in Fig. 2.2.

2.4.1 Removal of H_2S

H_2S is commonly found in natural gas, refinery gas, syngas, coke oven gas, associated gas, water gas, and Claus exhaust gas. H_2S should be handled with caution because it is toxic and corrosive with many negative consequences for the environment and human health. However, it is an important source of sulfur that can be recovered for the production of sulfur compounds. Thus, there is a growing interest in the removal of H_2S from waste gas across the world.

Currently, available technologies for the removal of H_2S from flue gas are classified into dry and wet flue gas desulphurization (FGD). The dry FGD utilizes a dry powdered sorbent (e.g., activated carbon and zinc oxide) that generally has low sulfur capacity, and the used sorbent is disposed of as waste because it is difficult to be

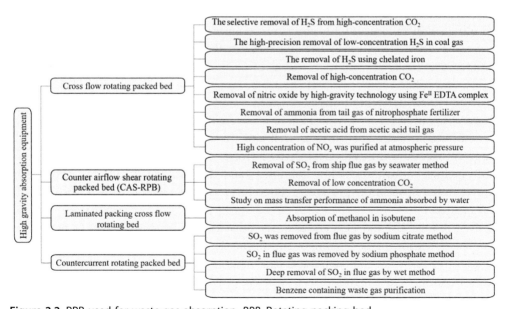

Figure 2.2 RPB used for waste gas absorption. *RPB*, Rotating packing bed.

regenerated. The major shortcomings of dry FGD are the high cost and potential environmental pollution caused by inappropriate disposal of used sorbent. The dry FGD is suitable for the removal of low-concentration H_2S, especially fine desulphurization. The wet FGD utilizes an alkaline solution as the absorbent to oxidize H_2S into elemental sulfur catalyzed by a catalyst. Common absorbents include sodium carbonate, ammonia water, and organic amine (organic base), and common catalysts include PDS (dinuclear cobalt-phthalocyanine sulfonate), TV (tannin extract), DDS (iron-alkali based complex), and ADA (sodium anthraquinone-2-sulfonate). Each has distinct characteristics that make it more suited to a particular problem. PDS is characterized by large sulfur capacity, high catalytic activity, low catalyst loading, low viscosity, low sulfur plugging, and simultaneous removal of organic sulfur. TV is characterized by the abundance of tannin extract, low cost, easy availability, and no sulfur plugging. DDS is often used for the removal of low-concentration H_2S.

Our team has combined high-gravity technology and wet FGD for the removal of H_2S [16,17]. In this section, we will introduce the selective removal of H_2S from high-concentration CO_2 using sodium carbonate solution as the desulfurizer and discuss the high-precision removal of low-concentration H_2S in coal gas using chelated iron.

2.4.1.1 Alkaline desulfurizers
2.4.1.1.1 Key technologies for selective removal of H_2S from high-concentration CO_2

The high-gravity technology has been widely used for selective removal of H_2S from high-concentration CO_2 emissions in ammonia synthesis in a group company in Shanxi Province, China. The exhaust gas flow rate is 21,000 m^3/h, which is composed of 98.97% (volume fraction) of CO_2, 0.68% of H_2S, and 0.35% of CH_4, C_2H_6, and other components. It should be noted that both CO_2 and H_2S are acid gases and their concentrations in waste gas emissions are very high. If an alkaline absorbent is used for removing H_2S in a conventional column, CO_2 will inevitably be removed in large quantities as well and the cost will be higher. Another challenge is that sodium bicarbonate derived from CO_2 can seriously block equipment and pipeline because of its low solubility, and this problem is difficult to solve in conventional absorption processes. Therefore, several technical breakthroughs are still needed.

First, both CO_2 and H_2S are acid gases that can undergo an acid-base neutralization reaction with the basic components in the desulfurizer. This reaction is known to be a fast reaction, but the chemical activity and rate of H_2S with the alkaline components are only slightly higher than that of CO_2. When a conventional column is used, a substantial amount of CO_2 would also be removed along with H_2S because of the long residence time of desulfurizer and gas in the column.

Second, the gas-liquid contact time in the RPB is generally less than 1 s, but it can be reduced to 0.1 s or even less if necessary simply by adjusting the parameters of the RPB or the rotation speed of the rotor. Previous studies have demonstrated that there are slight differences in chemical reactivity and rate between CO_2 and H_2S. However, we found that these two acid gases had very different reactions with the alkaline components in the RPB. In conclusion, the use of RPB makes it possible for selective absorption of H_2S from high-concentration CO_2.

Third, the H_2S concentration in CO_2 emissions is very high (10,600 mg/m^3), which is about 10 times that in the usual case, and one has the choice to either use a desulfurizer with higher sulfur capacity or increase the amount of desulfurizer in the desulphurization process.

In order to address these problems, we have developed an innovative technique for selective removal of H_2S from CO_2 emissions using RPB based on their differences in chemical reactivity and rate. A fundamental aspect of this technology is to ensure the selective absorption of H_2S within a very short period by controlling the gas-liquid contact time and increasing the absorption efficiency of H_2S, and CO_2 will leave the RPB before it can react with the desulfurizer. Consequently, the absorption of CO_2 is inhibited and H_2S is selectively removed from the gas stream containing a high concentration of CO_2. The PDS with a concentration of 120-480 mg/m^3 is used as a catalyst and the industrial-grade sodium carbonate is used as desulfurizer. The liquid-gas ratio is 5-20 L/m^3. Over 700 tons of sulfur can be recovered each year, the desulphurization rate is 98.5%, the removal rate of CO_2 is $\leq 0.5\%$, and energy consumption is reduced by 2.6×10^6 kW·h each year. This technique offers a promising option for the highly selective removal of H_2S and has brought significant economic and environmental benefits.

The selective removal process of H_2S by the high-gravity technology is schematically shown in Fig. 2.3. The gas streams containing CO_2 and H_2S are introduced into RPB through the gas pipeline and then flow upwards through the packing that rotates

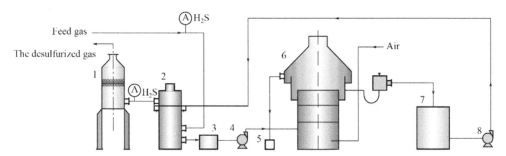

Figure 2.3 The high-gravity technology for selective removal of H_2S (1—demister, 2—RPB, 3—H_2S-rich solution tank, 4—H_2S-rich solution pump, 5—sulfur foam tank, 6—regeneration tank, 7—lean solvent storage tank, 8—lean solvent pump). *RPB*, Rotating packing bed.

at a high speed. The lean solvent is pumped from the lean solvent storage tank into RPB and then flows from the inner edge of the packing to the outer edge of the packing. The gas-liquid mass transfer is significantly enhanced because of high turbulence, strong mixing, and rapid interfacial renewal as a result of the high centrifugal force. Gas samples are collected using sampling tubes installed at both inlet and outlet for the measurement of H_2S concentrations. Finally, the desulfurized gas is discharged from the gas outlet, and the liquid mists and droplets entrained in the gas are removed using a demister. The H_2S-rich solution flows into the solution storage tank from the liquid outlet and is then pumped into the regeneration tank where it contacts with fresh air flowing countercurrent to the liquid and is regenerated under the catalysis of PDS. The regenerated lean solvent flows into the lean solvent storage tank for reuse. The sulfur foam generated in the regeneration process is transferred to the sulfur foam tank. The photograph for the selective removal of H_2S from high-concentration CO_2 by the high-gravity technology is shown in Fig. 2.4.

The operating conditions for the selective removal of H_2S from high-concentration CO_2 by the high-gravity technology are summarized as follows:

1. Liquid-gas ratio. Increasing the flow rate of the lean solvent can make the packing surface wetter and reduce the equilibrium partial pressure of H_2S in the solvent, which in turn can increase the driving force for the gas-liquid mass transfer and consequently the removal rate of H_2S. The removal rate of H_2S reaches 99% at a liquid-gas ratio of 8.5 L/m^3 (which is only 1/40-1/8 of that in conventional desulphurization processes), and increasing the liquid-gas ratio could increase the removal rate of H_2S but at the expense of high operating cost. Theoretically, it is

Figure 2.4 The photograph for selective removal of H_2S from high-concentration CO_2 by the high-gravity technology.

essential to select an appropriate liquid-gas ratio in order to ensure the total sulfur capacity of the desulfurizer is higher than the sulfur content in the gas. The stoichiometry of the chemical reaction between gas and liquid phases should also be satisfied. However, the actual liquid-gas ratio may be much higher than and even several times that of the theoretical liquid-gas ratio. The higher the mass transfer rate in the desulphurization process, the closer the actual liquid-gas ratio is to the theoretical one. In conclusion, the liquid-gas ratio is not only technically but also economically important, and an optimal liquid-gas ratio should be selected.

2. High-gravity factor. The removal rate of H_2S first increases and then decreases with the increase of high-gravity factor. A larger gas-liquid contact area is expected as the solvent can be sheared into smaller liquid droplets, filaments, and films by the packing that rotates at higher speeds, potentially increasing the removal rate of H_2S. However, the liquid has a shorter residence time in the packing because it will flow through the packing more rapidly at higher levels of high-gravity factor, potentially decreasing the removal rate of H_2S. In this regard, an optimal high-gravity factor should be selected to maximize the removal rate of H_2S. Finally, the optimal high-gravity factor is determined to be 106.2.

3. Alkali content. Because of the high concentration of CO_2 in the gas stream, the amount of desulphurization solution and Na_2CO_3 concentration should be carefully selected to ensure a sufficiently high sulfur capacity for effective absorption of H_2S and to reduce the absorption of CO_2 as much as possible. A high Na_2CO_3 concentration contributes to reducing the liquid-gas ratio and consequently the energy consumption for the transfer of the desulphurization solution. However, too high Na_2CO_3 concentrations can lead to the absorption of CO_2 and the production of salts, which makes it more difficult for the regeneration of the solvent and results in an increase in Na_2CO_3 consumption. Under optimal conditions (high-gravity factor >106.2 and liquid-gas ratio >8.5 L/m^3), the outlet H_2S concentration is <50 mg/m^3 at Na_2CO_3 concentrations of 10-14 g/L, 50-80 mg/m^3 at Na_2CO_3 concentrations of 7-9g/L, and >300 mg/m^3 at Na_2CO_3 concentrations of <7 g/L, respectively, which meets the environmental requirements. Thus, the alkalinity of the desulphurization solution should be kept at about 10 g/L.

4. PDS concentration. Increasing the PDS concentration from 120 to 360 mg/m^3 can increase the removal rate of H_2S by about 15%, while as the PDS concentration is further increased from 360 to 480 mg/m^3, the removal rate of H_2S is increased by less than 1%. Thus, the PDS concentration is preferably kept at 360 mg/m^3 in order to ensure high desulphurization efficiency and low operation cost.

5. Temperature. The removal rate of H_2S is lower at higher temperatures as the absorption of H_2S is an exothermic process. However, it should be noted that side reactions are likely to take place at higher temperatures, causing the generation of more salt by-products and consumption of more alkali; but too low

temperatures will adversely affect the regeneration of the solution and the flotation separation of sulfur foam. The gas phase temperature is kept at 30-35°C and the desulphurization solution temperature is kept at 35-40°C. As a result, the vapor in the gas would not condense into the solution to ensure water equilibrium in the system.

6. CO_2 content. The removal rate of CO_2 is $\leq 0.5\%$ and the selectivity is increased by 27%-29% compared to the technique used in the United States. The high-gravity technology has technical and economic advantages for selective removal of H_2S from high-concentration CO_2 emissions due to lower alkali consumption and operation costs.

2.4.1.1.2 Key technologies for fine removal of H_2S from coal gas

The ultra-low emission policy and rapid development of the coal chemical industry in China have placed unprecedented stress on the removal of H_2S from coal gas in coal gasification and coking. Our team has proposed innovative technologies and equipment based on high-gravity for deep desulphurization of coal gas, and the sulfur content can meet the requirements of fuel gas and syngas in the industries of ceramics, aluminum oxide, and kaolin.

The high-gravity technology has been successfully applied to fine desulphurization of coal gas using an RPB of ϕ1400 mm × 2500 mm in a ceramics company, where the coal gas flow rate is 13,000-15,000 m³/h, and the H_2S concentration is 300-500 mg/m³. Sodium carbonate solution (0.2-0.6 mol/L, pH = 7-10) is used as absorbent, and CoS is used as catalyst. Under conditions of high-gravity factor = 80 and liquid-gas ratio = 5-20 L/m³, the H_2S concentration in the purified gas is reduced to 5 mg/m³. The purified gas can be used as a substitute for liquefied petroleum gas for vehicles, which can help to solve the problems of transport and supply of liquefied petroleum gas for vehicles. More importantly, it lowers the operation cost by nearly 30% and the fuel cost by over 5 million RMB, implying that this technology is economically attractive.

This technology has also been successfully applied to the removal of sulfur from producer gas in a large-scale aluminum oxide company in Guangxi Province, China, and the photograph of the scene is shown in Fig. 2.5. The yield of the coal gas is 16.5×10^4 m³/h and the H_2S concentration is 2500 mg/m³. The desulphurization rate can reach above 98% by using six RPBs of ϕ2000 mm × 4600 mm. Sulfur foam is separated and the desulphurization solution is regenerated in the regeneration tank, filtered, and melted for the production of brimstone. Also, this technique is applicable to the nonferrous metal industry.

The Beijing University of Chemical Technology has cooperated with China National Offshore Oil Corporation to remove H_2S from natural gas at an offshore platform with high-gravity technology [18].

Figure 2.5 The photograph of the scene for the removal of H_2S from coal gas by the high-gravity technology.

The unique advantages of the high-gravity technology, such as high selectivity and deep desulphurization, make it applicable to the removal of H_2S from natural gas, syngas, coal gas, semi-water gas, coke oven gas, and shift gas.

2.4.1.2 Chelated iron desulfurization

Chelated iron desulfurization provides another solution for the removal of H_2S because of its advantages of high sulfur capacity, high desulphurization rate, easy recovery of sulfur, less side reaction, and environmental friendliness. In China, chelated iron desulfurization is still in its infancy and there is a need to import the technology from abroad. Another important problem is that the traditional column for chelated iron desulfurization has poor mass transfer efficiency, non-uniform gas–liquid flow, bulky device, and high energy consumption. We make some contributions to the development of technologies and devices for chelated iron desulfurization in China. Based on the advantages of chelated iron desulfurization (e.g., high sulfur capacity and desulphurization rate) and high-gravity technology (e.g., high mass transfer rate, short residence time, small equipment size, and low energy consumption), a novel chelated iron is used for removal of H_2S (9.11 g/m^3) from simulated gas in RPB (Fig. 2.6). The gas is introduced into the RPB from the gas inlet and then flows axially from the bottom to the top of the packing. The desulfurizer is pumped from the storage tank into the RPB and measured using a flowmeter. Then, it is uniformly spayed on the inner edge of the packing using a liquid distributor and flows radially from the inner edge to the outer edge of the packing, and comes into contact with the gas for absorption of H_2S. The gas is discharged into the absorbing device from the gas outlet, and

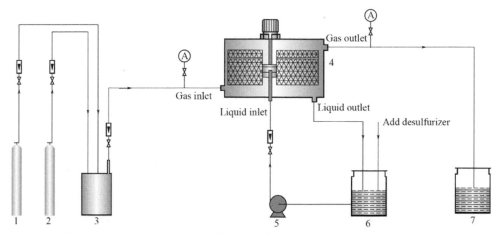

Figure 2.6 The experimental device and flowchart for the high-gravity chelated iron method for removal of H₂S (1—N₂ steel cylinder, 2—H₂S steel cylinder, 3—gas storage tank, 4—cross-flow RPB, 5—liquid pump, 6—desulfurizer storage tank, 7—absorbing device).

the H₂S-rich solution is discharged into the liquid storage tank from the liquid outlet and then regenerated for reuse. The H₂S concentration at the gas outlet is measured.

The effects of desulfurizer formulation, gas flow rate, liquid flow rate, liquid–gas ratio, and high-gravity factor on the desulphurization rate in a cross-flow RPB are investigated. The experimental results reveal that the working sulfur capacity of the desulfurizer with a $n(EDTA):n(HEDTA):n(Fe^{3+}):n(Fe^{2+})$ ratio = 1.2 : 6 : 5 : 1 reaches 4.25 kg/m³, which is much higher than that of traditional desulfurizers (0.1–0.3 kg/m³), and the desulphurization rate reaches 94% at a gas–liquid contact time of only 0.6 s. The desulphurization rate decreases with increasing gas flow rate but increases with increasing liquid flow rate, liquid–gas ratio, and high-gravity factor. The optimal conditions are as follows: liquid flow rate = 72 L/h, gas flow rate = 4 m³/h, and high-gravity factor = 35. Compared with conventional columns, the high-gravity technology may have better application prospects because of its remarkable advantages of high desulphurization rate, low liquid–gas ratio, small equipment size, high operating flexibility, and low consumption of desulphurization solution and energy [19,20].

2.4.2 Removal of SO₂

Sulfur dioxide (SO₂) is known as one of the most harmful atmospheric pollutants. In China, an ultra-low emission (ULE) policy was implemented in 2014 for renovating coal-fired power-generating units to limit SO₂ to 35 mg/m³. In recent years, concerns have been raised about flue gas emissions from coal-fired power plants, nonferrous smelting, fluid catalytic cracking, and the Clause process for sulfur production. Many techniques have been proposed for the removal of SO₂. The columns used in wet desulphurization are often criticized for their large size, high investment, low efficiency, high flue gas resistance, high operating cost, and difficult operation and maintenance. In comparison, the high-gravity technology has the advantages of small equipment size, low space requirement, high desulphurization rate, low system resistance, low operating cost, and low investment, which make it a promising FGD technique. For instance, the high-gravity technology was used for the removal of SO_2 in a sulfuric acid plant [21], where the gas flow rate was 3000 m^3/h and 0.244 mol/L NH_4HCO_3 solution was used as the desulfurizer. The SO_2 concentration was reduced from 5000 mg/m^3 before treatment to 100 mg/m^3 after treatment. In addition, the technology also has great advantages in terms of equipment investment, power consumption, and gas phase pressure drop.

We have previously used high-gravity technology for the removal of SO_2 with sodium citrate, sodium phosphate, sodium sulfite, and seawater as regeneratable desulfurizers. A variety of countercurrent-flow RPB, counter airflow shear RPB, and cross-flow RPB are devised, and novel desulfurization processes are also proposed, as summarized in Table 2.2.

2.4.2.1 Countercurrent-flow rotating packing bed

The countercurrent-flow RPB used for desulphurization is schematically shown in Fig. 2.7. The fume gas containing SO_2 is introduced into the RPB from the gas inlet, which flows radially from the outer edge to the inner edge of the packing and finally leaves the RPB from the gas outlet at the center of the bed. The solvent is pressurized by a diaphragm pump and sprayed on the inner edge of the packing through a liquid distributor. The flow rate is measured using a flowmeter. The liquid radially flows through the packing driven by the centrifugal force and comes into contact with the fume gas that flows countercurrent to the liquid. Then, SO_2 is absorbed by the desulfurizer, and the liquid is discharged into the solution tank from the liquid outlet at the bottom of the RPB. The liquid flowing through the rotating packing is sheared into small liquid films, filaments, and droplets several times, resulting in a large surface area and rapid absorption of SO_2. The countercurrent-flow mode combined with high-gravity can enhance the mass transfer and desulphurization efficiency.

2.4.2.1.1 Removal of SO_2 from flue gas using sodium citrate

The citric acid-sodium citrate buffer solution is used for the removal of SO_2 from fume gas in RPB [22]. The effects of high-gravity factor, liquid-gas ratio, gas flow rate, citric acid concentration, and pH on the removal rate of SO_2 and the gas-phase volumetric mass transfer coefficient (K_Ga) are investigated at inlet SO_2 concentrations of 1500-12,000 mg/m^3. It is found that both the removal rate of SO_2 and the gas-phase volumetric mass transfer coefficient increase with increasing high-gravity factor, liquid-gas ratio, citric acid concentration and pH. The optimal conditions are listed as

Table 2.2 Applications of the high-gravity technology for desulfurization.

No	Desulfurizer	Advantages	Disadvantages	RPB type
1	Citric acid–sodium citrate	It has a high desulphurization rate and is applicable to a wide range of SO_2 concentrations, especially high-concentration SO_2 It is not toxic and has a low vapor pressure, and it can be easily desorbed and regenerated There is low steam consumption and operating cost in the regeneration process, and high-purity SO_2 is recovered without causing secondary pollution	The hydroxyl groups of citric acid are vulnerable to degradation, leading to high consumption of citric acid and change in solution color to dark It is temperature sensitive The regeneration temperature is high and the equipment may be severely corroded Side reactions are likely to occur and uncontrollable, and sulfate crystallization and blockage occur easily	Countercurrent-flow RPB (rotating packing bed)
2	Phosphoric acid–sodium phosphate	It is not degradable because of stable physical and chemical properties and thus the consumption is low It is easily available and applicable to a wide range of SO_2 concentrations It leads to a high degree of purification and causes no liquid/solid wastes and secondary pollution	The solution can be regenerated for reuse only by heating or chemical reaction Oxidation may occur in the absorption and regeneration process, and thus extra alkaline solution will be consumed and secondary salt separation is needed It is time-consuming and costly	Countercurrent-flow RPB
3	Sodium sulfite	It is not toxic, low cost, and easily available Alkaline solution is used for recycled absorption, causing no blockage or corrosion to the equipment, pump, and pipeline because no solid products formed The equipment is unlikely to be corroded and it is easier for maintenance and long-term stable operation	Na_2SO_3 is oxidized to Na_2SO_4 in the desulphurization process It is difficult to be regenerated and more alkaline solution is consumed, leading to a higher operating cost The sulfur capacity is lower compared to above desulfurizers, and crystallization and precipitation are more likely to occur	Countercurrent-flow RPB

(Continued)

Table 2.2 (Continued)

No	Desulfurizer	Advantages	Disadvantages	RPB type
4	Na-Ca double alkali	Alkaline solution is used for recycled absorption, causing no blockage or corrosion to the equipment, pump, and pipeline The solution has strong alkalinity, fast reaction rate, and a desulphurization rate $\geq 95\%$	Na_2SO_3 is oxidized to Na_2SO_4 in the desulphurization process It is difficult to be regenerated and more alkaline solution is consumed, leading to a higher operating cost The gypsum quality is reduced due to the presence of Na_2SO_4, leading to low utilization rate of ash	Countercurrent-flow RPB
5	Seawater	It is cheap and easily available The process is simple and a high desulphurization rate is achieved at low investment and operating costs The operation is stable with no need for chemical reagents The used seawater can be discharged directly into the sea, causing no waste and secondary pollution	It is applicable only to coastal areas or other areas where seawater is easily available It is applicable to low-concentration SO_2	Counter airflow shear RPB

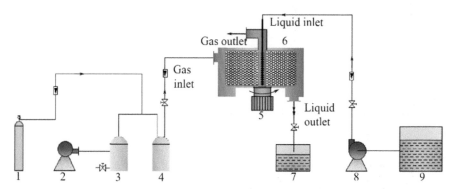

Figure 2.7 The desulphurization process in RPB (1—SO_2 steel cylinder, 2—Roots blower, 3—buffer tank, 4—mixing tank, 5—motor, 6—RPB, 7—rich solution storage tank, 8—diaphragm pump, 9—lean solvent storage tank). RPB, Rotating packing bed.

follows: citric acid concentration = 1.0 mol/L, initial pH = 4.5, liquid-gas ratio = 3-7 L/m³, and high-gravity factor = 54.53-90.14. The desulphurization rate is higher than 98% as the SO_2 concentration is reduced from 4500 to 80 mg/m³. The solution is regenerated for reuse and sulfur is recovered.

2.4.2.1.2 Removal of SO_2 from flue gas using sodium phosphate

The phosphoric acid-sodium phosphate buffer solution is used for removal of SO_2 from fume gas in RPB [13]. The effects of high-gravity factor, initial pH, liquid-gas ratio, gas flow rate, spraying density, phosphoric acid concentration, and absorbent temperature on the removal rate of SO_2 and the gas-phase volumetric mass transfer coefficient are investigated at inlet SO_2 concentrations of 1000-14,000 mg/m³ under concurrent and countercurrent flow conditions. The results reveal that the following points: (1) The removal rate of SO_2 and the gas-phase volumetric mass transfer coefficient increase at a decreasing rate with the increase of initial pH, liquid-gas ratio, high-gravity factor and phosphoric acid concentration under both conditions, but decrease with increasing absorbent temperature. At spraying densities lower than 4 m³/(m²·h), the removal rates of SO_2 decrease with increasing SO_2 concentration and gas flow rate, and the removal rate of SO_2 under concurrent flow conditions fluctuates more quickly; (2) The removal rate of SO_2 and the gas-phase volumetric mass transfer coefficient under concurrent flow conditions are lower than that under countercurrent flow conditions, but the differences become less significant with increasing liquid-gas ratio, high-gravity factor, and initial pH. The same desulphurization rate can be obtained by appropriately increasing the liquid-gas ratio, initial pH, or high-gravity factor; (3) Under similar conditions, the gas-phase volumetric mass transfer coefficient in RPB is more than 10 times higher than that in a high-efficiency packed column and the mass transfer efficiency is greatly improved; (4) The optimal conditions for concurrent flow are as follows: initial pH = 5.5-6.0, liquid-gas ratio = 2.0-3.0 L/m^3, temperature <50°C, high-gravity factor = 80-100, phosphoric acid concentration = 1.5 mol/L, gas flow rate = 0.3-1.2 m/s, and SO_2 concentration ⩽ 14 g/m^3. The removal rate of SO_2 is over 98% and the final SO_2 concentration is lower than 35 mg/m^3, which meets the emission standards. The optimal conditions for countercurrent flow are as follows: liquid-gas ratio = 1.5-2 L/m^3, high-gravity factor = 80, phosphoric acid concentration = 1-1.5 mol/L, others are the same as above. The removal rate of SO_2 is over 98% and the final SO_2 concentration is lower than 35 mg/m^3, which also meets the emission standards; (5) The SO_2-rich solution can be regenerated by means of steam stripping in both RPB and packed column, and the desorption rate decreases with increasing pH of the SO_2-rich solution, gas flow rate, and phosphoric acid concentration, but increases with the increasing preheating temperature of the SO_2-rich solution, SO_2 concentration, liquid flow rate, and high-gravity factor. The optimal conditions for regeneration are as follows: high-gravity factor = 70, liquid-gas ratio = 1.25 L/m^3, solution flow rate = 20 L/h, and SO_2 concentration = 5700 mg/m^3. The desulphurization rate and desorption rate stabilize at 98% and 81% after 5 h, respectively, and the outlet SO_2 concentration is lower than 100 mg/m^3. It is also important to note that both the absorption rate and the desorption rate of the phosphoric acid-sodium phosphate buffer solution remain constant after regeneration several times.

2.4.2.1.3 Removal of SO_2 from flue gas using sodium sulfite

The Na_2SO_3 solution is used for the removal of SO_2 from fume gas in a countercurrent-flow RPB [23], where the inlet SO_2 concentration is 1200-11,200 mg/m^3, and the effects of high-gravity factor, liquid-gas ratio, gas flow rate, spraying density, sodium sulfite concentration, and inlet SO_2 concentration on the removal rate of SO_2 are investigated. It is found that the removal rate of SO_2 increases with increasing liquid-gas ratio, high-gravity factor, and sodium sulfite concentration, but decreases with increasing gas flow rate and inlet SO_2 concentration. The optimal conditions are as follows: liquid-gas ratio = 2.5-4 L/m^3, high-gravity factor = 98-150, sodium sulfite concentration = 0.075-0.100 mol/L, inlet SO_2 concentration <6000 mg/m^3. The removal rate of SO_2 is over 98% and the final SO_2 concentration is lower than 15 mg/m^3, which meets the emission standards. As no CO_2 is absorbed, the desulfurizated fume gas can be used to supplement carbon for vegetable greenhouse crops.

2.4.2.1.4 Removal of SO_2 from flue gas using Na-Ca alkali solution

The Na-Ca alkali solution is used for the removal of SO_2 from fume gas in a multi-stage cross-flow RPB, and the effects of Na^+ concentration c_{Na^+}, inlet SO_2 concentration, high-gravity factor, liquid-gas ratio, and empty bed gas flow rate on the removal rate of SO_2 are investigated. The experimental results reveal that the removal rate of SO_2

increases with increasing Na^+ concentration c_{Na^+}, high-gravity factor, liquid-gas ratio and empty bed gas flow rate, but decreases slightly with the increase of inlet SO_2 concentration. The optimal conditions are as follows: c_{Na^+} = 0.6-1.0 mol/L, high-gravity factor = 55-86, liquid-gas ratio = 2L/m³, empty bed gas flow rate = 1.7-2m/s. At inlet SO_2 concentrations of 1714-2285 mg/m³, the removal rate of SO_2 reaches 98% and the final SO_2 concentration is lower than 35 mg/m³. For high-concentration SO_2, the final SO_2 concentration can still be kept below 35 mg/m³ by increasing the Na^+ concentration or the liquid-gas ratio in order to meet the ultra-low emission standards. Compared with the bubble column or absorption column, the liquid-gas ratio and Na^+ concentration are smaller in RPB. For instance, when the removal rate of SO_2 is 98.1% and the outlet SO_2 concentration is lower than 35 mg/m³, the liquid-gas ratio is only 2 L/m³ and the Na^+ concentration is only 1 mol/L, which can reduce alkali consumption and scaling, as well as the operation cost.

2.4.2.2 Counter airflow shear rotating packing bed

More SO_2 is emitted from ship diesel engines with the rapid development of the shipping industry in recent years. The Annex VI of MARPOL 73/78 sets very strict limits on the emissions of SO_x from ship exhausts in an effort to reduce oceanic pollution, and the Marine Environmental Protection Committee (MEPC) believes that FGD is a preferred option for this purpose. Given the limited space in most ships and the inability of diesel engines to withstand high back pressure, it is imperative to find new desulphurization methods and devices that are specifically designed for ships. Seawater is often used to absorb SO_2 emitted from ship diesel engines because seawater is a natural alkaline solution containing bicarbonate that assists in buffering the addition of acid and the absorbed SO_2 is transformed into sulfate ions that are natural constituents of seawater. Note that the adsorption process of SO_2 is a gas film-controlled process. RPB has the potential to intensify the liquid film-controlled mass transfer process, but the gas will rotate synchronously with the packing and flows through the packing in an intact manner because of the drag force of the rotating packing on the gas. Thus, the mass transfer in the gas film is not markedly increased as that in the liquid film in RPB. In order to overcome this problem, we have developed a novel RPB called the counter airflow shear rotating packed bed (CAS-RPB), in which a number of form-drag baffles are installed at the outer edge of each packing ring and adjacent packing rings rotate in opposite directions, causing high shear stress and deformation of the gas. Continuous vortex and strong turbulence are formed in the gaps of packing rings, so that the gas is no longer intact but broken, which accelerates the renewal of the interface and consequently intensifies the gas film-controlled mass transfer process.

Seawater is considered the most economical desulfurizer for absorption of low-concentration SO_2 because it is readily available. However, when the SO_2 concentration is high, alkaline substances (e.g., sodium carbonate and sodium hydroxide) need to be supplemented because of the low sulfur capacity and activity of seawater.

Zhang et al. [14] investigated the removal of SO_2 from simulated flue gas using seawater with or without NaOH solution as the absorbent in CAS-RPB, as shown in Fig. 2.8. The SO_2 gas stored in a steel cylinder is mixed with the air supplied by a Roots blower in a mixer. The flow rates are measured using gas rotameters before mixing. The mixture is introduced into CAS-RPB. The seawater is pumped from the liquid storage tank into CAS-RPB by a centrifugal pump and the flow rate is measured using a liquid rotameter. Then, the seawater is sprayed uniformly on the inner edge of the packing through a liquid distributor at the center of the rotor and flows outwardly in the radial direction driven by the centrifugal force. The seawater is highly dispersed and smashed into smaller elements by the rotating packing. The gas flows in the radial direction from the outer edge of the packing to the inner edge of the packing driven by gas pressure and continually sheared by the rotating packing. The mass transfer takes place more dramatically between gas and liquid that flow countercurrent to each other because of the high turbulence and rapid renewal of the gas film. Finally, the gas is discharged from the gas outlet and the liquid is discharged from the liquid outlet into the liquid storage tank. Samples are collected at gas inlet

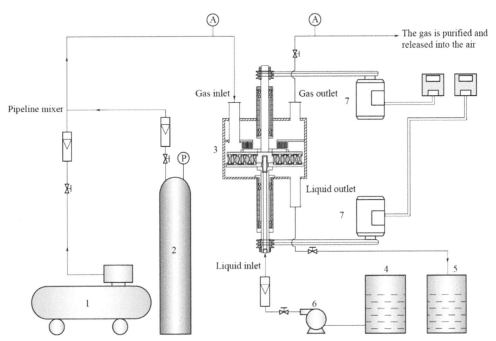

Figure 2.8 Removal of SO_2 by seawater (1—Roots blower, 2—SO_2 steel cylinder, 3—CAS-RPB, 4,5—liquid storage tank, 6—centrifugal pump, 7—frequency converter).

and outlet. The seawater in the liquid storage tank is changed with seawater supplemented with NaOH and experiments are performed under similar conditions.

Seawater is used for the removal of SO_2. The effects of alkalinity, liquid-gas ratio, high-gravity factor, SO_2 concentration, and gas flow rate on the desulphurization rate (η) and the gas-phase volumetric mass transfer coefficient ($K_G a$) under conditions of concurrent and counter-rotation of the two disks are investigated. The experimental results reveal that both the desulphurization rate and the gas-phase volumetric mass transfer coefficient increase with increasing alkalinity, liquid-gas ratio, and high-gravity factor, but they first increase and then decrease with increasing SO_2 concentration. Increasing the gas flow rate leads to a decrease in the desulphurization rate but an increase in the gas-phase volumetric mass transfer coefficient. It should be noted that the desulphurization rate and the gas-phase volumetric mass transfer coefficient under counter-rotation conditions are larger than that under concurrent rotation conditions. The optimal conditions are as follows: liquid-gas ratio should be 4 L/m³ and the high-gravity factor should be 66.3. SO_2 concentration is reduced from 610 to 50 mg/m³ with a desulphurization rate of over 92%, which meets the IMO standards. Compared with packed column operating under similar conditions, the removal rate is increased by 1.4 times and the gas-phase volumetric mass transfer coefficient is increased by 4.74 times, indicating that CAS-RPB can increase the removal rate of SO_2 from ship exhausts by increasing the gas-film controlled mass transfer process.

NaOH is added into seawater and the effects of alkalinity, liquid-gas ratio, high-gravity factor, SO_2 concentration, and gas flow rate on the desulphurization rate (η) and the gas-phase volumetric mass transfer coefficient ($K_G a$) under conditions of concurrent and counter rotation are investigated. It is found that both of them increase with increasing absorbent concentration, liquid-gas ratio, and high-gravity factor, but they first increase and then decrease with increasing SO_2 concentration. Increasing the gas flow rate leads to a decrease in the desulphurization rate but an increase in the gas-phase volumetric mass transfer coefficient, and they are larger under counter-rotation conditions than that under concurrent rotation conditions. The optimal conditions are as follows: absorbent concentration = 0.06 mol/L, liquid-gas ratio = 2.5 L/m³, and high-gravity factor = 70, under which the SO_2 concentration is reduced from 1000 to 20 mg/m³ with a desulphurization rate of over 98%, which also meet the IMO standards.

In conclusion, high-gravity technology has the advantages of a high desulphurization rate, high degree of purification, low liquid-gas ratio, low investment, and low operating cost. The RPB used for desulphurization is characterized by small size, low space requirement, low equipment height and weight, easy installation on the deck and less susceptible to environmental influences. What makes it particularly appealing is that desulphurization in a high-gravity environment is not affected by gravity, shaking and leaning, and thus it has great potential to be used for desulphurization of fume gas generated by ship diesel engines [14]. Fig. 2.9 shows the high-gravity technology for absorption of SO_2 from flue gas generated by ship diesel engines.

2.4.3 Removal of CO_2 from gas streams

CO_2 absorption is a gas-liquid mass transfer process, and the mass transfer performance of the absorbing device is related to the selectivity, removal rate, operational difficulties, energy consumption, and operating cost. The accelerated development of the petrochemical industry in China has brought about a dramatic increase in CO_2 emissions, and much effort is being made to prevent the emissions of large quantities of CO_2 into the atmosphere. Various physical methods (e.g., physical absorption, adsorption, and membrane separation) and chemical methods (e.g., absorption in activated MDEA and hot potassium carbonate solutions) have been proposed for this purpose. Each method has its own advantages: (1) Physical methods have higher absorption capacity at higher partial pressures, while chemical methods have higher absorption capacity at lower partial pressures; (2) Regeneration is achieved by means of vacuum flash evaporation for physical methods because of the relatively high desorption amount but by means of heating for chemical methods; (3) The partial pressure is high for physical methods but low for chemical methods when the dissolved quantity is very low. Thus, chemical methods are more suitable for CO_2 absorption and small-scale process.

However, most devices for decarburization are far from satisfactory due to their low absorption efficiency, complex operation, large size, and high energy consumption. How to devise new devices and absorbents and improve the current decarburization techniques have attracted much research interest. The high-gravity technology as a novel process intensification technology has promising applications in CO_2 capture

Figure 2.9 Field picture of high-gravity wet desulphurization.

because of its numerous advantages such as high decarburization efficiency, small size, and low energy consumption.

2.4.3.1 Removal of high-concentration CO_2

The N-methyldiethanolamine (MDEA) solution can be used for both physical absorption and chemical absorption with high absorption capacity (the ratio of the amount of CO_2 absorbed to the amount of absorbent used), low reaction heat, low desorption energy consumption, and high stability. However, DEA is a tertiary amine with a high steric hindrance and a slow absorption rate of CO_2. Triethylene tetramine (TETA) contains two primary amine nitrogen atoms and two tertiary amine nitrogen atoms, and the absorption rate of CO_2 is 2-3 times that of the MDEA solution. However, the poor desorption makes it difficult to be repeatedly used. The organic amine solution consisting of a small amount of TETA and MDEA can be used for the removal of CO_2 because of the high absorption rate of TETA and the low desorption energy consumption of MDEA. The reactions of CO_2 with the mixture of MDEA and TETA are described as follows.

$$CO_2 + H_2O \rightarrow H^+ + HCO_3^- \tag{2.2}$$

$$H^+ + R_3N \rightarrow R_3NH^+ \tag{2.3}$$

$$RNH_2 + CO_2 \rightarrow RNH_2^+ COO^- \tag{2.4}$$

$$RNH_2^+ COO^- + R_3N \rightarrow RNHCOO^- + R_3NH^+ \tag{2.5}$$

Reaction (2.2) is a slow liquid film-controlled reaction, reaction (2.4) is the control reaction for the absorption of CO_2 in primary and secondary amines, and reactions (2.3) and (2.5) are transient reactions. Thus, reactions (2.2) and (2.4) are the control reactions for the absorption of CO_2 in the mixed amine solution. Thus, increasing the gas-liquid mass transfer rate is crucial to increase the capture rate of CO_2. The high-gravity technology can enhance the mass transfer coefficient by 1-3 orders of magnitude because of its ability to greatly improve the gas-liquid interphase area. Thus, it facilitates the reactions (2.2) and (2.4).

We have used the high-gravity technology to remove CO_2 from the feed gas of synthetic ammonia (where the volume fraction of CO_2 is 4%) using mixed amine solution in RPB [24-27], as shown in Fig. 2.10. The high-pressure CO_2 in the CO_2 steel cylinder is decompressed by the pressure-reducing valve and then fed into the buffer tank, where it is mixed with the air supplied by the blower. The CO_2 concentration is controlled by changing the CO_2 or the gas flow rate. The gas mixture is

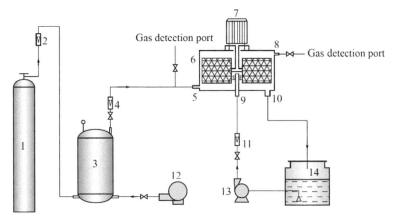

Figure 2.10 The high-gravity decarburization process (1—CO_2 steel cylinder, 2, 4—gas rotameter, 3—buffer tank, 5—gas inlet, 6—RPB, 7—variable frequency motor, 8—gas outlet, 9—liquid inlet, 10—liquid outlet, 11—liquid rotameter, 12—blower, 13—liquid pump, 14—liquid storage tank). *RPB*, Rotating packing bed.

introduced into RPB, and the absorbent is pumped from the liquid storage tank into RPB by a liquid pump and measured using a liquid rotameter. Then, the liquid is sprayed on the inner edge of the packing through a liquid distributor and flows radially through the packing driven by the centrifugal force. The gas and liquid are sufficiently contacted and react in the rotating packing. Finally, gas is discharged from the gas outlet at the top of the RPB, and liquid is thrown onto the interior wall of the casing by centrifugal force and then discharged from the liquid outlet at the bottom of the RPB under gravity.

The operating parameters are tuned (gas flow rate G = 5-15 m³/h, liquid flow rate L = 20-120 L/h, high-gravity factor β = 9-150) to investigate the effects of the formulation of the mixed amine solution, liquid flow rate, gas flow rate, liquid-gas ratio, high-gravity factor, circulating volume of solution, and temperature on the absorption efficiency. The main conclusions are as follows:

1. Effect of solution type

    Fig. 2.11 reveals that the removal rate of CO_2 by the mixed solution composed of 0.5 mol/L TETA and 1.0 mol/L MDEA is almost twice that of the MDEA solution.

2. Effect of CO_2 loading in the solution

    At room temperature, liquid-gas ratio was 8 L/m³, high-gravity factor was 120, experimental time was 100 min, and sampling interval was 10 min. The relationships between CO_2 loading and the removal rate of CO_2 for MDEA, TETA, and MDEA + TETA solutions are shown in Fig. 2.12. The removal rate of CO_2 decreases with increasing CO_2 loading in the solution in RPB. The lowest removal

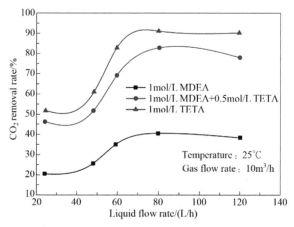

Figure 2.11 The effect of solution type on the removal rate of CO_2.

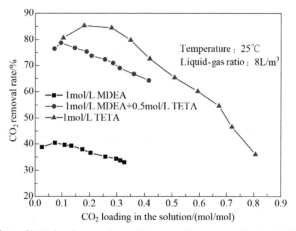

Figure 2.12 The effect of CO_2 loading in the solution on the removal rate of CO_2.

rate of CO_2 (33.5%) is observed for MDEA and the CO_2 loading is 0.335 at 100 min, while the highest removal rate of CO_2 is observed for TETA and the CO_2 loading reaches 0.8 at 100 min. TETA could not be used alone in the industry. Notably, there is a small variation in the CO_2 loading in the MDEA + TETA mixed solution, and the CO_2 loading at 100 min is 0.42, at which the removal rate of CO_2 is still 64.3%. Thus, the MDEA + TETA mixed solution has higher absorption capacity and can be easily regenerated for reuse.

3. Effect of high-gravity factor

The gas flow rate is fixed at 10 m³/h and the liquid (0.6 mol/L TETA + 1.4 mol/L MDEA) flow rate is fixed at 80 L/h, and the high-gravity factor is varied to investigate its effect on the removal rate of CO_2, as shown in Fig. 2.13. The removal rate of

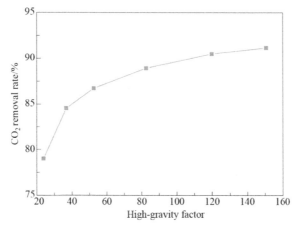

Figure 2.13 The effect of high-gravity factor on the removal rate of CO_2.

CO_2 increases at a decreasing rate with an increasing high-gravity factor. However, increasing the high-gravity factor can reduce the gas–liquid contact time because of the shorter residence time of the liquid in the rotating packing, which may reduce the total mass transfer efficiency.

The removal rate of CO_2 reaches a maximum of 82.64% at a high-gravity factor $\beta = 120$, liquid-gas ratio = 10 L/m^3, circulating volume of solution $W = 2$ L, and temperature = 30°C. Compared to traditional columns, RPB can significantly increase the mass transfer rate and absorption efficiency by increasing the total specific surface area. However, more studies are needed to better understand the desorption and regeneration processes of the solution.

The researchers at the PetroChina Daqing Chemical Research Center have also investigated the removal of CO_2 from shift gas in RPB using the mixture consisting of 60% (mass ratio) of H_2O, 28% of MDEA, and 12% of diethyl amine (R_2NH) [28]. The effects of four parameters (liquid flow rate = 50-110 L/h, rotation speed = 700-1300 r/min, reaction temperature = 50-110°C, gas flow rate = 1000-1600 L/h) are explored. It is found that (1) the optimal operating conditions are liquid flow rate of 80-90 L/h, rotation speed of 1200 r/min, reaction temperature of 80-90°C, and gas flow rate of 1200 L/h, and a removal rate of CO_2 can meet the requirements for ammonia synthesis (the volume fraction of CO_2 is 0.2%), (2) RPB has promising industrial applications because of high removal rate of CO_2, high processing capacity, low energy consumption, low operating cost, low space requirement, simple and convenient operation, and (3) at 30×10^4 t/a ammonia synthesis, the high-gravity technology can reduce the operating cost by 3.65 million RMB and the equipment investment by 19.40 million RMB compared to packed column. Thus, the high-gravity technology is economically more viable than traditional decarburization technologies.

2.4.3.2 Removal of low-concentration CO_2

The volume fraction of CO_2 is approximately 0.04% in ambient air and 0.07% in indoor air. However, the CO_2 concentration in densely populated areas, closed cabins, and confined spaces can be several times higher than the normal level. One will feel shortness of breath and palpitation at CO_2 concentrations higher than 1%, dizziness at CO_2 concentrations of 4%–5%, unconsciousness, gradual cessation of respiration, and even death at CO_2 concentrations higher than 6%. There is a need for the treatment of excess CO_2 in indoor air. We have used NaOH solution to remove low-concentration CO_2 ($\leq$1.2%), especially extremely low-concentration CO_2, in the air in a CAS-RPB [29], as shown in Fig. 2.14. It is found that at inlet CO_2 concentrations lower than 12,000 mg/m^3, the removal rate of CO_2 increases with increasing inlet CO_2 concentration, indicating that it is hard to remove low-concentration CO_2. The removal rate of CO_2 is 26.8%–30.2% at CO_2 concentrations of 1000–4000 mg/m^3, indicating that a high decarburization rate can still be obtained when the CO_2 concentration is slightly higher than that in the atmosphere. The partial pressure of CO_2 is low and the driving force for CO_2 absorption is weak when the CO_2 concentration is lower than or close to that in the atmosphere. The high decarburization rate is attributed to the ability of RPB to intensify the gas-liquid mass transfer process.

In practical applications, CO_2-containing gas can be subject to repeated decarburization processes in order to reach the requirement of air quality. This technique is applicable to air purification in densely populated areas, closed cabins, and confined spaces.

2.4.4 Removal of NO_x from flue gas

High NO_x emissions are generated in the production of explosives, nitric acid, nitrogenous fertilizers, and dyes. Unlike industrial boiler flue gas, the NO_x concentration can reach 10,000–560,000 mg/m^3. NO_x is called the "yellow dragon" in China because it is dark

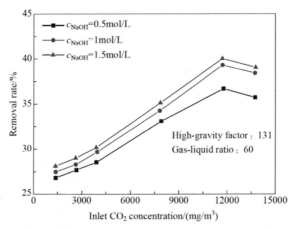

Figure 2.14 The effect of inlet CO_2 concentration on the removal rate of CO_2.

yellow in color. The compositions of NO_x emissions are very complex, mainly including N_2O, NO, NO_2, N_2O_4, and N_2O_5. Although NO_x emissions are low compared to industrial boiler flue gas, they may originate from multiple sources and diffuse very rapidly, which makes it extremely difficult to be removed. Generally, high-pressure absorption is used for the removal of nitric oxide. For instance, high-pressure column is used in PLINKE of Germany for removal of NO_x or reduction of nitric acid under pressurized conditions. A mixture of NO_x-containing gas and air is compressed from ambient pressure to 0.6-0.7 MPa using a four-stage compressor equipped with special internals for refrigeration and circulating cooling, and thus there are high requirements for the device and pipeline. As a result, the investment and operating costs are high. In China, high-pressure absorption is used only in the production of nitric acid, and wet absorption is used for the treatment of atmospheric pressure exhaust gas using packed column, bubble column, and sieve plate column. In general, a number of columns are connected in series. For instance, seven packed columns are connected in series in an explosives and powders enterprise for treatment of NO_x exhaust gas. However, nitric oxide is still present in the treated gas because of the low mass transfer of these columns and consequently low absorption rates of nitric oxide. The series connection makes the process very complex and leads to enormous investment and low efficiency. The wet absorption processes mainly include water absorption, acid absorption, alkali absorption, oxidation absorption, absorption reduction, and complex absorption, as listed in Table 2.3.

Hydration reaction is a fast diffusion-controlled reaction [e.g., $NO(g) \rightarrow NO(l)$, $NO_2(g) \rightarrow NO_2(l)$, $N_2O_3(g) \rightarrow N_2O_3(l)$], and the mass transfer rate from the gas phase to the

Table 2.3 Typical wet absorption techniques of nitric oxide.

Absorption techniques	Key technical points	Main shortcomings
Ozone/ oxidation absorption	Ozone and nitric oxide are mixed, and NO is oxidized to NO_2 and absorbed by the solution	Ozone is produced by high voltage ionization, which requires expensive equipment, high power consumption, and high cost
ClO_2/ oxidation absorption reduction	NO is oxidized to NO_2 by ClO_2 and then absorbed by the Na_2SO_3 solution to reduce NO_x into N_2	The device is prone to corrosion, and it is difficult to treat oxidant and absorbing solution
Absorption reduction	NO is reduced to N_2 using a reducing agent	The oxidation degree of NO_x has an effect on the absorption efficiency
Complex absorption	NO is fixed using $Fe^{II}EDTA$ complex and then reduced to N_2 using Na_2SO_3	The cost may be high because of the loss of complexant and regeneration

liquid phase is critical for the absorption efficiency. Currently, a major challenge is to design small devices that can significantly increase the mass transfer rate. Although there may be no NO in the flue gas, the absorbed NO_2 can be converted into NO, which is insoluble in water and thus leads to the presence of high-concentration NO in the final gas after multiple-stage absorption because of the low oxidation rate of NO. Many new absorption technologies are being explored, such as complexation and oxidation. The gas phase oxidants include O_2, O_3, Cl_2, ClO_2, etc, and the liquid phase oxidants include HNO_3, $KMnO_4$, $NaClO_2$, $NaClO$, H_2O_2, $KBrO_3$, $K_2Cr_2O_7$, Na_2CrO_4, $(NH_4)_2Cr_2O_7$, etc. The cost will be high for the treatment of high-concentration NO_x, which may make it less economically attractive in the industry. Thus, a two-stage absorption process is established based on the NO_x concentration. Specifically, the majority of NO_x is removed by the high-gravity technology that permits high gas-liquid mass transfer, followed by oxidation or complexation for absorption of low-concentration nitric oxide. Next, we describe the high-gravity technology and complexation for the removal of NO.

2.4.4.1 Removal of high-concentration nitric oxide by high-gravity technology at atmospheric pressure

We have investigated the absorption of high-concentration NO_x in water and dilute nitric acid by the high-gravity technology at atmospheric pressure [10,30], as shown in Fig. 2.15. The absorbing solution for the first stage RPB is cooled in a plate cooler in order to improve the absorption of nitric oxide at room temperature, and as the nitric acid concentration reaches a given value, the solution is transferred to the workshop for future use. The second-stage circulating fluid is transferred to the first-stage recycling tank for absorption of high-concentration nitric oxide, and process water is supplemented in order to ensure the liquid level of each recycling tank. Most nitric oxide is absorbed by the two-stage high-gravity device, leaving only a trace amount of nitric

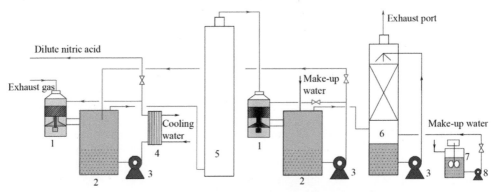

Figure 2.15 High-gravity technology for purification of exhaust gas (1—high-gravity device, 2—circulating tank, 3—circulating pump, 4—heat exchanger, 5—oxidation tank, 6—decomposing column, 7—decomposing agent preparation tank, 8—transfer pump).

oxide to be reduced into N_2 by the decomposing agent in the decomposing column. N_2O, NO, NO_2, N_2O_4, and N_2O_5 can be rapidly transferred from the gas phase to the liquid phase in the RPB so that the absorption rate and purification efficiency are significantly improved. The solution is highly dispersed in a high-gravity field, which can reduce the circulating quantity of the absorbent and the resistance of the gas phase flowing through the packing. The high-gravity technology can also reduce the power consumption for the circulating pump and blower.

Water and dilute nitric acid solution are used for the absorption of nitric oxide, and one NO molecule will be formed for every three NO_2 molecules consumed. As NO is insoluble in water and acid, an oxidation tank is placed between the two RPBs in order to prolong the residence time of the gas. NO can be oxidized to NO_2 by the oxygen presented in the exhaust gas or gaseous O_3 supplemented to improve the absorption and conversion efficiencies.

The experiments conducted in a chemical plant reveal that the absorption rate by a single RPB reaches 63.7% under appropriate conditions, which accounts for 95.5% of the theoretical absorption rate at equilibrium (66.7%) and is 23% higher than that in a single column when water is used as the absorbent. The total volumetric mass transfer coefficient is 480 times that of the sieve plate column. When A/additive (where A is the decomposing agent or oxidant) is used as the absorbent, the absorption rate reaches 85% in a single-stage RPB, indicating that the high-gravity technology can significantly improve the mass transfer efficiency of nitric oxide and thus lead to high-efficient treatment of high-concentration nitric oxide.

This technology has been successfully applied to the treatment of exhaust gas in a chemical plant in Shanxi Province, China. Two RPBs of $\phi 1200$ mm $\times$ 2800 mm are connected in series. The exhaust gas generated in the plant is first absorbed by dilute nitric acid in the first-stage RPB, and the resulting solution is recycled. When the nitric acid concentration in the solution reaches 50%, it is transferred to the acid recovery workshop and recovered for subsequent nitrification. The loss of solution is supplemented with the solution derived from the second-stage RPB. The exhaust gas is transferred to the oxidation tank for further oxidation of NO, after which it is transferred to the second-stage RPB. In case of no sufficient solution (or the solution is transferred to the first-stage RPB), water is supplemented. Finally, the exhaust gas flows into the decomposing column, where the concentration of nitric oxide is reduced. The photograph of the removal of high-concentration NO_x by the high-gravity technology at atmospheric pressure is shown in Fig. 2.16. The initial concentration of nitric oxide is 18,000 mg/m^3, which is reduced to 2000 mg/m^3 after absorption in the two RPBs and then to below 240 mg/m^3 after decomposition in the decomposing column. The final concentration of nitric oxide meets the national emission standards. The operating parameters are listed in Table 2.4. Compared to the original "three columns + absorption in alkaline solutions" technology, the total

Figure 2.16 The photograph for the absorption of high-concentration NO_x by the high-gravity technology at atmospheric pressure.

Table 2.4 The conditions and operating parameters for the absorption of high-concentration NO_x by the high-gravity technology.

Conditions	Operating parameters	Conditions	Operating parameters
Processing capacity/(m^3/h)	600–20,000	Gas flow rate/(m/s)	0.2–0.5
Inlet NO_x concentration/ (mg/m^3)	18,000–20,000	Residence time of absorbent/s	0.3
Outlet NO_x concentration/ (mg/m^3)	$\leqslant 2000$	Residence time of oxidant/s	60
Liquid-gas ratio/(L/m^3)	20	Hydraulic loading/ $[m^3/(m^2 \cdot h)]$	10–20

absorption rate is increased by 26%, 50% of dilute nitric acid is produced, and the operating cost is reduced by over 50%.

This technology can be used for the treatment of exhaust gas at room temperature and atmospheric pressure and it requires fewer devices, smaller space, and lower investment. The final treated gas meets the emission standards. Compared with high-pressure absorption, the equipment cost, investment, and operating cost are reduced by 30%, 75%, and 79%, respectively, which provide an economically attractive option for the absorption of high-concentration exhaust gas in a low-cost and high-efficient manner at atmospheric pressure. A higher driving force is expected for gas-liquid absorption under high-pressure conditions, which potentially leads to higher absorption efficiency. It should be noted that the high-gravity technology is applicable not only to absorption at atmospheric pressure but also to absorption at high pressure.

2.4.4.2 Removal of nitric oxide by high-gravity technology using Fe II EDTA complex

As described in previous sections, the low-concentration NO generated in the absorption process of nitric oxide is difficult to be oxidized, which represents a technical bottleneck for the absorption of nitric oxide in acidic or alkaline solutions. The oxidation rate of NO is proportional to the square of NO concentration ($r_{NO_2} = kc_{NO}^2$) [31]. Thus, a low oxidation rate of NO is expected at low NO concentrations and it is close to 0 at NO concentrations lower than 0.1%. Previous studies on the oxidation absorption of NO have focused on how to alter the reaction equilibrium (e.g., increasing the system pressure) and improve the reaction rate (e.g., adding catalyst or oxidant such as $NaClO_2$, H_2O_2, and $KMnO_4$). However, these methods may be of limited benefit in practice on account of the complex operating conditions and high absorbent cost. Researchers have found that NO can be removed by the complexation with transition metal complex [32-34]. At room temperature and atmospheric pressure, the complexation between Fe II EDTA complex and NO leads to the rapid absorption of NO because of the fast reaction rate and high absorption capacity. Importantly, Fe^{2+} is environmentally friendly without secondary pollution. Therefore, Fe II EDTA complex is expected to be a promising NO absorbent in the chemical industry.

Wang et al. [11] used Fe II EDTA complex as the absorbent for absorption of NO, as shown in Fig. 2.17. It is found that the absorption of NO by Fe II EDTA follows the pseudo-first-order kinetics. The effects of high-gravity factor β, initial absorbent

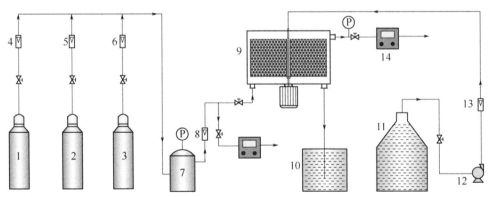

Figure 2.17 The flow chart of the high-gravity technology for the removal of NO using Fe II EDTA complex (1—NO steel cylinder, 2—N_2 steel cylinder, 3—O_2 steel cylinder, 4-6 , 8—gas flowmeter, 7—gas buffer tank, 9—RPB, 10—absorbed liquid storage tank, 11—absorbent storage tank, 12—liquid pump, 13—liquid flowmeter, 14—NO gas detector).

concentration, liquid-gas ratio L/G, gas flow rate G, inlet NO concentration, absorption temperature, and solution pH on the removal rate of η_{NO} and the gas-phase volumetric mass transfer coefficient K_Ga are investigated. The relationships between the total volumetric mass transfer coefficient and the influencing factors of interest are determined by regression.

$$K_Ga = \frac{2.625 c_B^{0.2798} \beta^{0.2730} G^{0.9144}}{c_A^{0.03201} R^{0.1427}} \qquad (2.6)$$

where K_Ga—total volumetric mass transfer coefficient, s^{-1};
c_A—NO concentration, mg/m^3;
c_B—absorbing solution concentration, mg/m^3;
β—high-gravity factor;
G—gas flow rate, m^3/h;
R—gas-liquid ratio.

The results reveal that increasing the high-gravity factor can effectively increase the absorption rate of NO and inhibit the reaction between oxygen and absorbent. Both the removal rate of NO and the total volumetric mass transfer coefficient increase with increasing high-gravity factor, liquid-gas ratio, and absorbent concentration, but decrease slowly with increasing inlet NO concentration. Increasing the gas flow rate increases the total volumetric mass transfer coefficient but reduces the removal rate of NO. However, NO absorption is inhibited at high temperatures or at too high or too low pH levels. The highest removal rate of NO ($>70\%$) is obtained when high-gravity factor is 80-90, liquid-gas ratio is 18 L/m^3, absorbent concentration is 0.02 mol/L, pH = 7.0, absorption temperature is equal to room temperature, and inlet NO concentration is 200-2000 mg/m^3 (volume concentration = 0.01%-0.1%). The absorption rate of NO is 50% higher and the total volumetric mass transfer coefficient is 147% higher compared with that in packed column, and the service life of the complex is longer in the presence of oxygen.

2.4.5 Removal of ammonia in the production of nitrophosphate fertilizers

Ammonia-containing exhaust gas is produced in large quantities in the production processes (e.g., acidolysis, filtration, neutralization, and conversion) of nitrophosphate fertilizers. The temperature of the exhaust gas can reach 70°C, and the compositions mainly include fluorine, ammonia, moisture, and nitric oxide [12], as shown in Table 2.5.

The exhaust gas generated in the production of nitrophosphate fertilizers has the following characteristics: (1) Extremely high ammonia concentration. According to the Emission Standard of Odor Pollutants (GB 14554—1993), the maximum allowable concentration is 739 mg/m^3 when the exhaust funnel height is higher than 60 m. This is achievable only when the absorption rate of ammonia is higher than 90.72%;

Table 2.5 The compositions of exhaust gas emissions in the production of nitrophosphate fertilizers by freezing method.

Location	Temperature /°C	NH_3 concentration/ (mg/m³)	NO_x concentration/ (mg/m³)	Fluorine concentration/ (mg/m³)	H_2O/%
Chimney	70	7960	751	62	20.6

(2) High moisture content—the volume fraction is 20.6% and the humidity is 0.1782 kg (water)/kg (dry gas), which is much higher than the saturated air. Thus, the exhaust gas once released into the air is like white smoke because of the sudden decrease in temperature. In order to meet the emission standards, the moisture content in the exhaust gas should be lower than the saturated moisture content in the air; (3) Presence of nitric oxide and fluorine, both of which, however, can easily meet the established standard. In conclusion, it is technically challenging for the exhaust gas generated in the production of nitrophosphate fertilizers to meet the emission standards.

The technique for separate removal of harmful components from the exhaust gas may not be desirable for most companies because of complex processing, long time, enormous investment, and operating cost. An integrated approach is needed to adsorb ammonia and moisture, the two main components of the exhaust gas. However, the absorption of ammonia in water or acidic solution in a column has the problems of low absorption efficiency, large equipment size, large space requirement, high construction and investment cost [12,35]. This is especially pronounced when there is a need to upgrade the technology or install new devices. Current dehumidification methods include cooling, adsorption, absorption, and membrane dehumidification [36]. As the exhaust gas generated in the production of nitrophosphate fertilizers is characterized by large emissions, high humidity, and high temperature, substantial heat exchange is required to cool the exhaust gas, and a key technical consideration is the cooling rate and efficiency of the exhaust gas. The concentrations of oxynitride and fluorine are low and they will be absorbed in water, implying that the emission standards can be easily met. Thus, of the utmost importance is to develop a novel technology for simultaneous deamination, denitrification, and dehumidification. Our team has proposed high-gravity technology for the absorption of exhaust gas generated in the production of nitrophosphate fertilizers, which has been successfully applied in the industry with significant economic and environmental benefits [12].

2.4.5.1 Parameter optimization and removal efficiency of the high-gravity technology

Ammonia is removed using the simulated wastewater in the production of nitrophosphate fertilizers as the absorbent in a cross-flow RPB. The experimental results have demonstrated the feasibility of the high-gravity technology for the absorption of ammonia. Then, the technical parameters are optimized. The empirical formula of the volumetric mass transfer coefficient is derived from the experimental results [12]. The mass transfer coefficient of the RPB is found to be 1734.2 times that of traditional columns.

$$k_y a_e = 6604 \frac{pDa_f}{RTR'} Re_G^{0.5} Re_L^{0.2118} \beta^{0.03354} \tag{2.7}$$

where $k_y a_e$—gas-phase volumetric mass transfer coefficient, mol/(m³·s);
p—system total pressure, kPa;
Re_G—gas Reynolds number, dimensionless;
Re_L—liquid Reynolds number, dimensionless;
D—diffusion coefficient, m²/s;
a_f—packing specific surface area, m⁻¹;
T—temperature, K;
R—molar gas constant, kJ/(kmol·K);
R'—equivalent diameter, m;
β—high-gravity factor, dimensionless.

The volumetric heat transfer coefficient is 89.9 times that of the column, indicating that the high-gravity technology has the potential to significantly improve the heat transfer rate. At a gas-liquid ratio of 1000, the removal rate of ammonia and moisture shows an increasing trend with increasing high-gravity factor from 30 to 90. At liquid flow rates of 0.2-0.8 m³/h, the removal rate of ammonia and moisture increases with increasing liquid flow rate, as shown in Figs. 2.18 and 2.19. However, the removal rate remains stable once the high-gravity factor or the liquid flow rate exceeds a given limit. Under conditions of initial ammonia concentration = 7960 mg/m³, high-gravity factor = 90, and gas inflow rate = 500 m³/h, the removal rate of ammonia can reach 91% and that of moisture can reach 62.4%, which is 92.7% of the theoretical value. Meanwhile, the outlet ammonia concentration is 716.4 mg/m³, which meets the emission standards.

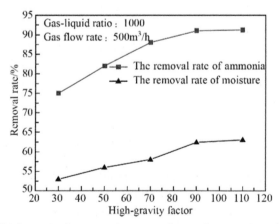

Figure 2.18 Effect of high-gravity factor on the removal rate of ammonia and moisture.

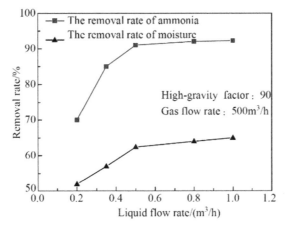

Figure 2.19 Effect of liquid flow rate on the removal rate of ammonia and moisture.

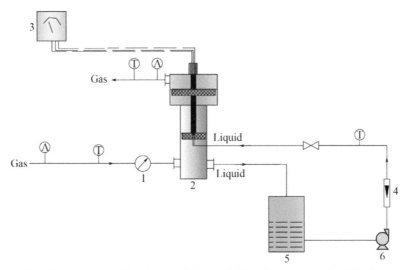

Figure 2.20 Deamination, denitrification, and dehumidification process in RPB (1—gas flowmeter; 2—RPB; 3—frequency converter; 4—liquid flowmeter; 5—liquid circulating tank; 6—liquid circulating pump). *RPB*, Rotating packing bed.

2.4.5.2 The High-gravity technology for deamination, denitrification, and dehumidification

Here, we describe our patented technology (No. ZL200710139291.2) for simultaneous deamination, denitrification, and dehumidification of the exhaust gas generated in the production of nitrophosphate fertilizers, as shown in Fig. 2.20. The RPB with two packing layers that are coaxial and of different diameters is used as the absorbing device, and the acid wastewater generated in the production of nitrophosphate fertilizers is used for cooling, deamination, denitrification, and dehumidification. Finally, ammonium salts that are derived from ammonium, nitric acid compounds that are derived from NO_x, and water that is derived from dehumidification are transferred to the acid wastewater for reuse.

The RPB with two coaxial packing layers of different diameters is schematically shown in Fig. 2.21 [37]. Note that the lower packing layer is wet, the upper packing layer is dry, and the segment between them is expanded in order to reduce the gas flow velocity. The acid wastewater is fed into the lower packing from the liquid inlet and the exhaust gas generated in the production of nitrophosphate fertilizers is fed into the lower packing from the bottom of the RPB, leading to gas-liquid contact in the rotating packing. The chemical reaction takes place between the acidic substances in the wastewater and the ammonia in the exhaust gas, and the direct contact heat exchange between gas and liquid leads to a rapid decrease in the temperature of the exhaust gas and subsequently the condensation of moisture into wastewater. NO_x is absorbed in water. As the gas flows from the lower packing to the upper packing through the expanded segment, the flow velocity is reduced. The shear of the rotating packing and strong collision leads to rapid condensation of liquid entrained in the gas and consequently a lower moisture content in the gas. Thus, simultaneous deamination, denitrification, and dehumidification can be realized in RPB.

2.4.5.3 A case for the application of the high-gravity technology

The high-gravity technology has been successfully used for deamination, denitrification, and dehumidification of exhaust gas in a company in Shanxi Province in 2009. As there is limited space for the installation of traditional absorbing devices, RPB is

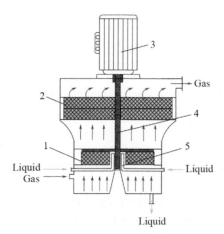

Figure 2.21 A schematic diagram of the RPB with two coaxial packing layers of different diameters (1—lower packing, 2—upper packing, 3—motor, 4—rotor shaft, 5—liquid distributor). *RPB*, Rotating packing bed.

the preferred option because of the small size and low space requirement. The processing capacity of a single RPB is 5.5×10^4 m^3/h. The acid wastewater generated in the production of nitrophosphate fertilizers is used as the absorbing solution. The removal rate of ammonia is 92% and approximately 3220 tons of ammonia could be recovered annually. The removal rate of moisture is 60.8% and approximately 43,500 tons of water is recovered annually. The pressure drop of the system is less than 1000 Pa, implying that there is no need for additional blowers. The high-gravity technology allows simultaneous deamination, denitrification, and dehumidification, and it is characterized by low space requirement, high removal rates of ammonia, NO$_x$ and moisture, convenient installation and maintenance, and low investment and operating cost. It has also been successfully applied in other companies in Shandong Province and contributes greatly to reducing exhaust gas emissions and bringing about substantial technical progress in the production of nitrophosphate fertilizers.

2.4.6 Purification of volatile organic compounds

Volatile organic compounds (VOCs) are defined by the World Health Organization (WHO) as any organic compound with a boiling point in the range of 50-260°C at room temperature [38]. In China, VOCs are defined as those organic compounds having a saturated vapor pressure of >70 Pa at room temperature and a boiling point of <260°C at atmospheric pressure, or those that are volatile and have a vapor pressure of ≥ 10 Pa at 20°C. VOCs are toxic and foul-smelling and have strong carcinogenic, teratogenic, and mutagenic potential to human [39]. Some VOCs can form photochemical smog and haze, and chlorofluorocarbon (CFC) compounds can even deplete the ozone layer. Thus, many countries have enacted laws to reduce VOC emissions into the atmosphere. In China, VOCs have been listed as one of the priority contaminants in 2011. Industrial VOCs are generally derived from solvent manufacturers and users, such as pharmaceutical production, interior decoration, spraying industry, packing and printing, petrochemical engineering, fuel combustion, vehicles, and planes. Typical VOCs emissions in some industries in China are listed in Table 2.6 [40].

In China, much progress has been made in the technology and equipment for the treatment of industrial VOCs, such as condensation, adsorption, absorption, catalysis, and membrane separation. Some novel techniques that are environmentally friendly and energy efficient and cause no secondary pollution (such as low-temperature plasma processing, photocatalysis, biodegradation, and high-gravity technology) have been developed. Absorption is one of the most commonly used methods for the treatment of industrial VOCs. The absorbed VOCs can be desorbed and recovered for future use, and the absorbent is of low cost, easily available, safe, and stable. Wastewater treatment unit is needed if water is used as the absorbent. Similar to other absorption

Table 2.6 Typical VOCs emissions in some industries in China.

Industries	Typical VOCs	Industries	Typical VOCs
Oil refining and petrochemical engineering	BTEX, 1,3-butadiene, hexane	Synthetic fiber	Formaldehyde, benzene
Storage, transport, and marketing of oil	Hexane, benzene, methylbenzene, xylene	Textile industry	Formaldehyde
Semiconductor and electronic industry	BTEX, formaldehyde, trichloroethylene, methylene chloride	Coking and steel industry	Benzene, methylbenzene, ethylbenzene
Printing and inking	Methylbenzene, xylene, ethylene glycol	Artificial board	Formaldehyde
Synthetic rubber	Methylbenzene, xylene, Benzo [a] pyrene	Architectural coating	Methylbenzene, formaldehyde, acetaldehyde
Painting	Methylbenzene, xylene, methylene chloride, ethylene glycol	Glass fiber-reinforced plastics	Styrene
Adhesives	Formaldehyde, methylbenzene, xylene, carbon tetrachloride	Waterproof coil	BTEX, Benzo [a] pyrene

processes, absorption of VOCs should also take into account the coupling between absorbent and RPB. In the following section, we will discuss the application and efficiency of high-gravity technology for absorption of VOCs using different absorbents.

2.4.6.1 Recovery of acetic acid in the production of energetic compounds

In the production of energetic compounds, the high temperature in the nitrification process leads to heat accumulation and consequently volatilization of acetic acid from the reaction vessel. Because the operation is intermittent, the exhaust gas emissions and acetic acid concentrations vary considerably over time. However, no high-efficient, low-cost method is available, and the exhaust gas has to be discharged directly into the air, causing a waste of acetic acid on the one hand and potential adverse consequences for the environment and workers on the other. The inherent advantages of the high-gravity technology, such as high absorption efficiency, small equipment size, rapid start-up and shut-down, and high operating flexibility, make it very suitable for the treatment of exhaust gas generated in intermittent operation. For instance, the rapid start-up and shut-down of the RPB can ensure an immediate response to the start-up and shut-down of the production line; and the high absorption efficiency and operating flexibility can ensure stable absorption in case of intermittent exhaust gas emissions.

RPB is used for the absorption of exhaust gas containing high-concentration (60 g/m^3) of acetic acid in a company, as shown in Fig. 2.22. The gas inlet of the RPB is directly connected to the gas outlet of the workshop, and the treated gas is discharged from the original exhaust funnel. The processing capacity is 4500 m^3/h. In the RPB, the stainless steel porous corrugated panel modified hydrophilic packing is used, the high-gravity factor is 50-90, and water is used as the absorbent. When the acetic acid concentration in the solution reaches about 70%, it is transferred to the rectification column for recovery of acetic acid. Note that the purity of the recovered acetic acid is over 99% and a total of 527 tons of acetic acid could be recovered every year. The high-gravity technology provides a promising solution for the recovery of solvent.

2.4.6.2 Absorption of volatile methanol

The high-purity isobutylene derived from the pyrolysis of methyl tertiary butyl ether (MTBE) is an important raw material for the production of gasoline alkylate. It is estimated that over 70% of isobutylene is produced from the pyrolysis of MTBE in the world. As MTBE can be cracked into isobutylene (which is insoluble in water) and methanol (which is highly soluble in water), water can be used for absorption of methanol in order to obtain high-purity isobutylene.

Figure 2.22 Recovery of acetic acid generated in the production of energetic compounds by the high-gravity technology.

Currently, packed or plate columns are often used for the removal of methanol. Since the gas-liquid mass transfer rate is low for a single column, a number of columns are connected in series, but this brings about new problems to be addressed, such as an increase in gas phase resistance, high quantity of absorbing solution, high energy consumption, and high production cost. In order to facilitate the absorption process of methanol that is gas film controlled, Du et al. [41] developed a novel type of RPB, multilevel cross-flow rotating packed bed (MC-RPB), which is shown in Fig. 2.23. MC-RPB is composed of two rotors and one stator. Specifically, the upper and lower packing are rotatable (rotor), and the middle packing is fixed (stator) and spaced apart from the upper and lower packing. The gas flows axially through the three packing layers and the gaps between them, and the liquid flows radially through the upper and lower packing driven by the centrifugal force and comes into contact with the gas passing through the packing. Such a configuration can significantly improve the gas-liquid slip velocity and contact time. Notably, as the gas flows from the lower rotating packing to the fixed packing and then to the upper rotating packing, the flow direction changes suddenly and dramatically because of the shear effect of the rotating packing, leading to high turbulence. As a result, the mass transfer rate between gas and liquid is increased and subsequently, the gas film-controlled absorption process of methanol is enhanced.

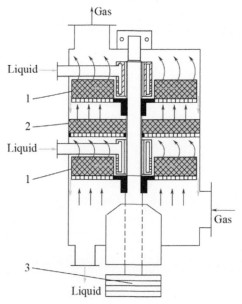

Figure 2.23 A schematic diagram of the multiple-stage cross-flow RPB (1—rotor, 2—stator, 3—motor). *RPB*, Rotating packing bed.

In order to compare the gas phase turbulence between stator and rotor, the total volumetric mass transfer coefficient (K_Ga) and the removal rate of methanol in the presence or absence of a stator are determined, as shown in Figs. 2.24 and 2.25, respectively. Under the same conditions, the total volumetric mass transfer coefficient and the removal rate of methanol are increased by 1.33-2.36 times and 1.15-1.54 times in the presence of a stator, respectively. This is because the static packing facilitates the redistribution of the gas, increases the disturbance and dispersion of the gas and its slip velocity relative to the packing, and reduces the gas side resistance to significantly increase the mass transfer efficiency of the gas phase.

Under similar conditions, the total volumetric mass transfer coefficient in the MC-RPB using Pall ring as the packing is 1.1-3.9 times that in the countercurrent plate packed bed [42] and 2.0-7.7 times that in the cross-flow plate packed bed [43], and the liquid-gas ratio is 1/3 of that in the cross-flow plate packed bed. These results have demonstrated that MC-RPB can effectively intensify the mass transfer process of

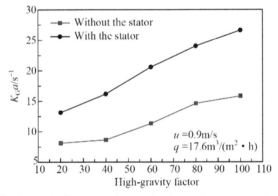

Figure 2.24 Effect of high-gravity factor on the total volumetric mass transfer coefficient.

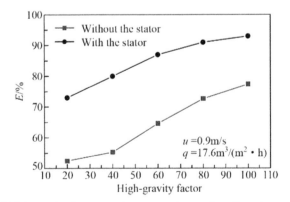

Figure 2.25 Effect of high-gravity factor on the removal rate of methanol.

methanol and reduce the quantity of the absorbing solution and power consumption. The empty bed gas flow rate is 3-10 times that in the cross-flow plate-packed bed, implying that the processing capacity is significantly improved.

For MC-RPB using structured metal gauze as the packing, the amount of the absorbing solution is reduced by 57.0% compared to the scrubber under similar conditions. When the inlet mass fraction of methanol is 1.4%, MC-RPB can take the place of one scrubber, and the height is reduced to 1/10, which makes it easier for industrial scale-up. It is applicable to a wide range of gas velocities compared to traditional columns and avoids liquid dumping. The radial size of the packing is smaller. Thus, MC-RPB has promising applications in the absorption of methanol from the gas mixture of methanol and isobutylene [44].

2.4.6.3 Absorption of other volatile gases

Chen and Liu [6] investigated the absorption of isopropanol, acetone, and ethyl acetate in water in a countercurrent-flow RPB. The experimental setup is shown in Fig. 2.26. It is found that the high centrifugal force in the countercurrent-flow RPB increases the effective interfacial area between gas and liquid phases, which in turn increases the gas-phase volumetric mass transfer coefficient and the mass transfer between gas and liquid phases. The empirical formula for the gas-phase volumetric mass transfer coefficient in the countercurrent-flow RPB is derived.

$$\frac{K_G a H_y^{0.27} RT}{D_G a_t^2} = 0.077 \, Re_G^{0.323} Re_L^{0.328} Gr_G^{0.18} \tag{2.8}$$

where a_t—total packing specific surface area, m²/m³;

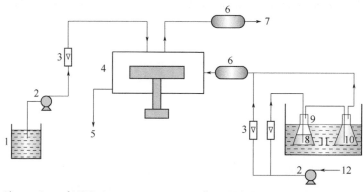

Figure 2.26 Absorption of VOCs in a countercurrent-flow RPB (1—water storage tank, 2—centrifugal pump, 3—flowmeter, 4—RPB, 5—liquid outlet, 6—gas detection, 7—gas outlet, 8—VOC solution, 9—bubbler, 10—buffer tank, 11—water bath, 12—air inlet). *RPB*, rotating packing bed; *VOCs*, volatile organic compounds.

D_G—gas diffusion coefficient, m^2/s;
H_y—dimensionless Henry constant;
Gr_G—gas Grashof number.

In their pilot test, Lin et al. [45] investigated the absorption of isopropanol from exhaust gas in a cross-flow RPB and compared the effects of rotation speed, gas flow rate, and liquid flow rate on the gas-phase volumetric mass transfer coefficient with that in a countercurrent-flow RPB. The results reveal that the mass transfer unit height of the cross-flow RPB is higher than that of the countercurrent-flow RPB, and the gas-phase volumetric mass transfer coefficient is increased by 13-77 times. The removal rate reaches 95% under conditions of gas flow rate = 150-300 m^3/h, gas-phase volumetric mass transfer coefficient = 81-165 s^{-1}, mass transfer unit height = 2.5-3.6 cm, and initial isopropanol concentration = 100-300 mg/m^3.

Chiang et al. [46] investigated the absorption of hydrophobic organic compounds (methylbenzene and xylene) in silicone oil, an organic solvent with a high boiling point, in a cross-flow RPB, where the inner diameter, outer diameter, and height of the packing is 1.3, 5.45, and 10.4 cm, respectively. It is found that the removal rate of methylbenzene and xylene waste reaches 98% when the flow rate of silicone oil is 500 mL/min, the flow rate of methylbenzene and xylene is 10.4 L/min, the initial concentration is 1200 mg/m^3, and the rotation speed is 1600 r/min, indicating that even a small-sized RPB can have excellent absorption efficiency.

Lin et al. [47] investigated the mass transfer process of VOCs (ethanol, acetone, and ethyl acetate) in water in an RPB with blade packing and the effects of gas flow rate, liquid flow rate, and rotation speed on the gas-phase volumetric mass transfer coefficient. The results show that the gas-phase volumetric mass transfer coefficient increases with increasing rotation speed, gas flow rate, and liquid flow rate. Rotation speed has no effect on the gas-phase volumetric mass transfer coefficient, but the gas-liquid effective contact area is higher at higher rotation speeds and it increases as a function of the 0.55 power of the rotation speed. Thus, the high rotation speed is the main reason why RPB is capable of intensifying the mass transfer process of VOCs.

Lin et al. also investigated the absorption of methanol, *n*-butyl alcohol, and the mixture of methanol and *n*-butyl alcohol in water in an RPB with blade packing [48,49]. They found that the removal rate of methanol and *n*-butyl alcohol increases with increasing rotation speed and liquid flow rate, but decreases with increasing gas flow rate. Their total volumetric mass transfer coefficients increase with increasing rotation speed, gas flow rate, and liquid flow rate, but the gas flow rate has a more pronounced effect than the liquid flow rate. The absorption of methanol and *n*-butyl alcohol is controlled by the gas phase mass transfer. Finally, the mass transfer unit height for absorption of methanol and *n*-butyl alcohol decreases with increasing rotation speed and liquid flow rate but increases with increasing gas flow rate.

2.5 Future perspectives

The most notable advantage of high-gravity absorption processes is the high mass transfer rate, which implies that the same task can be accomplished using a smaller device in an energy- and cost-effective way. The RPB can decrease the liquid transfer height to 3-4 m and the energy consumption to 70%. It can also lower the gas pressure drop and needs no additional blowers to pressurize the exhaust gas. The gas is directly introduced into the RPB for subsequent treatment, which avoids some unnecessary investment in equipment and operation cost. RPB will operate smoothly for several minutes after startup, and it is also easy to shut down. It requires minimal maintenance because the high-speed rotation prevents the packing from blockage and enables the packing to self-clean, leading to an extended maintenance time. RPB has a wide range of applications and it is particularly suitable in special circumstances with a limited space because of low space requirements. Despite a growing interest and recognition of high-gravity technology as a promising process intensification technology, more studies are needed to promote the industrialization and application ranges. The high-gravity technology can be coupled with other techniques according to the requirements of different systems to make full use of their distinct advantages and maximize their technical and economic benefits.

References

[1] Liu YZ. High-gravity Chemical Process and Technology. Beijing: National Defence Industry Press, 2009.
[2] Burns JR, Ramshaw C. Process intensification: visual study of liquid maldistribution in rotating packed beds. Chemical Engineering Science, 1995,51(8):1347-1352.
[3] Jiao WZ, Liu YZ, Qi GS. Gas pressure drop and mass transfer characteristics in cross flow rotating packed bed with porous plate packing. Industrial & Engineering Chemistry Research, 2010,49 (8):3732-3740.
[4] Chen ZQ, Xiong SX, Wu JG. Helical rotating absorber. CIESC Journal, 1995,46(3):388-391.
[5] Wang DC, Jia Y, Du YG. Measurement of gas-liquid interfacial area of the plate rotating column by chemical methods. Chemical Engineering (China), 1985,2:30-35.
[6] Chen YS, Liu HS. Absorption of VOCs in a rotating packed bed. Industrial & Engineering Chemistry Research, 2002,41(6):1583-1588.
[7] Chen HH, Deng XH, Zhang JJ. The effective gas-liquid interfacial area and volumetric mass-transfer coefficient measured by chemical absorption methods in rotating packed bed with multiple spraying under centrifugal force. Chemical Reaction Engineering and Technology, 1999,15(1):97-103.
[8] Zhao HH. The Fluid Mechanics and Mass Transfer of Cross-flow Multistage Rotating Packed Bed. Taiyuan: North University of China, 2004.
[9] Shi J, Wang JD, Yu GC. Chemical Engineering Handbook. Beijing: Chemical Industry Press, 1996.
[10] Li P. Treatment of High-concentration Nitrogen Oxides by High-gravity Technology. Taiyuan: North University of China, 2007.
[11] Wang F. Research of the Nitric Oxide Complexing Absorbed by $Fe^{II}EDTA$. Taiyuan: North University of China, 2015.
[12] Meng XL. Foundation Study on Ammonia Removing and Dehumidifying from Nitro-Phosphate Tail Gas by High-gravity Technology. Taiyuan: North University of China, 2008.

[13] Yuan ZG. Study on Regenerable Flue Gas Desulfurization by Sodium Phosphate Buffer Solution in a Rotating Packed Bed. Taiyuan: Taiyuan University of Technology, 2014.
[14] Zhang FF. Removal of SO_2 in Ship Flue Gas in the Counter Airflow Shear Rotating Packed Bed. Taiyuan: North University of China, 2016.
[15] Gu DY. The Effect of Form-Drag Baffle on the Mass Transfer Performance of Counter Airflow Shear Rotating Packed Bed RPB. Taiyuan: North University of China, 2015.
[16] Jiao WZ, Yang PZ, Qi GS, et al. Selective absorption of H_2S with high CO_2 concentration in mixture in a rotating packed bed. Chemical Engineering & Processing: Process Intensification, 2018,12 (7):142-147.
[17] Qiu SH, Liu YZ, Han JZ. Selective removal of hydrogen sulfide from gas in high-gravity environment. Natural Gas Chemical Industry, 2011,36(1):30-33.
[18] Zou HK, Sheng MP, Sun XF, et al. Removal of hydrogen sulfide from coke oven gas by catalytic oxidative absorption in a rotating packed bed. Fuel, 2017,204(9):47-53.
[19] Zhu ZF, Liu YZ, Luo Y. Hydrogen sulfide removal by iron chelate based method in a cross-flow rotating packed bed. Natural Gas Chemical Industry, 2014,39(1):77-81.
[20] Qi GS, Liu YZ, Jiao WZ. Study on industrial application of hydrogen sulfide removal by wet oxidation method with high-gravity technology. China Petroleum Processing & Petrochemical Technology, 2011,13(4):29-34.
[21] Zhao XD, Sun WF, Xu YM. The application of high-gravity technology in disposal of hazardous chemicals gas. Safety Health & Environment, 2016,16(7):1-4.
[22] Jiang XP. Flue Gas Desulfurization by Sodium Citrate in High-gravity Rotating Packed Bed. Taiyuan: North University of China, 2011.
[23] Yu NN. Deep Removal of Sulfur Dioxide in Flue Gas by High-gravity Wet Method. Taiyuan: North University of China, 2013.
[24] Xing YQ, Liu YZ, Wang QC. Experimental study on the removal of carbon dioxide from feed gas for ammonia synthesis by high-gravity method. Natural Gas Chemical Industry, 2007,33(1):29-33.
[25] Xing YQ, Liu YZ, Wang QC. Experimental study on the removal of carbon dioxide from feed gas for ammonia synthesis in a rotating packed bed. M-Sized Nitrogenous Fertilizer Progress, 2008,3:55-58.
[26] Zhang J. The Absorption of Carbon Dioxide in Ammonia Solution to Form Ammonium Bicarbonate Crystal in a High-gravity Field. Huainan: Anhui University of Science and Technology, 2006.
[27] Liu YZ, Shen HY. Carbon Dioxide Emission Reduction Process and Technology—Solvent Absorption. Beijing: Chemical Industry Press, 2013.
[28] Zeng QY, Bai YJ, Yang CJ. Removal of carbon dioxide from shift gas with high-gravity technology and its application prospects. Natural Gas Chemical Industry, 2010,35(1):23-54.
[29] Shi XJ, Liu YZ, Qi GS. Experimental research on high-gravity method treatment of excessive indoor CO_2. Chemical Industry and Engineering Progress, 2014,33(4):1050-1053.
[30] Li Y, Liu YZ, Zhang LY, et al. Absorption of NO_x into nitric acid solution in rotating packed bed. Chinese Journal of Chemical Engineering, 2010,18(2):244-248.
[31] Hüpen B, Kenig EY. Rigorous modelling of absorption in tray and packed columns. Chemical Engineering Science, 2005,60(22):6462-6471.
[32] Weisweiler W, Blumhofer R, Westermann T. Absorption of nitrogen monoxide in aqueous solutions containing sulfite and transition-metal chelates such as Fe (Ⅱ)-EDTA, Fe (Ⅱ)- NTA, Co (Ⅱ)-Trien and Co (Ⅱ)-Treten. Chemical Engineering and Processing: Process Intensification, 1986,20(3):155-166.
[33] Chien TW, Hsueh HT, Chu BY. Absorption kinetics of NO from simulated flue gas using Fe (Ⅱ) EDTA solutions. Process Safety & Environmental Protection, 2009,87(5):300-306.
[34] Sada E, Kumazawa H, Hikosaka H. A kinetic study of absorption of nitrogen oxide (NO) into aqueous solutions of sodium sulfite with added iron(Ⅱ)-EDTA chelate. Industrial & Engineering Chemistry Fundamentals, 1986,25(3):386-390.
[35] Meng XL, Liu YZ, Jiao WZ. Research on treatment of ammonia in phosphate tail gas by rotating packed bed. Chemical Industry and Engineering Progress, 2008,27(2):308-310.

[36] Jiao WZ, Meng XL, Liu YZ. Study on dehumidification of gas by rotating packed bed. Natural Gas Chemical Industry, 2011,36(3):17-20.
[37] Liu YZ, Jiao WZ, Qi GS. Technology and device for deamination, denitrification and dehumidification of tail gas in the production nitrophosphate fertilizers. CN 200710139291.2, 2009.
[38] Lu ZX. Purification Techniques of Organic Emissions. Beijing: Chemical Industry Press, 2011.
[39] Liu P, Zhou XM. Recycle of VOC and its treatment technology. Environmental Protection in Petrochemical Industry, 2004,24(3):39-42.
[40] Chen Y, Li LN, Yang CQ. Countermeasures for priority control of toxic VOC pollution. Environmental Science, 2011,32(12):3469-3475.
[41] Du J. Mass Transfer Performance of Methanol Absorption in Layered Packing Cross-Flow Rotating Beds. Taiyuan: North University of China, 2017.
[42] Hsu LJ, Lin CC. Binary VOCs absorption in a rotating packed bed with blade packings. Journal of Environmental Management, 2012,98:175-182.
[43] Chen YS, Hsu LJ, Lin CC, et al. Volatile organic compounds absorption in a cross-flow rotating packed bed. Environmental Science & Technology, 2008,42(7):2631-2636.
[44] Zhou JD, Liu M. Optimization of demethanol separation process for isobutylene preparation process. Petrochemical Design, 2002,19(4):18-20.
[45] Lin CC, Wei TY, Hsu SK, et al. Performance of a pilot-scale cross-flow rotating packed bed in removing VOCs from waste gas streams. Separation & Purification Technology, 2006,52(2):274-279.
[46] Chiang CY, Liu YY, Chen YS, et al. Absorption of hydrophobic volatile organic compounds by a rotating packed bed. Industrial & Engineering Chemistry Research, 2012,51(27):9441-9445.
[47] Lin CC, Chien KS. Mass-transfer performance of rotating packed beds equipped with blade packings in VOCs absorption into water. Separation & Purification Technology, 2008,63(1):138-144.
[48] Lin CC, Lin YC, Chien KS. VOCs absorption in rotating packed beds equipped with blade packings. Journal of Industrial & Engineering Chemistry, 2009,15(6):813-818.
[49] Lin CC, Lin YC, Chen SC, et al. Evaluation of a rotating packed bed equipped with blade packings for methanol and 1-butanol removal. Journal of Industrial & Engineering Chemistry, 2010,16(6):1033-1039.

CHAPTER 3

Desorption

Contents

3.1 Intensification of heat and mass transfer in thermal desorption by high gravity	76
3.1.1 Theory of simultaneous heat and mass transfer	76
3.1.2 Characteristics of simultaneous heat and mass transfer	79
3.2 Key technologies	80
3.2.1 Matching of desorption and high-gravity technology	80
3.2.2 Development of rotating packing bed	81
3.3 Application examples	82
3.3.1 Stripping of ammonia in the denitration of flue gas in power plants	82
3.3.2 High-gravity technology for stripping of ammonia nitrogen wastewater	84
3.3.3 Devolatilization of polymers	87
3.3.4 Extraction of bromine from brine	91
3.3.5 Recovery of iodine in wet-process phosphoric acid	93
3.3.6 Stripping of acrylonitrile in wastewater	95
3.3.7 Removal of nitrobenzene from high-concentration wastewater	97
3.3.8 Water deoxygenation	98
References	99

Desorption is the inverse process of absorption whereby the solute previously adsorbed by the adsorbent is released from the adsorbent. It is essentially a mass transfer process of the solute from the liquid phase to the gas phase in contact with the liquid. Desorption is commonly used for recovery of high-purity gaseous solute and regeneration of the absorbent for reuse, and it can be performed either immediately after absorption that forms a continuous absorption-desorption process or as a stand-alone process, such as stripping of ammonia nitrogen in wastewater. Desorption can be categorized into physical desorption without chemical reaction and chemical desorption with chemical reaction. As the desorption process is totally opposite to the adsorption process, any factor that facilitates the adsorption process is expected to be a limiting factor of the desorption process, and vice versa. Typical desorption methods include low-pressure desorption, thermal desorption, and stripping (air or vapor). They manage to make the equilibrium partial pressure of the solute in the liquid phase greater than that in the gas phase in order to desorb the solute from the liquid. Thermal desorption is usually performed in combination with stripping using water vapor as the desorption agent. The vapor flows upwards from the bottom to the top of the column, and the liquid flows downwards from the top to the bottom of the column and comes into countercurrent contact with the vapor. The regenerated absorbent is collected at the bottom, and vapor and solute are withdrawn from the top. If the solute is insoluble in water, the high-purity solute can be obtained by vapor condensation; but if the absorbent is also volatile, a distillation process is required to separate the solute.

In the desorption process, gas and liquid are brought into contact for mass transfer, and the limit of the thermodynamic and chemical equilibrium should be taken into consideration. How to improve the mass transfer is of utmost importance for desorption. In the chemical industry, desorption often takes place in a column and the mass transfer rate between gas and liquid phases, column size, energy consumption, and operating cost are far from satisfactory.

The high-gravity technology allows chemical processes to be carried out in a high-gravity field rather than in a normal gravity field so that the mass transfer coefficient will be increased by 1-3 orders of magnitude. This is because the liquid is sheared into micro/nano elements and the gas-liquid interface is continuously renewed in a high-gravity field [1]. In Chapter 2, we have described various absorption processes in a high-gravity field. In this chapter, we will focus on desorption, the inverse process of absorption, in a high-gravity field. It is suggested that the essential difference between desorption and absorption is the opposite mass transfer direction. The gas-liquid absorption process is an exothermic process and requires lower temperatures, while the desorption process is endothermic and requires higher temperatures. Both heat and mass transfer occur in absorption and desorption processes. Heat transfer is negligible in absorption and hence not discussed here, but it plays a crucial role in desorption and should be considered seriously. In this chapter, we first describe the mechanism of how high gravity enhances heat and mass transfer in desorption, and then introduce various desorption processes, such as stripping of ammonia nitrogen, polymer devolatilization, and extraction of bromine, in a high gravity field.

3.1 Intensification of heat and mass transfer in thermal desorption by high gravity

High gravity can enhance the heat and mass transfer required in thermal desorption. In this section, we discuss the principles and characteristics of simultaneous intensification of heat and mass transfer in a high-gravity field using hot air stripping of ammonia as an example.

3.1.1 Theory of simultaneous heat and mass transfer

Intensification of heat transfer in a high gravity field will not change the mass transfer limit, which is still dependent on the dissolution equilibrium. The driving force for mass transfer is shown in Fig. 3.1.

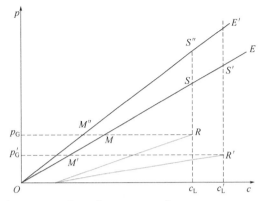

Figure 3.1 Schematic of the driving force for mass transfer.

Table 3.1 Relationship between ionization equilibrium constant of free ammonia and temperature.

T/°C	0	10	20	25	30	40	50
$K_b \times 10^5$	1.37	1.57	1.76	1.80	1.84	1.94	2.02

The straight line OE represents the equilibrium that can be described by Henry's law ($p^* = c_L/H$). At a given cross-section of a desorption device, the components of gas (p_G) and liquid (c_L) that are in contact with each other are represented by the point R (p_G, c_L). It is clear that the farther the point R is from the line OE, the greater the driving force for mass transfer, and then the higher the mass transfer rate will be. This can be achieved by increasing the slope of the line OE (increasing Henry's coefficient H by heating or reducing the pressure) and moving point R toward R' (p'_G, c'_L) in the lower right corner (increasing the free ammonia concentration c_L in the liquid phase or reducing the partial pressure p_G of ammonia in the gas phase). The dissolution equilibrium involves the desorption equilibrium of the air-ammonia-water system and the dissociation equilibrium of free ammonia and ammonium ions in the solution.

3.1.1.1 Thermodynamic analysis

There are two equilibrium states for free ammonia: ionization equilibrium and gas-liquid equilibrium, which are described by formulas (3.1) and (3.2), respectively.

$$NH_3 + H_2O \rightleftharpoons NH_4^+ + OH^- \tag{3.1}$$

$$NH_3(l) \rightleftharpoons NH_3(g) \tag{3.2}$$

For ionization equilibrium, the ionization equilibrium constants of ammonia (K_b) at different temperatures are listed in Table 3.1. The higher the temperature is, the higher the value of K_b will be. Increasing the temperature shifts the ionization equilibrium (3.1) toward the left (formation of more free ammonia).

For gas–liquid equilibrium, the variation of the solubility of ammonia in water with temperature is shown in Fig. 3.2. Ammonia is highly soluble in water (52 g of NH_3 will be dissolved in 100 g of water at atmospheric pressure and 20°C), and its solubility in water shows a decreasing trend with increasing temperature. Thus, increasing temperature enhances the transfer of free ammonia from the liquid phase to the gas phase.

3.1.1.2 Intensification of heat transfer by high-gravity

The heat transfer process in high-gravity intensified thermal desorption is essentially a direct-contact heat transfer process in which heat and mass transfer occur simultaneously, but their directions are dependent on their respective temperature and partial pressure. In a rotating packing bed (RPB), the hot gas flows from the periphery of the rotor into the packing driven by pressure and then passes through the packing in a radial direction. After that, it enters the inner cavity of the rotor and is discharged from the gas outlet. The liquid is introduced into the inner cavity of the rotor from the liquid inlet and then sprayed on the inner edge of the packing. After that, it passes through the packing driven by the centrifugal force and comes into the gas for simultaneous heat and mass transfer. The liquid flowing through the packing is sheared into very tiny liquid films, filaments, and droplets and it is subjected to repeated dispersion-coagulation cycles in the rotating packing. Finally, the liquid is discharged from the liquid outlet at the bottom of the RPB. Note that both heat and mass transfer are greatly intensified due to the increase of the interfacial area for heat and mass transfer and the decrease of the distance and resistance of heat and mass transfer. The gas and

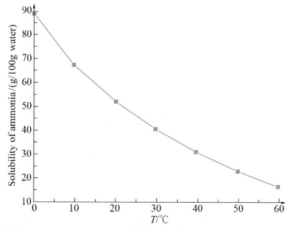

Figure 3.2 Relationship between ammonia solubility and temperature.

liquid phases are dispersed and aggregated several times in the packing, resulting in strong disturbance and a significant reduction in interphase transfer resistance. Consequently, both heat and mass transfer rates would be increased by several orders of magnitude.

The heat and mass transfer directions in thermal desorption in a cross-flow RPB are shown in Fig. 3.3. Heat is transferred from the gas phase with a higher temperature to the liquid phase with a lower temperature, increasing liquid temperature and vaporization of a small volume of liquid. The sensible heat gained by the liquid is transferred as latent heat to the gas phase with the vaporization of the solvent, resulting in a slight increase of the liquid phase temperature at a rate much lower than that of the gas phase. Simultaneously, mass is transferred in an opposite direction from the liquid phase to the gas phase, especially volatile solute [2].

3.1.2 Characteristics of simultaneous heat and mass transfer

The hot air stripping of ammonia is a typical thermal desorption process with simultaneous heat and mass transfer. Here, we summarize the characteristics of simultaneous heat and mass transfer in thermal desorption using hot air stripping of ammonia in an RPB as an example (Figs. 3.3 and 3.4). Ambient air is pumped into the buffer tank by a roots blower and then

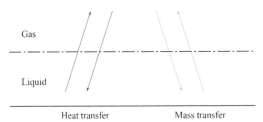

Figure 3.3 Heat and mass transfer direction in thermal desorption in a cross-flow RPB. *RPB*, Rotating packing bed.

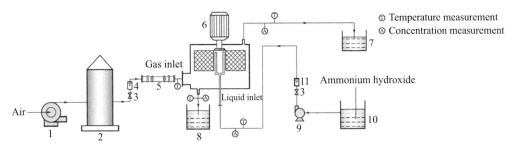

Figure 3.4 Flowchart for high-gravity intensified desorption of ammonia (1—roots blower; 2—air buffer tank; 3—valve; 4—rotameter; 5—air heater; 6—cross-flow RPB; 7—ammonia recovery tank; 8—liquid storage tank; 9—centrifugal pump; 10—ammonia storage tank; 11—liquid flowmeter).

supplied to the air heater. The flow rate is regulated by a valve and measured by a rotameter. When the air is heated to the required temperature, it is introduced into the RPB. As the hot air flows through the packing, the ammonia water is pumped from the liquid storage tank to the RPB by a centrifugal pump. The flow rate is regulated by a valve and measured by a liquid flowmeter. The liquid is uniformly sprayed on the inner edge of the packing using a liquid distributor at the center of the rotor. As the liquid flows outwards in the radial direction under the high centrifugal force, it will be sheared into micro/nano pieces, such as liquid droplets, filaments, and films, which provide a larger interfacial area for simultaneous heat and mass transfer. This along with rapid interface renewal has greatly improved the direct-contact heat exchange between hot air and ammonia and the desorption rate of ammonia in the RPB. Therefore, free ammonia is transferred more rapidly from the liquid phase to the gas phase. Ammonia can be used directly for denitration or recovery for other applications. The desorption liquid is discharged into the liquid storage tank. The temperatures of liquid and gas phases are measured in real-time by temperature sensors at the inlet and outlet.

The effects of several operating parameters on the gas phase volumetric mass transfer coefficient and the total heat transfer coefficient are investigated, and the main conclusions are summarized as follows:

1. High gravity can significantly enhance the heat transfer, and the higher temperature of inlet gas can further increase the heat transfer between gas and liquid phases.
2. The total mass transfer coefficient and heat transfer coefficient increase with increasing high-gravity factor, but the increasing rate is decreasing.
3. The total mass transfer coefficient and heat transfer coefficient increase with increasing gas flow rate. However, the increasing rate is dependent on the packing type, and the total mass transfer coefficient and heat transfer coefficient are higher for the wire mesh packing compared to randomly arranged packing.
4. Increasing the liquid spraying density decreases the total mass transfer coefficient, but increases the total heat transfer coefficient. Again, the wire mesh packing favors the mass and heat transfer in thermal desorption.

3.2 Key technologies

In order to ensure sufficient contact and mixing of gas and liquid phases and subsequently the efficient transfer of a solute from the liquid phase to the gas phase, it's important that RPB used for desorption should take advantage of high gravity technology.

3.2.1 Matching of desorption and high-gravity technology

The high gravity technology can simultaneously enhance the heat and mass transfer in the desorption process with the use of RPB which is much smaller than most conventional desorption columns. The supporting facilities of RPB should also be small and the desorption process should be compatible with the high-gravity technology, such as selecting the appropriate pump and blower, adjusting the spatial arrangement, improving the compatibility between high-gravity device and technology, and recovering solute for other applications. A key challenge is to optimize the operating parameters of RPB based on the unique characteristics of RPB and gas/liquid phases involved in desorption (e.g., short residence time and rapid heat and mass transfer). For high-viscosity materials, it is necessary to increase the temperature and reduce the viscosity in order to ensure high desorption rate because their residence time is very short and heat and mass transfer occur almost instantaneously in the RPB. For heat-sensitive materials, it is necessary to appropriately reduce the temperature and increase the flow rate in order to avoid decomposition because the mass and heat transfer rate are high and the temperature gradient is low in the RPB. For low-pressure desorption requiring high vacuum, the advantages of the high-gravity technology in terms of high mass transfer efficiency and low-pressure drop make it possible to reduce the vacuum requirement and minimize energy consumption.

3.2.2 Development of rotating packing bed

As described in previous sections, RPB falls into three distinct categories according to gas-liquid contact modes, including countercurrent-flow RPB, concurrent-flow RPB, and cross-flow RPB. The choice of the most appropriate gas-liquid contact mode depends largely on the desorption requirement and liquid-gas ratio. For instance, countercurrent contact can maximize the driving force for mass transfer in the desorption process, but the gas phase resistance is also high. For cross-flow contact, the gas phase resistance is low, but the driving force for mass transfer is also reduced. However, irrespective of the gas-liquid contact mode, the liquid always flows in a radial direction from the inner edge of the packing at the center of the rotor to the outer edge of the packing, and it is subjected to repeated dispersion-coalescence-dispersion processes that allow the liquid to be highly dispersed and sheared into micro/nano pieces. As a consequence, the interfacial area is greatly increased and the surface is rapidly renewed, both of which contribute to the diffusion of solute from the liquid phase. More emphasis should be placed on the structure of the packing. For high-viscosity liquid, the packing would be more effective if pores and channels are regularly arranged and a streamlined shape is adopted to ensure high dispersion and coalescence and reduce flow resistance. For thermal desorption, the structure should have the potential to simultaneously increase heat and mass transfer, such as regular wire mesh. For oily materials, lipophilic packing is preferred; while for water-based materials, hydrophilic packing is preferably used to ensure high dispersion in the liquid. For the desorption of non-Newtonian fluid, it is necessary for the rotor structure to be suitable for high-speed shear of the liquid in order to reduce the apparent viscosity of the non-Newtonian fluid and increase the heat and mass transfer rate. Unlike absorption, desorption is concerned with the treatment of liquid. Thus, the diameter and axial thickness of the rotor should be considered in scale-up based on the liquid flow rate and desorption requirement.

3.3 Application examples

High-gravity desorption is advantageous compared to many traditional desorption processes due to its high efficiency, small size, low cost, and easy release of solute, and it has been successfully applied to the stripping of ammonia, treatment of ammonia nitrogen wastewater, removal of formaldehyde in urea-formaldehyde adhesive, extraction of bromine from brine, and recovery of iodine.

3.3.1 Stripping of ammonia in the denitration of flue gas in power plants

Selective non-catalytic reduction (SNCR) of nitrogen oxides (NO_x) is currently one of the most widely used in-furnace denitration methods in power plants. NO_x-containing reducing agent (e.g., ammonia gas, ammonia water, and urea) is injected into the furnace or flue wherethetemperatureis900-1100°C using a spray gun in the region where the superheater is located. The agent is rapidly decomposed into NH_3, which subsequently undergoes SNCR with NO_x in the flue gas to produce N_2 and H_2O [3]. This technology is characterized by high denitration rate, low ammonia escape rate, no secondary pollution, small device size, and high stability. The ammonia gas is mainly derived from the gasification of liquid ammonia followed by air dilution, and the dilute ammonia content is about 5%. However, much concern has been expressed about the possible consequences resulting from the explosion and leakage of liquid ammonia in the storage, transportation, and application processes. In order to eliminate these risks, ammonia stripping is used instead of liquid ammonia. However, new problems will arise if traditional stripping tanks (aeration tanks) and columns are used because of the low mass transfer rate, low efficiency, large device size, high power consumption and operating cost, and more importantly, difficulty in controlling the ammonia concentration in the gas phase.

The use of high-gravity technology for stripping ammonia water leads to high heat and mass transfer rate, high stripping efficiency, small device size, low energy consumption, wide operation range, and precise control of ammonia concentration required for denitration by adjusting the liquid or gas flow rate. The mechanism is described in Section 3.1.

3.3.1.1 High-gravity stripping of ammonia water

The flowchart for high-gravity stripping of ammonia water is similar to that for high-gravity desorption, as shown in Figs. 3.3 and 3.4. The main difference is that the flue gas used does not need to be heated, and the desorbed flue gas containing ammonia and fresh flue gas is diluted to a given concentration and introduced into the denitration system. The dynamic matching between the total supply and demand of the amount is controllable to ensure the denitration efficiency of flue gas and to prevent the escape of ammonia.

3.3.1.2 Parameter optimization

1. Inlet gas temperature: With all other conditions being equal, the ammonia concentration in the gas phase increases with increasing inlet gas temperature. The stripping rate reaches a maximum of 47% at ammonia concentrations of 21.5%–31.3% and an inlet gas temperature of 120°C. Under conditions of inlet gas temperature = 30°C and ammonia concentration >10%, the use of the flue gas can meet the requirement.
2. High-gravity factor: The stripping rate and the ammonia concentration in the gas phase increase and finally stabilize with an increasing high-gravity factor. The stripping rate reaches 47% when the ammonia concentration is 24.3%–31.5% and the high-gravity factor is 58. Taking into account the energy consumption, the optimal high-gravity factor ranges from 35 to 58.
3. Gas–liquid ratio: Increasing the gas–liquid ratio leads to a linear increase in the stripping rate but a linear decrease in the ammonia concentration in the gas phase. At ammonia concentrations of 24.4%–33.3% and a gas–liquid ratio of 1000, the stripping rate reaches 71.4%. The ammonia concentration in the gas phase and the total amount of ammonia can be precisely controlled by regulating the flow rate of flue gas.
4. Evaporation of ammonia water: Under appropriate conditions (high-gravity factor = 58, gas–liquid ratio = 500, inlet gas temperature = 120°C, inlet liquid temperature = 20°C), the outlet liquid temperature is 33°C, the stripping rate is 47.3%, and the ammonia concentration in the gas phase is 31.5%. It is known from the heat and material balance that the concentration of ammonia derived from evaporation is 0.28%, which is much lower than that derived from stripping. Stripping has a more significant effect on ammonia concentration than evaporation, and ammonia is mainly derived from stripping rather than from evaporation. It is concluded that high-gravity stripping of ammonia can overcome the problems encountered in evaporation of ammonia water, such as low efficiency and high energy consumption.

3.3.1.3 Stripping efficiency

We have used the high-gravity technology for stripping ammonia in denitration of flue gas generated in circulating fluidized bed boilers in a coal gangue power plant in Shanxi Province, China, as shown in Figs. 3.3–3.5. The diameter of the RPB is only 800 mm and the height is 2.4 m. The flow rate of ammonia water is 0.1–0.8 m^3/h

Figure 3.5 The photograph for stripping of ammonia by the high-gravity technology.

and that of flue gas is 1000-2400 m³/h. The stripping rate of ammonia can reach over 90%. The NO_x concentration in the flue gas is lower than 100 mg/m³ (and even 50 mg/m³ sometimes), and the concentration of escaped ammonia is lower than 8 mg/m³. The denitration rate is increased by over 30% compared to the direct injection of ammonia water. Less ammonia is escaped, reducing the secondary pollution. This project has demonstrated the unique advantages of the high-gravity technology for stripping ammonia, such as small device size, low production cost, and high safety.

3.3.2 High-gravity technology for stripping of ammonia nitrogen wastewater

The ammonia nitrogen concentration in wastewater is much lower than that in ammonia water and the driving force for gas-liquid mass transfer is smaller. In view of this, it is extremely challenging to meet the ultra-low emission standard.

3.3.2.1 Technological principles

There is an equilibrium between ammonium ions (NH_4^+) and free ammonia (NH_3) in the wastewater ($NH_3 + H_2O = NH_4^+ + OH^-$), which is related to pH and temperature. At pH > 7, the equilibrium will shift to the left as the pH or temperature increases or as the gas phase partial pressure of ammonia decreases, which leads to the formation of more free ammonia and thus has a positive effect on the stripping. The high-gravity technology can facilitate the stripping of ammonia nitrogen from wastewater under alkaline conditions, because the ammonia nitrogen is converted to free ammonia and brought into more intimate contact with a large volume of air (to decrease the partial pressure of ammonia) in the rotating packing, and ammonia is transferred from the liquid phase to the gas phase for the removal of ammonia nitrogen.

3.3.2.2 Technological processes

The wastewater quality and quantity are adjusted in the regulating pond and the pH is adjusted to 10-11 with alkali before it is introduced into the RPB. The wastewater is uniformly sprayed on the inner edge of the packing though a liquid distributor, and it is sheared into small droplets, filaments, and films under the effect of high gravity as it flows radially through the rotating packing. Simultaneously, the air is introduced into the RPB from the gas inlet and flows in a radial direction from the outer cavity of the rotor to the inner cavity of the rotor driven by the gas pressure gradient. In this case, the mass transfer is enhanced because of high turbulence and large gas-liquid contact area, and then free ammonia is stripped out. A small amount of liquid entrained in the gas is removed by the demister and the gas is discharged after deamination in the absorbing device. Stripping is often used in combination with biochemical methods (Figs. 3.3-3.6). After stripping in the RPB, the mass concentration of ammonia nitrogen in the wastewater is reduced below 300 mg/L. When the input standard for biochemical treatment is met (multiple-stage stripping is performed if necessary), the wastewater is injected into the biochemical treatment system.

3.3.2.3 Technological characteristics

The experimental results have demonstrated that the optimal conditions for the high-gravity stripping of ammonia nitrogen wastewater are pH = 10.5-11.0, liquid-gas ratio = 0.83 L/m^3, high-gravity factor = 100, and temperature = 35-40°C [4-7]. The comparison between high-gravity and traditional stripping is shown in Table 3.2.

Notably, the high-gravity technology for stripping ammonia nitrogen wastewater has the following characteristics:

1. Low power consumption: The air supply is reduced by about 75% because the gas-liquid ratio is only 1/4 of that of the traditional stripping, and the pressure drop is also very small (<1200 Pa) in the RPB.

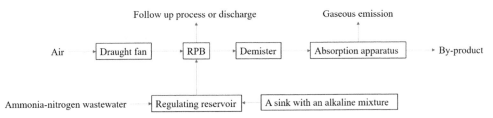

Figure 3.6 Flowchart for high-gravity stripping of ammonia nitrogen wastewater.

Table 3.2 Comparison between high-gravity and traditional stripping.

Method	Liquid-gas ratio/(L/m³)	Stripping rate/%	Bed bottom pressure drop/Pa	Device size/mm	Blower power/kW
High-gravity	0.83-1.67	75-85	850-1500	1200 × 3500	15
Traditional	0.21-0.42	20-55	3000-4600	3500 × 28000	45

Note: the flow rate of ammonia nitrogen wastewater is kept at 15 m³/h.

2. High stripping efficiency: The stripping rate is 2-3 times that obtained by traditional stripping. Under the most favorable conditions, the single-stage stripping rate is higher than 85%. At concentrations <2000 mg/L, the input standard for biochemical treatment is met after the primary treatment (300 mg/L).
3. Small size, low weight, low space requirement, and low construction cost: The RPB can be easily started up and shut down within a few minutes.
4. The gas-liquid ratio is only 1/4 of that of the traditional stripping. After stripping, the ammonia concentration in the gas is increased by 3 times, which is favorable for the recovery of ammonia. The removal rate is also high for volatile oily components and COD.
5. Self-cleaning: It is difficult for dirt, aerobe, and alga to deposit and block the RPB because of the high centrifugal force, which overcomes the problems of scaling and blockage of the packing that are often encountered in traditional columns.
6. High operating flexibility: The high-gravity device is not sensitive to changes in gas-liquid ratio and it still operates normally even under changing conditions. However, this is difficult to achieve for any other columns.
7. High gas-liquid mass transfer efficiency: RPB is able to increase the dissolution rate of oxygen and thus the concentration of dissolved oxygen in the water to provide sufficient oxygen for subsequent biochemical treatment.

The stripping rate reaches over 85% for ammonia nitrogen wastewater containing low amounts of oily substances, such as wastewater discharged in the production of rare earth, and over 75% for coking ammonia nitrogen wastewater that is difficult to treat. Note that the high-gravity technology has no special requirement for the wastewater (e.g., pH and temperature), making it easier to upgrade the traditional stripping technologies.

In summary, the high-gravity stripping technology has the advantages of high stripping rare, low-pressure drop, low energy consumption, low operating cost, low space requirement, simple operation, and ease of industrial scale-up, and it is an economically and technologically feasible approach for stripping of ammonia nitrogen wastewater in the industry.

3.3.3 Devolatilization of polymers

In the polymer manufacturing process, the polymer generally consists of approximately 10%-80% of low molecular weight volatile matter, such as monomers, solvents, and other by-products. The presence of volatile matter has an impact not only on the quality, physical and chemical properties, and applications of the polymer but also on users' health and environment. Polymer devolatilization is the process of reducing volatile matter from the polymer and it plays a key role in the production of polymer [8]. In view of the high safety, environment, and health requirements, the content of volatile matter is expected to be lower than 5% and even the level of mg/m^3. The energy consumption involved in polymer devolatilization often accounts for over 60% of the total energy consumption in the production of polymer. All these have brought about additional challenges in the devolatilization of the polymer.

The phase equilibrium principle suggests that the driving force for devolatilization can be increased by increasing the temperature or decreasing the partial pressure of the volatile matter, and the resistance for devolatilization can be decreased by increasing the heat transfer rate or the surface area and renewal rate of the polymer. It is also noted that the polymer will become more viscous in the case of high heat and mass transfer and low volatile content, which makes it more difficult for devolatilization. Flash evaporation, bubble devolatilization, and diffusion devolatilization are used for polymers with low to high volatile contents, respectively, and typical devolatilizers include flash tank devolatilizer, falling-strand devolatilizer, slit devolatilizer, film devolatilizer, single-screw devolatilizer, and twin-screw devolatilizer. However, these devolatilization techniques have low devolatilization rate, poor efficiency, low processing capacity, high energy consumption, and limited application. The surface is rapidly renewed with the high-speed rotation of the screw devolatilizer, which can accelerate bubble nucleation, growth, and rupture. Thus, it is particularly applicable to the removal of volatile matter with very high viscosity. Because of negative pressure operation, venting up occurs at the air outlet and the devolatilization rate is generally low. The residence time is long because of the high viscosity, which can easily cause the degeneration of the polymer. The falling-strand devolatilizer is simple, reliable, and cost-efficient, but it is only applicable to the polymer with very low viscosity and high mobility because it relies on the gravity for surface renewal and removal of volatile. Thus, it is large in size and shows a devolatilization rate of only 30%-40%.

Compared to conventional desorption processes, the main difficulties for polymer devolatilization are as follows: (1) High viscosity—the viscosity becomes remarkably higher as the devolatilization continues, which in turn makes devolatilization more difficult; (2) Low mass transfer coefficient—this means that a larger device is needed and a longer residence time of the polymer is expected; (3) High temperature—a high temperature is required to improve the devolatilization efficiency, which may lead to color changes and degradation of the polymer. The high-gravity technology is beneficial for polymer devolatilization in such a way that (1) the high-speed rotating packing shears the polymer into micro/nano liquid droplets, filaments, and films and increases the gas-liquid interfacial area and consequently the heat and mass transfer rate; (2) due to the short residence time of the polymer in the RPB, the removal rate can be increased by increasing the temperature to reduce its viscosity; and (3) the high centrifugal force increases the heat and mass transfer as it can reduce the apparent viscosity of the polymer.

Here, we take the devolatilization of urea-formaldehyde adhesive as an example to illustrate how high-gravity technology improves polymer devolatilization.

3.3.3.1 Technological principles

The urea-formaldehyde adhesive has found many applications in the manufacture of wood-based boards (e.g., fiberboard and plywood) and wood processing industry because of low cost, ease of use, and high bonding strength, and it accounts for approximately 80% of wood adhesives. In general, the urea-formaldehyde adhesive contains 1.5% of free formaldehyde that can be released into the atmosphere during service life [9,10] and cause indoor air pollution. The high-gravity technology has been demonstrated to be effective in removing formaldehyde from the urea-formaldehyde adhesive. Inert gas (e.g., air, CO_2, and N_2) is fed into the RPB as the foaming agent. Formaldehyde volatilizes easily from the adhesive because of its low boiling point (19.5°C) and is mixed with the inert gas, and mass and heat transfer occurs between the gas mixture and the adhesive solution. The addition of inert gas can also reduce the partial pressure of formaldehyde and accelerate its diffusion in the adhesive solution. When heated, the high-viscosity adhesive solution exhibits low apparent viscosity and rapid surface renewal because of the high shear force in the RPB, which increases the gas-liquid mass transfer coefficient and the volatilization of formaldehyde from the adhesive solution. The stripping with inert gas leads to the evaporation of some moisture from the heated adhesive solution and thus formaldehyde is released into the atmosphere. Heating can also lead to the release of formaldehyde from the thermal decomposition of free formaldehyde and unstable structures (e.g., hydroxymethyl and ether bonds) in the urea-formaldehyde adhesive.

Because of the high-speed rotation of the packing, the adhesive solution is dispersed into tinny liquid droplets, filaments, and films, as well as the gas-liquid contact area is greatly increased. According to the surface renewal theory, a thin liquid film is formed on the rotating packing and the surface renewal rate and mass and heat transfer coefficient are greatly increased, and consequently, a higher devolatilization rate is obtained. We have developed a series of RPBs for the effective removal of formaldehyde in the urea-formaldehyde adhesive [11,12].

3.3.3.2 Technological processes
The removal of formaldehyde in the urea-formaldehyde adhesive containing 1.5% of free formaldehyde by the high-gravity technology is shown in Fig. 3.7.

3.3.3.3 Operating conditions
1. Preheating temperature: Increasing the preheating temperature can reduce the viscosity and density of the urea-formaldehyde adhesive and increase the mass transfer coefficient. It also accelerates the formation, growth, and coagulation of formaldehyde monomer bubbles and the diffusion of formaldehyde monomers in the solution. Thus, the removal rate of formaldehyde increases as the temperature increases, but it shows less changes at too high temperatures.
2. High-gravity factor: The removal rate of formaldehyde increases with increasing high-gravity factor, but it remains unchanged when the high-gravity factor is greater than 75. A large mass transfer area and rapid surface renewal are required to accelerate the diffusion of free formaldehyde in the adhesive because of the high viscosity of the adhesive [13]. As RPB is capable of improving film formation and surface renewal, increasing the high-gravity factor favors the removal of volatile matter in materials with high viscosity.
3. Feed loading: All other conditions being equal, the removal rate of formaldehyde first increases and then decreases as the feed loading increases. The feed loading is a determining factor of film formation when the rotation speed of the packing is kept constant. The packing will not be sufficiently wetted at low loading, leading to poor film formation and low surface renewal; while the film will be too thick at high loading, leading to a reduction of the residence time in the packing. Thus,

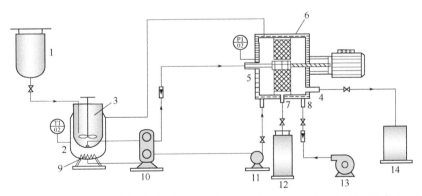

Figure 3.7 Removal of formaldehyde in the urea-formaldehyde adhesive by the high-gravity technology (1—raw material tank; 2—raw material preheating tank; 3—stirrer; 4—gas outlet; 5—liquid inlet; 6—jacket; 7—liquid outlet; 8—gas inlet; 9—heating wire; 10—gear pump; 11—hot-water pump; 12—product tank; 13—air blower; 14—absorption liquid tank).

too low or too high loading is not good for the diffusion of formaldehyde monomers in the adhesive.
4. Air flow rate: The removal rate of formaldehyde first increases and then decreases with the air flow rate when the liquid flow rate is kept constant. Increasing the air flow rate will decrease the partial pressure of formaldehyde vapor and increase its diffusion velocity, leading to an increase in the removal rate. As the air flow rate further increases, the gas-liquid ratio is increased, causing entrainment and consequently lower mass transfer of formaldehyde. Thus, the airflow rate should be controlled within an appropriate range in order to avoid entrainment.

3.3.3.4 Removal efficiency

The optimal conditions for the removal of free formaldehyde in urea-formaldehyde adhesive are preheating temperature = 80°C, devolatilization temperature = 85°C, gas-liquid ratio = 100-300, and high-gravity factor = 75-100. The transient high temperature would have no adverse effect on the properties of the urea-formaldehyde adhesive. After treatment at a transient high temperature, the formaldehyde resulting from the decomposition of hydroxymethyl and methylene ether bonds is removed from the adhesive, the resulting adhesive can be stored stably, and the time needed for dehydration is significantly reduced. The averages of various operating parameters obtained under optimal conditions are listed in Table 3.3. It is seen that the average content of free formaldehyde is 0.23% after devolatilization, which is lower than the national standard (0.3%). Other properties are also significantly improved.

3.3.3.5 Technological advantages

The high-gravity technology can be applied for the removal of free formaldehyde in the urea-formaldehyde adhesive. For the urea-formaldehyde adhesive containing 1.5% of formaldehyde, the final formaldehyde concentration is lower than the national standard without affecting the properties of the adhesive. The devolatilization rate is above

Table 3.3 Comparison of the properties of the urea-formaldehyde adhesive before and after devolatilization.

Properties	Before devolatilization	After devolatilization	Properties	Before devolatilization	After devolatilization
pH	7.92	8.01	Solid content/%	52.3	62.5
Free formaldehyde content/%	1.48	0.23	Solidification time/s	55.5	48.5
Viscosity/mPa · s	380.5	419.8	Working life/h	4.5	10

85%, the gas phase pressure drop is ⩽1000 Pa, and the residence time is ⩽1 s. Compared with other devices, RPB has the advantages of high devolatilization rate, low energy consumption, short residence time, and low investment and operating cost.

3.3.4 Extraction of bromine from brine

Bromine is an important raw material for a wide variety of high-value-added inorganic and organic chemicals used in flame retardant, medicine, pesticide, printing and dyeing, oilfield exploitation, and military field [14-16]. Bromine is traditionally recovered from the brine by vapor distillation or air blowing, and some new methods are being used, such as resin adsorption, gas membrane, and emulsion liquid membrane [17-20]. In China, the application of vapor distillation is limited because of large steam consumption and high cost, and over 90% of bromine is recovered by air blowing in a packed column. In order to recover as much bromine as possible, there is a need to increase the air amount and flow rate. However, this is restricted by liquid flooding. Another possible problem is that the column is prone to blockage because of the presence of impurities (e.g., mud and sand) in the brine and the formation of calcium sulfate crystals. Thus, air blowing may lead to a low removal rate and high energy consumption.

This problem can be solved with the use of high-gravity technology. A high-gravity field is created by the high centrifugal force resulting from the high-speed rotation of the packing, so that the bromine-containing brine can be dispersed into very small liquid droplets, filaments and films. Because of this, the specific surface area is increased and the gas–liquid transfer resistance is reduced, which greatly contributes to the mass transfer and subsequently the desorption of bromine. The high-gravity technology for air blowing of bromine in brine is shown in Fig. 3.8, which includes acidification and oxidation of the brine, blowing, absorption, and distillation. A novel feature of this technology is to use air to blow free bromine in the acidified and

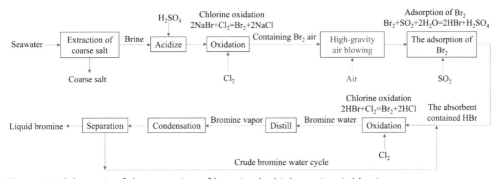

Figure 3.8 Schematic of the extraction of bromine by high-gravity air blowing.

oxidized brine in the RPB, and it is absorbed by the absorbent (e.g., sulfurous acid or alkaline solution) in the absorption column.

3.3.4.1 Extraction of bromine by high-gravity air blowing
The extraction of bromine by high-gravity air blowing is shown in Fig. 3.8.

The parameter optimization and extraction efficiency are discussed below:
1. The stripping rate of free bromine in the oxidation solution shows an increasing trend with the decrease of the liquid-gas ratio. Specifically, it increases rapidly as the liquid-gas ratio decreases from 16.7 to 8.3 L/m^3, but gradually levels off at liquid-gas ratios higher than 8.3 L/m^3. For this, the optimal liquid-gas ratio is set to 8.3 L/m^3.
2. The stripping rate increases rapidly from 78.6% to 90.1% as the spraying density increases from 6.3 m^3/(m$^2 \cdot$ h) to 10.5 m^3/(m$^2 \cdot$ h), after which it decreases slowly with the further increase of the spraying density. Thus, the most appropriate spraying density is determined to be 10.5 m^3/(m$^2 \cdot$ h).
3. The stripping rate increases with the increase of the high-gravity factor, but the increasing rate is decreasing when the high-gravity factor is higher than 42. Thus, the high-gravity factor of 42 is selected. It is noted that the variation of the stripping rate is more pronounced in high-concentration oxidation solution (2000 mg/L) than in low-concentration oxidation solution (50-450 mg/L). When the high-gravity factor is higher than 19, the stripping rate is higher for the high-concentration oxidation solution, implying that a stronger high-gravity field has a more significant effect on the stripping efficiency of free bromine in the high-concentration oxidation solution.
4. As the pH is decreased from 4.5 to 3.5, the stripping rate is increased from 79.75% to 88.36%. At pH <3.5, most bromine exists in the form of free bromine, and the increasing rate is reduced. Considering the blowing efficiency and the effect of SO_2 on the subsequent salt production process, pH = 3.5 is selected for the oxidation solution.
5. The stripping rate increases with the increase of the total bromine concentration. However, it is noted that the increasing rate is higher at bromine concentrations of 50-250 mg/L but lower at bromine concentrations greater than 250 mg/L.
6. The most favorable conditions for one-stage stripping are temperature = 20-25°C, oxidation solution pH = 3.5, high-gravity factor = 43, and liquid-gas ratio = 8.3 L/m^3. The second-stage stripping rate shows a similar pattern to that of the first-stage stripping rate. As the total bromine concentration is increased from 250 to 450 mg/L, the second-stage stripping rate is increased only slightly from 91.6% to 93%, and the third-stage stripping rate is increased from 93.3% to 95.3%. It is concluded that the increase trend of stripping rate of the third stage to the second is not as significant as that of the second to the single stage, and the contribution to the stripping rate is relatively lower.

3.3.4.2 Technological advantages

Compared to other technologies, the high-gravity technology for extraction of bromine from the brine has the following advantages:

1. Compared to the packed column, it can greatly improve the stripping rate of free bromine. The mass transfer coefficient of free bromine in the oxidation solution is $0.1238\ s^{-1}$, which is 170 times that of the packed column ($0.000728\ s^{-1}$).
2. It can improve the oxidation of bromine ions, resulting in an increase in the utilization rate of chlorine and a decrease in chlorine consumption.
3. It can reduce air consumption and consequently the operating cost is reduced by 31%.
4. It can improve the stripping rate of free bromine, especially in high-concentration oxidation solution.
5. Compared to the packed column, the gas phase resistance is reduced by at least 30%. As a result, the blower pressure is reduced and therefore investment and cost are also reduced.
6. It has high operating flexibility, low cost, and convenient installation and maintenance. The device size is only 1/28 of that of traditional columns.

3.3.5 Recovery of iodine in wet-process phosphoric acid

Iodine is a naturally occurring element commonly present in seawater. Its main producers include Japan, Chile, and the United States. It is mainly extracted from the water produced by natural gas wells and brine in Japan, from the production of sodium nitrate from saltpeter in Chile, and from the water produced by petroleum wells in the United States. In China, iodine is mainly extracted from laminaria extract and well brine with an annual production of only several hundred tons. This falls far short of the demand, contributing to the high price of iodine in China. Each year, a considerable sum of money is spent to import iodine. Thus, there has been a growing interest in the recovery of iodine in wet-process phosphoric acid.

3.3.5.1 Recovery of iodine in wet-process phosphoric acid using high-gravity air blowing

A packed column is commonly used for the recovery of iodine in wet-process phosphoric acid. The liquid flows slowly to form a thick fluid layer. In this case, the interphase mass transfer is limited because of the small interfacial area and the low surface renewal frequency. A feasible way to increase the mass transfer is to change the flow pattern of the liquid phase by increasing the gas flow rate. However, there is not much room to increase the gas flow rate because of liquid dumping. Thus, a high concentration of iodine (>0.03% in the feed solution) is required for the air-blowing method and the energy consumption is high. The presence of impurities (i.e., mud, sand, and viscous materials) in the feed solution leads to long-term accumulation of deposits, which makes the channel narrower and the gas-liquid contact area smaller and thus causes complete blockage of the packing. As a result, the pressure drop in the column is increased, leading to a sudden increase in energy consumption for gas supply. In some serious cases, the device may be blocked and need to be cleaned. Thus, the packed column requires high concentrations of iodine in the feed solution and it is prone to blockage. In addition, the energy consumption is also high.

The air-blowing method for recovery of iodine in wet-process phosphoric acid in the RPB is shown in Fig. 3.9. First, chlorine (or chlorine water) is introduced into the feed solution to oxidize iodine ions into free iodine ($2NaI + Cl_2 \rightarrow I_2 + 2NaCl$). The feed solution containing free iodine is introduced into RPB and brought into contact with the hot air that flows countercurrent to the liquid for recovery of iodine. The iodine-containing air is introduced into the absorption column, and iodine is absorbed by the SO_2 solution sprayed from the top of the column and then reduced to hydroiodic acid ($SO_2 + I_2 + 2H_2O \rightarrow H_2SO_4 + 2HI$). The solution is recycled several times, and when the mass concentration of iodine reaches 150 g/L, it is introduced into the iodine extractor. Then, chlorine is introduced into the extractor to oxidize HI ($2HI + Cl_2 \rightarrow 2HCl + I_2$) for the precipitation of solid iodine. After separation, crude iodine is obtained. The mother liquid that contains a large amount of free acid and iodine is transferred to the storage tank to acidify and oxidize the feed solution. The crude iodine is liberated by adding concentrated sulfuric acid and cooled for crystallization to obtain pure iodine.

The high-gravity technology uses RPB instead of conventional columns. The iodine-containing water oxidized by chlorine is introduced into the RPB from the liquid inlet and uniformly sprayed on the inner edge of the high-speed rotating packing. As the liquid flows through the packing in a radial direction driven by centrifugal force, it is sheared into micro/nano liquid forms (e.g., droplets, filaments, and films) and brought into contact with the hot air that flows countercurrent to the liquid phase in the packing. Because of the large specific surface area of the liquid and the rapid surface renewal, the heat and mass transfer are greatly increased and the free iodine in water can be rapidly transferred to the gas phase.

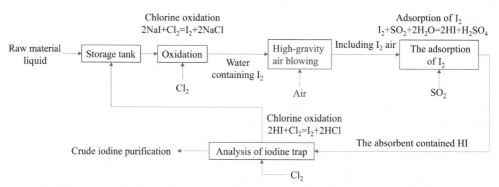

Figure 3.9 The schematic of the high-gravity air-blowing method for extraction of iodine.

3.3.5.2 Parameter optimization and removal efficiency

Iodine was recovered from dilute phosphoric acid for the first time by a company in Guizhou Province, China. Because of the presence of impurities in the dilute phosphoric acid and the impact of temperature, solid phases are continuously precipitated, causing serious scaling and blockage of the column. As a result, the rate of normal operation is less than 50%. In order to overcome these problems, the high-gravity technology is used.

1. The stripping rate reaches 91.27% for the feed solution with iodine concentrations of 35.2-45.1 mg/L, which is 2.80 times that obtained using conventional columns. The traditional air-blowing method is applicable only to high iodine concentrations (>300 mg/m^3), and the high-gravity technology is applicable to a wider range of iodine concentrations with higher removal efficiency.
2. The gas-liquid ratio is improved from 13 under the original conditions to about 34 (air consumption is reduced by 2/3), and the stripping rate is maintained at above 58%. The residual iodine in the feed solution is 8 mg/m^3. As less air is used and the final iodine concentration is increased, which is favorable for subsequent absorption of iodine.
3. The RPB has self-cleaning capacity and it is less prone to scaling and blockage, which can improve the operation rate of the device.
4. Iodine can be recovered from dilute phosphoric acid with an iodine concentration of only 60 mg/m^3, which provides a promising means of recovering halogen resources.

3.3.6 Stripping of acrylonitrile in wastewater

Acrylonitrile (AN) has a wide range of industrial and domestic uses since it is an important organic monomer in the manufacture of synthetic fiber, rubber, and plastics. In recent years, the annual production of AN has been increasing, and with it comes an increase in the discharge of AN wastewater. It is estimated that approximately 1.5 tons of AN wastewater is generated for each ton of AN produced. In 2013, about 1.9 million tons of AN wastewater were generated in China [21], which was mainly derived from the production of AN, acrylic fibers, and acrylonitrile-butadiene-styrene (ABS) plastics [22]. AN is highly toxic and difficult to degrade. The AN wastewater, if not disposed of in a proper manner, will destroy the ecosystem and impact human health [23]. For this reason, AN wastewater can be discharged only after proper treatment.

Incineration is the most commonly used method for the treatment of AN wastewater [24,25], such as the case in the Acrylon Chemical Plant of PetroChina Fushun Petrochemical Company, Sinopec Shanghai Petrochemical Co. LTD, Sinopec Anqing Company, CNPC Auspicious Chemical Group Co, and Sinopec Qilu Petrochemical

Company. Although incineration is a simple yet effective wastewater treatment method, the flue gas derived from the incineration of wastewater is likely to cause scaling, blockage, and corrosion of the boiler. The flue gas released into the atmosphere also causes secondary pollution and a wide range of environmental problems, such as greenhouse gas emissions and photochemical smog pollution. A substantial amount of auxiliary fuel is required for incineration of wastewater. Thus, there is a need to find a more efficient and cost-effective method for treatment of AN wastewater. Stripping is simple, cost efficient, and highly efficient for the treatment of industrial wastewater, especially wastewater containing high concentrations of volatile organics. Common stripping devices include packed column, plate column, and aeration tank, all of which have low gas–liquid transfer rates and thus require enormous investment and operating costs. This is because the concentration of AN is low in the wastewater. The closer the AN is to the top of the column, the closer the gas phase concentration is to the equilibrium concentration, and the lower the driving force for the gas–liquid mass transfer will be. The mass transfer resistance is higher because the liquid droplets are large and the liquid film is thick. Thus, the low gas–liquid mass transfer in traditional stripping devices leads to a low (<20%) removal rate of AN.

Next, we will describe the use of high-gravity stripping for the treatment of AN wastewater at room temperature [26-28].

3.3.6.1 High-gravity stripping for treatment of acrylonitrile wastewater

The treatment process of AN wastewater using high-gravity stripping is similar to that described in Fig. 3.7. One major difference is that no preheating of wastewater is needed. AN wastewater is simulated by dissolving AN (analytically pure) in deionized water to mass concentrations of 2000–4000 mg/L, and air is used as the desorption gas. The operating parameters are varied [high-gravity factor = 35–70, liquid-gas ratio = 0.7–2.5 L/m^3, liquid spraying density = 0.9–2.3 m^3/(m$^2 \cdot$ h)] to investigate their effects on the total volumetric mass transfer coefficient ($K_x a$), mass transfer unit height (H_{OL}), and removal rate of AN (η).

3.3.6.2 Parameter optimization and removal efficiency

The gas–liquid ratio has the most significant effect on the total volumetric mass transfer coefficient and the removal rate of AN, followed by liquid spraying density, high-gravity factor, and initial concentration. Both the total volumetric mass transfer coefficient and the removal rate of AN, increase with the increase of high-gravity factor, gas–liquid ratio, and spraying density, and both of them first increase and then decrease with increasing initial concentration. The mass transfer unit height decreases with the increase of high-gravity factor and gas–liquid ratio. However, it first increases and then decreases with increasing liquid spraying density, and an opposite trend is observed with increasing initial concentration. At room temperature and atmospheric pressure, the optimal conditions are high-gravity factor = 50, liquid–gas ratio = 0.77 L/m^3, and liquid spraying density = 1.60 m^3/(m$^2 \cdot$ h).

At initial AN concentrations of (3000 ± 100) mg/L, the removal rate of AN under the optimal conditions is 69.7%, 88.8%, and 97.1% for the single-, second-, and third-stage stripping, respectively. The final AN concentration in the wastewater is 192 mg/L with a difference of only 2.9% from the equilibrium concentration. The total volumetric mass transfer coefficient is 0.86 kmol/(m³·s), and the mass transfer unit height is 3.06 cm.

The empirical relationships among $K_x a$, H_{OL}, and η can be expressed as follows:

$$K_x a = 4.2340 \times 10^{-4} \beta^{0.3307} R^{0.6215} U^{0.6801} c_m^{0.1812}$$

$$H_{OL} = 37.2240 \beta^{-0.2948} R^{-0.5960} U^{0.1578} c_m^{-0.2085}$$

$$\eta = 0.9243 \beta^{0.1884} R^{0.3698} U^{-0.0494} c_m^{0.1127}$$

3.3.6.3 Comparison with other stripping devices

The removal rate of AN is 2.6 and 13 times that obtained in a column and stirrer respectively. The total volumetric mass transfer coefficient is increased by a factor of 8, the mass transfer unit height is reduced to 1/63, and the desorption gas resistance is only 1/10 of that obtained by stripping in a column. After desorption, the AN concentration is close to the theoretical equilibrium concentration, the mass transfer is improved and less energy is consumed. Thus, high-gravity stripping has promising industrial applications for the treatment of AN wastewater.

3.3.7 Removal of nitrobenzene from high-concentration wastewater

Nitrobenzene (NB) is an important raw chemical widely used in various industries such as petroleum chemistry, dyes, materials, and coking [29,30]. Notably, NB is highly volatile and chemically stable and it is difficult to biologically degrade. It also has high carcinogenic, teratogenic, and mutagenic potential and thus it is listed as a priority contaminant [31,32]. Current methods for the treatment of NB wastewater mainly include biological methods, physical methods and chemical methods. Stripping is often used for pre-treatment of wastewater with high concentrations of volatile organics [33]. As the removal of NB is a mass transfer controlled process, high-gravity technology has the potential to greatly improve the stripping rate of NB because of its ability to increase the gas-liquid mass transfer rate.

3.3.7.1 Technological process

The treatment process of NB wastewater using high-gravity stripping is similar to the one illustrated in Fig. 3.7. The simulated wastewater (550 mg/L) with an initial pH of 7.5 is prepared from NB and deionized water. High-gravity stripping is carried out at room temperature $(20 \pm 3)°C$, and the effects of high-gravity factor (40-150), liquid-gas ratio (0.8-4.0 L/m³), and spraying density [1.0-3.5 m³/(m² · h)] on the stripping rate of NB are investigated.

3.3.7.2 Parameter optimization and removal efficiency
The stripping rate of NB increases with the increase of high-gravity factor and liquid-gas ratio, but it first increases and then decreases with the increase of spraying density. The most favorable conditions are high-gravity factor = 80, liquid-gas ratio = 1.0 L/m³, and spraying density = 1.5 m³/(m²·h), and the single-stage stripping rate is 17.4%. After 10 cycles, the total stripping rate reaches 85.2%, and the final NB concentration in the wastewater is reduced to 90 mg/L, which meets the requirement of biochemical treatment.

3.3.8 Water deoxygenation
The removal of dissolved oxygen in water has important implications for the industry. An exceptionally high concentration of dissolved oxygen in water is likely to cause serious corrosion of iron pipelines and devices, as well as degradation of water quality because of the breeding of microorganisms and an increase in suspended solids in the water. Especially, the presence of dissolved oxygen in boiler feed water may induce oxygen depolarization corrosion of the boiler and water supply system. It is estimated that about 40% of boiler failures are related to oxygen corrosion and about 70% of boilers show different extents of corrosion, which will cause substantial economic losses. Thus, it is necessary to properly control oxygen concentration in water. Thermal deoxygenation is often used to remove the dissolved oxygen in the feed water of high-pressure boilers, and the removal efficiency depends largely on the structure and operating conditions of the deaerator. The most favorable conditions can be obtained when the water is heated to boiling and supplied at a constant rate to the deaerator, and the oxygen, vapor, and other gases flowing out of the deaerator can be readily discharged. However, a low mass transfer rate is expected because of the limited gas-liquid contact in the deaerator, which may cause poor deoxygenation, large device size, and high energy consumption. Also, the oxygen concentration in the water is much higher than the equilibrium concentration, which falls short of the national standard for boiler feed water.

High-gravity technology has been demonstrated to be able to improve the mass and heat transfer in thermal deoxygenation. The gas-liquid contact area and surface renewal in the thermal deoxygenation process can be increased by several orders of magnitude with the use of RPB, and the final oxygen concentration is reduced to 7 μg/L, which is close to the theoretical equilibrium concentration. Researchers at the Beijing University of Chemical Technology have used high-gravity technology in combination with low-pressure vapor deoxygenation to remove oxygen from 70 t/h feed water of high-pressure boilers in a power plant. The results reveal that the oxygen concentration and deoxygenation rate at the outlet meet the national standard, and a large amount of high-quality stream is saved with substantial economic profit. The high-gravity technology can also be combined with vapor thermal deoxygenation and vacuum deoxygenation depending on the field and operating conditions. No chemical agents are required and the deoxygenation efficiency is better than the traditional deaerator. It is also characterized by small device size, low cost, and advantages in terms of inlet water temperature, pressure, and investment. In conclusion, it is a promising deoxygenation method [34–36].

High-gravity technology is particularly suitable for thermal desorption because of its potential to improve mass and heat transfer. The desorption efficiency is high and the energy consumption is low. The final solute concentration in the liquid is close to the theoretical equilibrium concentration, and the standard can be met without the use of any other chemical agents. However, the application of high-gravity desorption in actual projects is still in its infancy, and efforts are being made to make the high-gravity device better suited to desorption processes and to improve industrial scale. A better understanding of how to enhance the mass and heat transfer and industrial application is essential to develop novel high-gravity devices and packing. To this end, continued effort is needed to explore the potential of the high-gravity technology, so that desorption can be achieved in a highly efficient but low-cost way.

References

[1] Liu YZ. High-gravity Chemical Process and Technology. Beijing: National Defence Industry Press, 2009.
[2] Li Y. A study on the heat transfer in the high gravity rotary field. Taiyuan: North University of China, 2007.
[3] Liu LM, Han BQ, Han ZH. Common problems appeared in SNCR denitrification system of coal-fired boilers and countermeasures thereof. Thermal Power Generation, 2010,39(6):65-67,70.
[4] Jiao WZ, Liu YZ, Liu JW, et al. Pilot study on coking ammonia-nitrogen wastewater by high-gravity rotary bed. Modern Chemical Industry, 2005,S1:257-259.
[5] Qi GS, Liu YZ, Wang JW, et al. Stripping ammonia-nitrogen wastewater by high gravity method. Coal Chemical Industry, 2007,1:61-63.
[6] Gu DY, Liu YZ, Qi GS, et al. Stripping ammonia nitrogen wastewater in a novel rotating packed bed. Natural Gas Chemical Industry, 2014,39(4):1-4.
[7] Liang XX, Liu YZ, Jiao WZ, et al. Application of high gravity technology in the industrial wastewater treatment. Paper Science & Technology, 2013,32(1):95-98.
[8] Ramon JA. Polymer Devolatilization. Zhao XT, Gong GB, editors. Beijing: Chemical Industry Press, 2005:105-110.
[9] Yan SY, Gu LL. Present situation and research prospect of the urea-formaldehyde resin. Science & Technology in Chemical Industry, 2005,13(4):50-54.
[10] Xu HY, Zhang JJ, Li HX. A review on development of bromic fine chemicals. Journal of North China University of Science and Technology (Natural Science Edition), 2006,28(3):88.
[11] Liu YZ, Liu ZH. Devices for removal of free formaldehyde from urea formaldehyde resin. CN 2892854Y, 2007-04-25.

[12] Liu YZ, Liu ZH. Methods and devices for removal of free formaldehyde from urea formaldehyde resin. CN 100427522C, 2008-10-22.
[13] Xie JJ, Pan QM, Pan ZR. Progress on polymer devolatilization. China Synthetic Rubber Industry, 1998,21(3):135-141.
[14] Sun PX, Zhang WF. Current situations and development of bromine products in China. Journal of Salt Science and Chemical Industry, 2000,30(1):7-8.
[15] Wang GQ, Feng HJ, Zhang FY. Developmental prospect on multipurpose utilization of seawater chemical resources. Journal of Ocean Technology, 2002,21(4):62-63.
[16] Wang SW, Lu BN. Current situation and prospect of elementary bromine production. China Well and Rock Salt, 2004,35(2):12.
[17] Li HM, Chen HD, Zhang QY. Evaluation of the technologies of comprehensive utilization and exploitation salt resource. Journal of Salt Lake Research, 2003,11(3):63-64.
[18] Yan SW, An LY, Tang ML, et al. Technological development of collecting bromine from brine by ion exchange. Journal of Salt Science and Chemical Industry, 1996,23(6):14-16.
[19] Zhang LJ, Wang W, Wang XL. Introduction to technique of manufacturing bromine and prospecting development trend in bromine products. Marine Sciences, 1998,5:20.
[20] Zhu CL, Kou JJ. Extraction of bromine from brine by RIP method. Multipurpose Utilization of Mineral Resources, 2003,10(5):14.
[21] Huang JX, Lu SL, Ji LC. Market analysis of acrylonitrile production for 2013. Chemical Industry, 2014,32(4):36-40.
[22] Wang K, Shen Z, Zhang M. A review of the methods applied to acrylonitrile wastewater treatment. Technology of Water Treatment, 2014,40(2):8-14.
[23] Labour Health and Occupational Diseases Teaching and Research Group, Beijing Medical College. Prevention of Acrylonitrile Poisoning. Beijing: Petroleum Chemical Industry Press, 1976:4-9.
[24] Kan HY. Brief introduction of waste water incinerator used in SECCO acrylonitrile plant. Process Equipment & Piping, 2007,44(5):24-27.
[25] Sun YM. Treatment of acrylonitrile wastewater. Contemporary Chemical Industry, 1996,4:48-50.
[26] Xue CF, Liu YZ, Jiao WZ. Mass transfer of acrylonitrile wastewater treatment by high gravity air stripping technology. Desalination and Water Treatment, 2016,57(27):12424-12432.
[27] Xue CF, Liu YZ, Jiao WZ. Experimental study on treatment of acrylonitrile waste water by high gravity air stripping technology. Chemical Industry and Engineering Progress, 2014,33(9):2501-2505.
[28] Xue CF, Liu YZ, Jiao WZ. Air stripping wastewater in rotating packed bed with different packings. Chemical Engineering (China), 2015,43(5):16-19.
[29] Jiao WZ, Liu YZ, Liu WL, et al. Degradation of nitrobenzene-containing wastewater with O_3 and H_2O_2 by high gravity technology. China Petroleum Processing & Petrochemical Technology, 2013,15(1):85-94.
[30] Guo L, Jiao WZ, Liu YZ, et al. Research progresses in treatment of wastewater containing nitrobenzene compounds. Environmental Protection of Chemical Industry, 2013,33(4):299-303.
[31] Carlos L, Nichela D, Triszcz JM, et al. Nitration of nitrobenzene in Fenton's processes. Chemosphere, 2010,80(3):340-345.
[32] Nichela DA, Berkovic AM, Costante MR, et al. Nitrobenzene degradation in Fenton-like systems using Cu(II) as catalyst—Comparison between Cu(II)- and Fe(III)-based systems. Chemical Engineering Journal, 2013,228:1148-1157.
[33] Quan XJ, Cheng ZL, Xu F, et al. Structural optimization of the porous section in a water sparged aerocyclone reactor to enhance the air-stripping efficiency of ammonia. Journal of Environmental Chemical Engineering, 2014,2(2):1199-1206.
[34] High gravity technology and devices for deoxidation of boiler water. Chemical Technology Market, 2003,26(3):69.
[35] Liu HX, Liu M. Progress on study of boiler water deoxidation. Corrosion Science and Protection Technology, 2007,19(6):432-434.
[36] Chen JM, Song YH. Deoxygenation of boiler feed water by high gravity technology. Chemical Industry and Engineering Progress, 2002,21(6):414-416.

CHAPTER 4

Distillation

Contents

4.1 Overview	101
4.2 Principles of high-gravity distillation	103
4.3 Key technologies	104
4.3.1 Liquid distributor	104
4.3.2 Packing	108
4.4 Characteristics of high-gravity distillation	111
4.4.1 Characteristics of high-gravity distillation in a rotating zigzag bed	111
4.4.2 Characteristics of high-gravity distillation in rotating packed bed	128
4.4.3 Comparison between high-gravity distillation equipment and traditional distillation columns	145
4.5 Application examples	146
4.5.1 Ordinary distillation	146
4.5.2 Special distillation	148
4.5.3 Distillation of high-viscosity and heat-sensitive materials	149
4.5.4 Distributed distillation	151
4.5.5 Recovery of ammonia wastewater	151
4.6 Prospects	153
References	154

4.1 Overview

The high-gravity technology (HiGee) has been widely used for distillation since its introduction in the 1970s [1]. The origin of the rotating packed bed (RPB) could be traced back to Colin Ramshaw's work on process intensification in the 1970s and 1980s at Imperial Chemical Industries (ICI), most notably in trying to find an alternative to conventional distillation. In 1979, ICI built two 0.762 m-diameter RPBs for the distillation of methanol–isopropanol mixture, whose separation capacity was equivalent to 30.48 m–high and 0.91 m–diameter packed columns. In 1983, ICI reported the successful use of industrial-scale RPB for the separation of ethanol/isopropanol and benzene/cyclohexane mixtures for several thousands of hours, which demonstrated the engineering and technological feasibility of this new technology. Most impressive is that the mass transfer unit height was only 10-30 mm, which was two orders of magnitude shorter than traditional packed column. Ramshaw used average centrifugal accelerations of up to 1000 g and found that the dumping rate of the liquid [15 kg/($m^2 \cdot s$)] was about 70% of that for safe operation and the theoretical plate height was 10-20 mm [2]. Richard [3] investigated the separation of isopropanol/ethanol mixture using two 800 mm-diameter RPBs connected in series, and a 40 theoretical plate column was obtained at rotation speeds of 1500-3000 r/min and average centrifugal accelerations of 200-1000 g.

In China, Chen and his colleagues at Zhejiang University [4] applied high-gravity technology to the distillation of ethanol/water mixture in 1989. The RPB used in their work had an outer diameter of 300 mm and a height of 40 mm, and it was filled with stainless steel θ ring packing of $\phi 6$ mm × 6 mm × 0.1 mm. Under conditions of atmospheric pressure and total reflux, the packing thickness in the radial direction that was equivalent to a theoretical plate was 3.93 cm at a reflux rate of 6.9 t/(m^2·h) and a rotation speed of 900 r/min, that is, the packing of one meter in thickness was equivalent to 25 theoretical plates in terms of separation efficiency.

In 1992, Martin and Martelli used RPB with metal mesh packing for distillation of cyclohexane/n-heptane mixture, where the rotor inner diameter was 175 mm, the rotor outer diameter was 600 mm, the specific surface area was 2500 m^2/m^3, and the porosity was 0.92. The theoretical plate number was in the range of 4-6, which was equivalent to a theoretical plate height of 30-50 mm.

In 1996, Trevour et al. [5] investigated the distillation of cyclohexane/n-heptane mixture using a high-gravity vapor-liquid contactor under total reflux conditions, where the diameter of the equipment was 914 mm, the rotor inner diameter was 87.5 mm, the rotor outer diameter was 300 mm, the axial direction height was 150 mm, the total volume was 0.0388 m^3, the packing specific surface area was 2500 m^2/m^3, and the porosity was 0.92. At operating pressures of 166-414 kPa and rotation speeds of 400-1200 r/min, 6 theoretical plates were obtained corresponding to a theoretical plate height of 30-90 mm.

Since 2001, Ji and his colleagues at Zhejiang University of Technology [6-35] have used rotating zigzag bed for the distillation of a wide variety of solvents, such as methanol, ethanol, DMF, and residual solvents in high-viscosity heat-sensitive materials. Different types of packing (e.g., Pall ring and triangular spiral packing) were used, and the effects of many structural parameters (e.g., pore distribution and plate spacing) on the fluid mechanics and distillation efficiency were investigated. The mass transfer unit height ranged from 30 to 40 mm.

Liu and his colleagues at the North University of China [36-67] have studied the distillation of ethanol/water and methanol/water mixture in a high-gravity environment, as well as the fundamental principles of high-gravity distillation, distillation efficiency, and fluid mechanics. Several empirical models have been established to calculate the separation efficiency and gas phase pressure drop in the RPB and provide experimental and theoretical evidence for industrial applications of the equipment. A series of liquid distributors, packings (e.g., screw axis packing and fin baffle packing), and high-efficient RPBs have been developed for high-gravity distillation, and their mass transfer and fluid mechanics have been investigated. The fin baffle packing was found to have the best mass transfer performance with a mass transfer unit height of 4.8-11.9 mm.

In 2002, Lin et al. [68] presented a novel RPB for distillation of methanol/ethanol mixture at atmospheric pressure and total reflux conditions. The rotor speeds ranged from 600 to 1600 r/min, which provided 42-298 equivalent gravitational force. The results indicated that 1-3 theoretical plates were achieved with a packing thickness of 8.6 cm, and consequently the height equivalent of a theoretical plate was about 3-9 cm in the RPB and it varied with centrifugal force to the 0.23-0.26 power.

In 2010, Chen and his colleagues at the Beijing University of Chemical Technology [69,70] developed a novel multi-stage counter-current RPB (MSCC-RPB) taking advantage of the conventional RPB and the rotating zigzag bed and investigated its performance in the continuous distillation of ethanol-water mixture at atmospheric pressure. The effects of rotation speed (n), feed concentration (x_F), thermal conditions of the feed (q), and reflux ratio (R) on the theoretical plate number (N_T) of MSCC-RPB were investigated. The theoretical plate number of MSCC-RPB increases with increasing rotation speed at $N < 800$ r/min but decreases with increasing rotation speed at $N > 800$ r/min. However, it remains largely constant with increasing feed concentration but increases with increasing thermal conditions of the feed and reflux ratio. The optimal rotation speed was 800 r/min, and the theoretical plate height was in the range of 19.5-31.4 mm.

4.2 Principles of high-gravity distillation

The RPB used for high-gravity distillation is schematically shown in Fig. 4.1. The liquid mixture to be separated is vaporized in a reboiler, and the gas is supplied into the external cavity of the RPB from the gas inlet and flows in the radial direction from the outer edge of the rotating packing to the inner edge of the packing driven by pressure gradient. Finally, the gas is gathered in the central cavity and exits the RPB from the gas outlet. The condensed liquid is introduced into the distributor from the liquid inlet, and then uniformly dispersed by the nozzle on the inner edge of the packing. The high gravity drives the liquid to flow radially from the inner edge to the outer edge of the packing.

The liquid flowing through the rotating packing (e.g., metal corrugated packing) is driven by both high gravitational force and frictional force and is sheared into micro- and even nanometer-scale pieces (e.g., droplets, filaments, and films). The gas-liquid interfacial area is increased and the liquid is rapidly renewed, increasing the effective specific surface area of the packing. Countercurrent contact takes place between the two phases in the zigzag channel at a very large relative velocity in a context of high dispersion, thorough mixing, strong turbulence, and rapid interface renewal, leading to high-efficient

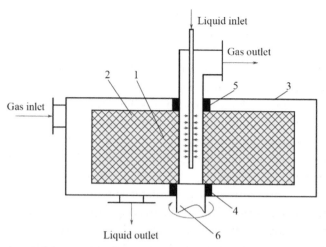

Figure 4.1 A schematic of the RPB used for high-gravity distillation (1—packing; 2—rotor; 3—casing; 4, 5—seals; 6—shaft). *RPB*, Rotating packed bed.

mass and heat transfer and repeated condensation and vaporization as the gas and liquid phases pass through each layer of packing. Phase equilibrium can be achieved rapidly, and the mixture can be separated and purified within a very short period of time.

The mechanism of high-gravity distillation is essentially the same as that of common distillation processes, both of which involve repeated condensation and vaporization based on the differences in the volatility or boiling point of components of the solution, but the separation efficiency is much higher in the high gravity field. In common distillation processes, gas-liquid mass transfer occurs in bubbles; while in high-gravity distillation processes, the liquid exists in the form of droplets, filaments, and films rather than bubbles and it is brought into more intimate contact with the gas in the packing.

4.3 Key technologies

The high-gravity distillation process involves simultaneous heat and mass transfer between gas and liquid phases, which is affected not only by the flow patterns of the two phases but also by the effective contact between them. For this reason, it is imperative to ensure a uniform distribution of the liquid in the rotating packing and to make the structural and technological parameters of the packing meet the requirement for efficient gas-liquid contact and high heat and mass transfer.

4.3.1 Liquid distributor

The liquid should be distributed on the packing surface in a form favorable for heat and mass transfer. In a typical configuration, the liquid distributor is installed at the center of the packing with a gap between them in order to ensure that all the liquid can be sprayed on the inner edge of the packing. However, it should be noted that the inner edge of the packing forms a cylinder, which suggests the need to consider the distribution of the liquid on the inner edge of the packing in both circumferential and axial directions.

4.3.1.1 Liquid distributor specific for rotating packed bed

Generally, 2-4 liquid delivery tubes are installed in the circumferential direction on the central liquid inlet tube, and a number of orifices are made for each tube at different heights to act as liquid distributors. The liquid is sprayed on the inner edge of the rotating packing in the circumferential direction. There are substantial differences in the flow velocity and flow rate of the liquid supplied from different orifices because of differences in liquid pressure. This is especially pronounced between top and bottom orifices in the axial direction. The liquid distributors have the advantages of simple structure and ease of replacement, installation, and maintenance. As high-gravity distillation is concerned with the treatment of liquid, the design of liquid distributors plays an important role in effective gas-liquid contact, repeated partial vaporization, and condensation.

Li et al. [54] proposed a novel liquid distributor consisting of an upper disk, a lower disk, and a number of liquid delivery tubes. The liquid distributor is fixed to the rotor shaft, and the liquid inlet is set at the upper disk. A number of semicircular cylinder liquid delivery tubes with different heights are installed at the periphery of the lower disk that has several axial and radial grooves. An opening is set at the top of each liquid delivery tube corresponding to the inner edge of the packing. All liquid delivery tubes are fixed to the lower disk but spaced at different distances from the upper disk because their heights are different. The overall structure of the liquid distributor is like a cylinder. As the liquid distributor rotates driven by the rotor shaft, the liquid in the liquid distributor is in a high gravity field and flows in a "U-N" pathway from the liquid distributor to the inner edge of the packing. Specifically, the liquid is introduced into the inner cavity from the liquid inlet at the upper disk and then flows into liquid delivery tubes in an "N" pathway along the axial and radial grooves on the lower disk. Subsequently, the liquid flows from the bottom to the top of the tube driven by the pressure gradient, and it also rotates at a high speed driven by the rotor shaft. Finally, the liquid is sprayed from the opening at the top of each liquid delivery tube to the inner edge of the packing at different heights. The overall flow pathway of the liquid in the liquid distributor is U shaped. As it is subject to both centrifugal force and pressure, thin liquid films would be formed at the opening of each liquid delivery tube, as shown in Figs. 4.2-4.4.

Such a design allows the liquid to have a fully developed initial velocity distribution in the liquid distributor and to flow along the preset path in the packing, which leads to longer flow path and residence time of the liquid, a more uniform liquid distribution, and a larger gas-liquid interfacial area in the packing. To sum up, this unique design increases the mass and heat transfer efficiency of the packing.

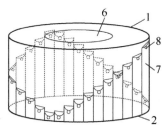

Figure 4.2 The three-dimensional structure of the liquid distributor for RPB. *RPB*, Rotating packed bed.

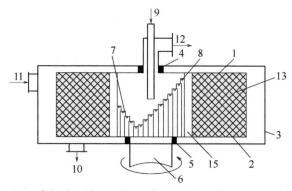

Figure 4.3 The front view of the liquid distributor for RPB. *RPB*, Rotating packed bed.

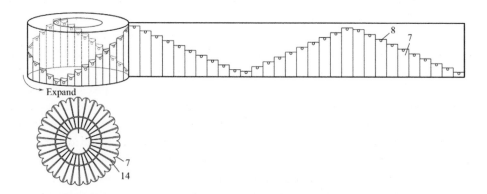

Figure 4.4 The expanded view of the liquid distributor for RPB. In Figs. 4.2-4.4: 1—upper disk; 2—lower disk with grooves; 3—casing; 4—mechanical seals; 5—dynamic seals; 6—rotor shaft; 7—liquid delivery tubes; 8—liquid openings; 9—liquid inlet; 10—liquid outlet; 11—gas inlet; 12—gas outlet;

4.3.1.2 Liquid distributor and redistributor for m-stage rotating packed bed [55]

The liquid distributor and redistributor for multi-stage RPB consist of a tubular or cone-shaped liquid distributor for static liquid distribution and a U- or disk-shaped liquid distributor for dynamic liquid distribution. There are N sinks at the front end of the liquid inlet tube and N spiral sinks at the cone-shaped liquid receiver and distributor. The U-shaped liquid distributor comprises a cylinder, a hollow annular plate at the lower edge of the lateral side of the cylinder, and an annular U-shaped sink at the upper edge of the lateral side of the cylinder. Small orifices are made at the bottom of the U-shaped sink. The U-shaped liquid distributors of different diameters and axial heights are arranged in a concentric fashion in order from the smallest to the largest diameter of the cylinder, as shown in Figs. 4.5–4.7.

A unique advantage of this liquid distributor is the ingenious combination of many different structures for different purposes, which can not only enable the liquid to have uniform initial radial, axial, and circumferential distribution in the distributor but also uniform distribution and redistribution in the RPB by increasing its residence time by means of the tortuous flow of the liquid in the distributor and redistributor.

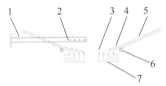

Figure 4.5 The front view of the liquid distributor and redistributor for multi-stage RPB. *RPB*, Rotating packed bed.

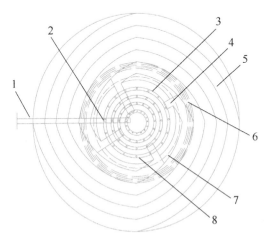

Figure 4.6 The top view of the liquid distributor and redistributor for multi-stage RPB. In Figs. 4.5 and 4.6: 1—liquid inlet tube; 2—opening; 3—cone-shaped spiral liquid receiver; 4—U-shaped liquid separating

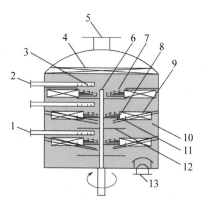

Figure 4.7 The schematic of the structure of the liquid distributor for multi-stage RPB (1—liquid inlet tube; 2—reflux inlet; 3—liquid distribution sink; 4—demister; 5—gas outlet; 6—shaft; 7—cone-shaped spiral liquid receiver; 8—U-shaped liquid separating sink; 9—maze; 10—packing; 11—liquid distribution orifice;

The mass transfer efficiency and space utilization are greatly improved, and the upper and lower packing is gas sealed. The inherent advantages of the liquid distributor and redistributor, such as simple structure, convenient installation, and multiple functions, make it applicable to various single-stage or multi-stage countercurrent-flow RPBs to intensify the gas–liquid transfer process and improve the transfer efficiency.

4.3.2 Packing

The equipment used for distillation mainly includes RPB and plate rotating bed. The structure, specific surface area, porosity, and installation mode of the packing would have important effects on fluid flow and gas–liquid contact efficiency. The packing falls primarily into three categories: random packing, quasi-structured packing, and structured packing. It is worth noting that structured packing is more effective in facilitating the uniform gas–liquid contact because of the ordered and uniform distribution of voids in the packing, and consequently, it is also more effective in intensifying mass and heat transfer and accelerating the renewal of the gas–liquid interface. The residence time (contact time) of gas and liquid phases in the rotating packing is prolonged, which also contributes to increasing the overall transfer efficiency. These factors have to be taken into consideration in the design of the packing.

4.3.2.1 Structured spiral packing [56]

The structured spiral packing consists of a hollow cylinder and two disks attached to the upper and lower ends of the cylinder respectively. A circular opening is made at the center of the upper disk, and a number of liquid distribution tubes are assembled on the lower disk around the cylinder with an opening to the interior of the cylinder. These tubes are wound around the outside of the cylinder between the lower and upper disk in a spiral manner. The lower disk will rotate together with the shaft, which creates a high centrifugal force field in the packing. The liquid is introduced into the interior of the cylinder from the opening on the upper disk and then flows into liquid distribution tubes. After that, it flows in an S (spiral) path in the liquid distribution tube and then flows from the inner edge to the outer edge of the packing in an N path. The liquid flowing in the liquid distribution tube is subject to the centrifugal force and the confinement stress of the tube, thus forming very thin liquid films. The gas can be supplied into the liquid distribution tube from the opening either on the external (countercurrent contact in the packing) or the internal (concurrent contact in the packing) side of the packing. The liquid distribution tubes are filled with gas and it flows in an S-N path from one side to the other side of the packing driven by the pressure gradient (see Figs. 4.8—4.10 for more details). This configuration can

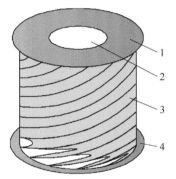

Figure 4.8 The overall structure of the structured spiral packing (1—upper disk; 2—cylinder; 3—liquid distribution tube; 4—lower disk).

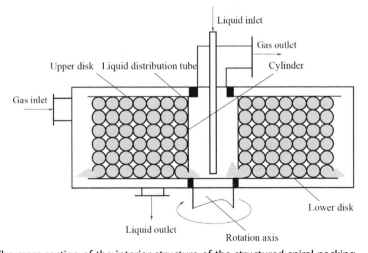

Figure 4.9 The cross-section of the interior structure of the structured spiral packing.

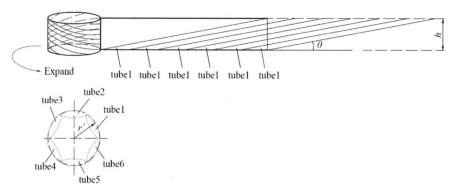

Figure 4.10 The expand view of a single liquid distribution tube of the structured spiral packing.

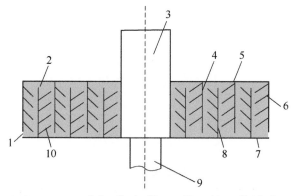

Figure 4.11 The internal structure of the fin baffle packing (1,2—fluid channel; 3—central tube; 4—upper baffler; 5—upper disk; 6,10—fans; 7—lower disk; 8—lower baffler; 9—rotor shaft).

improve space utilization of the rotating bed packing and allow the gas and liquid phases to flow along the preset path in the packing, which leads to longer flow path and residence time of the liquid, a more uniform liquid distribution, and a larger gas-liquid interfacial area in the packing. Therefore, the mass and heat transfer efficiency is significantly improved.

4.3.2.2 Structured fin baffle packing [57]

The fin baffle packing consists of an upper disk and a lower disk with a number of columnar bafflers that are of different diameters and arranged in an alternative way. There is a gap between two bafflers and between each baffler and the upper or the lower disk, for the passage of fluid. A number of fins that are inclined downwards and upwards are arranged on each baffler attached to the upper and lower disk, respectively (see Figs. 4.11 and 4.12 for more details).

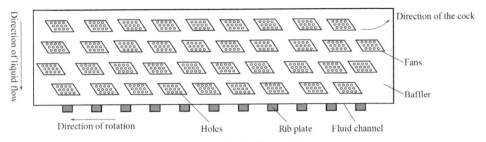

Figure 4.12 The expanded view of the upper fin baffle packing.

The columnar bafflers of different diameters are rigidly fixed to the upper or the lower disk with bolts in an alternative fashion and spaced apart at equal distances, and the high-speed rotation of these bafflers with the shaft creates a high centrifugal force field. The liquid flows along the bafflers fixed to the upper and lower disks alternatively from the innermost to the outermost baffler, forming an "N" shaped channel in the axial direction; while the gas flows in the opposite direction from the outermost to the innermost baffler. The surface tension of the liquid is overcome as a result of the high centrifugal force and thus a very thin liquid film is formed, which can greatly increase the gas–liquid contact area and subsequently the mass transfer between the two phases. Based on the liquid flow direction and baffler rotation direction, parallelogram fans are installed in an alternative fashion on the inside of the baffler with a fixed angle between the fan and the baffler, which will force the liquid to flow in a ≳ path between two adjacent bafflers. More specifically, as the liquid leaving the fan is subject to both inertial force and centrifugal force, a portion of the liquid will collide with the outside of the baffler that is opposite to the baffler on which the fin is installed, which makes the baffler wetter and thus increases the interfacial contact area and the surface utilization of the packing; the remaining liquid is forced onto the baffler on which the fin is installed, thus resulting in a ≳ flow path between two adjacent bafflers. This increases not only the gas–liquid contact area but also the flow path and residence time of the liquid, and as a result, the mass and heat transfer efficiency is improved. Several small holes are uniformly arranged on the fan in a triangular or parallelogram manner. The liquid will flow as a form of liquid film of several micrometers thick because of the strong centrifugal force, and it will flow forward rather than through these holes to the baffler on which the fin is installed. In this way, the gas–liquid contact area on each fan is greatly increased, leading to the intensification of the mass transfer process.

4.4 Characteristics of high-gravity distillation

4.4.1 Characteristics of high-gravity distillation in a rotating zigzag bed

4.4.1.1 Structure of rotating zigzag bed

The structure of the rotating zigzag bed is schematically shown in Fig. 4.13. Its core components include the rotor, which consists of a rotational disk and a set of baffles

112 HiGee Chemical Separation Engineering

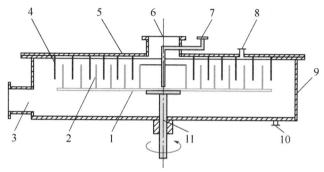

Figure 4.13 A schematic of the structure of the rotating zigzag bed (1—rotational disk; 2—rotational baffle; 3—gas inlet; 4—stationary baffle; 5—stationary disk; 6—gas outlet; 7—liquid inlet; 8—middle feed pipe; 9—casing; 10—liquid outlet; 11—rotor shaft).

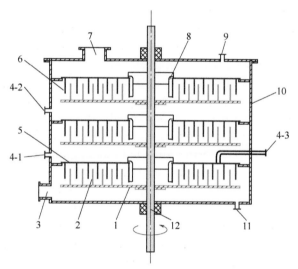

Figure 4.14 A schematic of the structure of the three-layer rotating zigzag bed (1—rotational disk; 2—rotational baffle; 3—gas inlet; 4-1,2,3—liquid inlet; 5—stationary disk; 6—stationary baffle; 7—gas outlet; 8—liquid delivery tubes; 9—reflux tube; 10—casing; 11—liquid outlet; 12—rotor shaft).

that are fixed to the disk and thus can rotate with the disk, and the stator, which consists of a stationary disk and a set of baffles that are fixed to the disk. Those rotational and stationary baffles are arranged in an alternative fashion and spaced apart at equal distances. The gaps between two bafflers and those between each baffler and the rotational or the stationary disk serve as the channels of the fluid.

The three-layer rotating zigzag bed is schematically shown in Fig. 4.14. An apparent difference between three-layer and single-layer rotating zigzag beds is the liquid delivery tubes at the center of the bed that are intended to introduce the liquid collected at the outer edge of the above rotor to the center of the lower rotor. Several liquid inlet tubes are also installed between two rotors (e.g., Fig. 4.7) or in a given rotor (e.g., Fig. 4.14) for distillation processes with different sources of feed in one rotating zigzag bed.

For the single-layer rotating zigzag bed, the gas is introduced into the bed through the gas inlet tube in the tangential direction and then flows from the outer edge to the inner edge of the rotor in an S-shaped channel formed by the annular space between rotational and stationary baffles and the gap between the disk and baffles, and finally, it exits the rotor through the gas outlet. The liquid is introduced into the rotor through the liquid inlet tube and then flows to the center of the rotational disk. Then, it is injected onto the stationary baffle through the small holes as a result of the high-speed rotation of the rotational baffles. Then it is collected on the liquid rotational disk under the effect of gravity and thrown again by the high-speed rotating baffles, and so forth. Finally, the liquid is collected at the casing and then exits the bed through the liquid outlet tube.

For the multi-layer rotating zigzag bed, the gas will flow from the center of the lower rotor to the outer edge of the upper rotor and then flow in the radial direction from the outer edge to the inner edge of the rotor driven by the pressure difference, and so forth. The liquid will flow from the outer edge of the upper rotor to the stationary disk of the lower rotor and then be introduced onto the rotational disk through liquid delivery tubes. After that, it flows in the radial direction to the outer edge of the rotor. In this way, the gas flows from the lower rotor to the upper rotor and then from the outer edge to the center in the upper rotor; the liquid flows from the upper rotor to the lower rotor and then from the center to the outer edge of the lower rotor, leading to countercurrent contact and an increase of the separation performance by several times.

4.4.1.2 Mass transfer performance of rotating zigzag bed

The rotor and stator are the core components responsible for the formation of the high-gravity field. Both gas and liquid are forced to flow along the predetermined path in the rotor and stator. The liquid is continuously dispersed and aggregated and the surface is renewed rapidly, and obvious end effects occur in the bed. The alternative arrangement of rotational and stationary baffles leads to a uniform distribution of the liquid and a low amplification effect. In the rotor, the liquid will go through repeated dispersion–aggregation processes, and it is dispersed into tiny droplets by the small holes on the rotational baffle, which in turn are thrown onto the stationary baffle at a very high velocity as a result of the high-speed rotation of the rotational baffle. The liquid films spreading on the stationary baffle fall on the rotational disk by gravity and then flow to the next rotational baffle driven by the centrifugal force. The above process is repeated until the liquid reaches the outer edge of the rotor. The gas flows from the outer edge to the inner edge of the rotor in an S-shaped channel formed by the annular space between rotational and stationary baffles and the gap between the disk and baffles driven by the pressure difference. The flow and contact of gas and liquid phases in the rotating zigzag bed are shown in Fig. 4.15. It is seen that the two phases flow in opposite directions and a mass transfer unit is formed between rotational and stationary baffles. The gas–liquid contact and mass transfer process in each unit can be divided into three stages. In the first stage, the liquid droplets are pulled away from the rotational baffle and brought into cross-flow contact with the gas flowing in the axial direction. These liquid droplets are small (<0.3 mm in diameter) and renewed almost instantaneously, leading to a very large specific surface area. In the second step, the liquid accumulated on the stationary baffle will rotate while falling because of the driving force resulting from the rotation of gas and the gravitational force, and it is brought into countercurrent contact with the gas that flows upwards in the annular space between rotational and stationary baffles. The liquid will also rotate on the baffle and then be continuously crashed and squeezed with the liquid thrown away from the rotational baffle. Surface renewal occurs instantaneously and the mass transfer rate is very high. The stationary baffles can be considered a group of wet walls. In the third step, the liquid leaves the stationary baffle and then flows to the rotational disk, and it is dispersed by the gas flowing in the radial direction, forming new liquid droplets for gas–liquid contact and mass transfer. The radial and circumferential velocity of the gas, as well as the slip velocity, are very high, resulting in a high mass transfer rate. It is concluded that the mass transfer rate is high in the unit, resulting in high overall mass transfer efficiency.

The experimental setup for testing the mass transfer performance of the two-layer rotating zigzag bed is schematically shown in Fig. 4.16. The bed is 800 mm in diameter and 550 mm in height, and it has two identical rotors with an outer diameter of 630 mm, an inner diameter of 200 mm, and a height of 80 mm. The upper rotor consisting of 10 stationary baffles and 9 rotational baffles is used for distillation, and the lower rotor consisting of 9 stationary baffles and 8 rotational baffles is used for

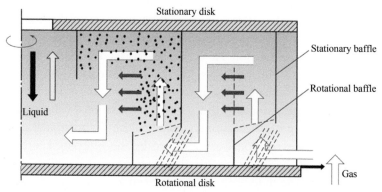

Figure 4.15 The flow and contact of gas and liquid phases in a rotating zigzag bed.

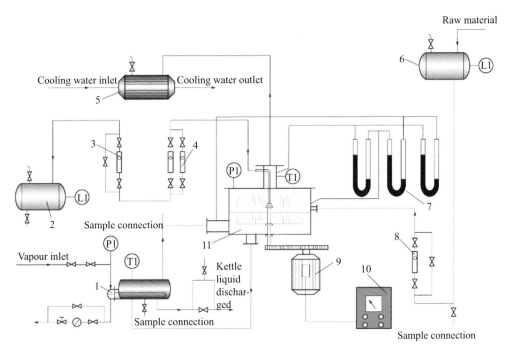

Figure 4.16 Experimental setup for testing the mass transfer performance of the rotating zigzag bed (1—reboiler; 2—product storage tank; 3—product rotameter; 4—reflux rotameter; 5—condenser; 6—head tank; 7—U-shaped differential pressure gauge; 8—feed rotameter; 9—adjustable-speed motor; 10—frequency modulator; 11—two-layer rotating zigzag bed).

stripping. The ethanol-water vapor generated from the reboiler is introduced into the lower rotor in the tangential direction and then flows through the high-speed rotating packing in an S-shaped channel. Subsequently, it is transferred from the lower rotor to the upper rotor through the central duct and then flows through the high-speed rotating packing. Finally, it exits the upper rotor to the condenser through the central duct. Normally, a portion of the condensate will be used as the reflux and its flow rate is measured using a rotameter, which is introduced into the central duct of the upper rotor and then uniformly sprayed on the first rotational baffle through the liquid distributor. As it flows through the rotating packing, it comes into countercurrent contact with the vapor. The liquid is collected and then flows into the central duct of the lower rotor. Finally, it is transferred to the reboiler. The rest condensate will be stored in the storage tank and its flow rate is measured using a rotameter. The feed solution is supplied from the head tank to the bed through the pipe between the upper and lower rotors and its flow rate is measured using a rotameter. The residual liquid spills over automatically from the reboiler. The rotation speed is adjusted by regulating the motor frequency converter. The mass transfer performance is evaluated by the theoretical plate number. The ethanol concentrations in samples collected at the condenser, reboiler, and feed are measured. The theoretical plate number is calculated by plate-by-plate calculation, and the reboiler is considered as a theoretical plate.

The mass transfer performance of the two-layer rotating zigzag bed is shown in Fig. 4.17, where the gas f factor is determined from the annular area between rotational and stationary baffles. As the radial distance between rotational and stationary baffles can be adjusted to ensure the same area for the flow of the gas from the outer edge to the inner edge of the rotor, the value of the gas f factor at the outer edge is equal to that at the inner edge. It is found in Fig. 4.17 that the separation efficiency of up to 25 theoretical plates/m can be obtained, and it increases dramatically with increasing rotation speed at rotation speeds of 800–1200 r/min, after which it remains almost unchanged. At a gas f factor greater than 15 $Pa^{0.5}$, the separation efficiency is low at low rotation speeds. In the case of large fluxes, a higher rotation speed is required to obtain high separation efficiency. This suggests the need to select an appropriate rotation speed based on the flux. The pressure drop as a function of rotation speed under different flux conditions is shown in Fig. 4.18. It is found that a higher rotation speed leads to a higher mass transfer rate but also a higher pressure drop. The pressure drop per theoretical plate first decreases and then increases with increasing rotation speed.

4.4.1.3 Fluid mechanics of the rotating zigzag bed

Unlike other high-gravity rotating beds, both gas and liquid phases flow in a complex zigzag channel in opposite directions in the rotating zigzag bed, and thus unique fluid mechanics is expected. The gas flow velocity vector distribution obtained by computational fluid dynamics (CFD) simulation in the absence of liquid in the rotor is shown in Fig. 4.19, Fig. 4.20. A large number of vortices appear in the channel between the stationary

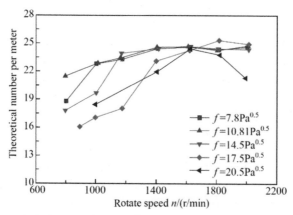

Figure 4.17 The theoretical plate number as a function of rotation speed.

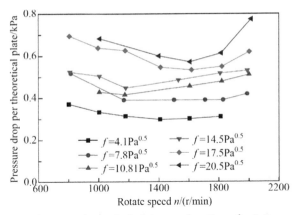

Figure 4.18 The pressure drop per theoretical plate as a function of rotation speed.

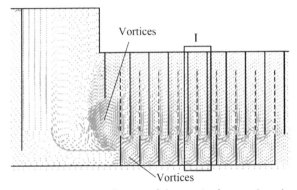

Figure 4.19 The flow velocity vector distribution of the gas in the rotating zigzag bed.

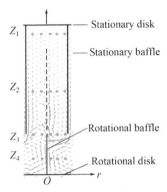

Figure 4.20 An enlarged view of the region indicated in Fig. 4.19.

baffle and the rotational disk, and the resistance here accounts for 55%-73% of the total gas flow resistance. Fig. 4.21 shows the ratios of gas pressure drop in the axial direction and at the turning. The gas flows in the channel between rotational and stationary baffles and its velocity can be decomposed into radial, axial, and tangential components. The radial and axial velocities can be calculated from the flow area, and the tangential velocity is determined by the rotation speed of the rotor and its shear force acting on the gas. A CFD simulation is carried out, where the casing diameter is 450 mm, the inner and outer diameter of the rotor is 101 and 284 mm, and there are 6, 7, 8, and 9 rotational baffles, respectively. The results of the gas phase in the rotating zigzag bed are shown in Fig. 4.22. The tangential velocity of the gas phase in the cavity rapidly reaches the rotation speed of the rotor at the outer edge, and that in the rotor is greater than the linear velocity at the same diameter, which leads to a large

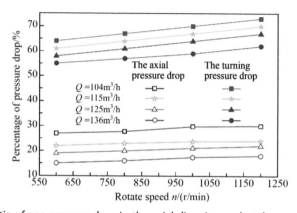

Figure 4.21 The ratio of gas pressure drop in the axial direction and at the turning.

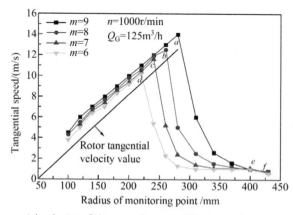

Figure 4.22 The tangential velocity of the gas phase at different radial positions in the rotor.

slip velocity between gas and liquid phases and thus improves the mass transfer process.

The total pressure drop Δp_T of the gas phase in the rotating zigzag bed consists of the pressure drop at the inlet, rotor, and outlet. The pressure drop in the rotor (Δp_R) can be determined by mathematical modeling and it consists of the pressure drop caused by the contraction of the cross-sectional area (Δp_m), centrifugation (Δp_c), and friction (Δp_f):

$$\Delta p_m = \rho_G \int_{r_i}^{r_o} \frac{v_r^2}{r} dr = \frac{\rho_G}{2} \left(\frac{G}{2\pi H \varphi}\right)^2 \left(\frac{1}{r_i^2} - \frac{1}{r_o^2}\right) \quad (4.1a)$$

$$\Delta p_c = \rho_G \int_{r_i}^{r_o} \frac{v_\theta^2}{r} dr \quad (4.1b)$$

$$\Delta p_f = \rho_G \int_{r_i}^{r_o} f_r \frac{v_r^2}{2d_h} dr = \frac{\rho_G}{2\varphi^2 d_h} \left(\frac{G}{2\pi H}\right)^2 \left[\alpha \frac{2\pi H \mu_G}{G d_h \rho_G} \ln \frac{r_o}{r_i} + \beta \left(\frac{1}{r_i} - \frac{1}{r_o}\right)\right] \quad (4.1c)$$

where φ—aperture ratio of the gas flow cross-section;
d_h—hydrodynamic diameter of the rotational baffle, m;
r—radial position, m;
r_i—rotor inner diameter, m;
r_o—rotor outer diameter, m;
H—rotor height, m;
v_θ—gas tangential velocity, m/s;
v_r—gas radial velocity, m/s;
ρ_G—gas density, kg/m³;
μ_G—gas viscosity, Pa · s;
G—gas flow rate, m³/s;
α, β—coefficient.

At a given rotor dimension, the gas tangential velocity is a function of the gas flow rate, liquid flow rate, rotation speed, and radial position; while coefficients α and β are the function of the liquid flow rate and rotation speed. The gas tangential velocity and coefficients α and β can be obtained by experiments. A single-layer rotating zigzag bed is used, as shown in Fig. 4.23, where the casing diameter is 700 mm, the rotor is 486 mm in outer diameter, 214 mm in inner diameter, and 104 mm in height, and there are 4 rotational baffles and 4 stationary baffles. The rotation speed of the rotor is in the range of 600-1200 r/min and the centrifugal acceleration is in the range of 141-563 g. The air-water system is used. The total pressure drop and rotor pressure drop are measured using a U-shaped differential pressure gauge. The

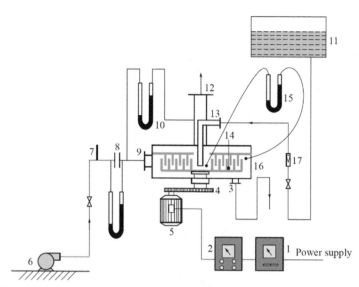

Figure 4.23 The experimental setup for the fluid mechanics in a rotating zigzag bed (1—three-phase watt-hour meter; 2—frequency modulator; 3—liquid outlet; 4—transmission belt; 5—adjustable-speed motor; 6—vortex gas pump; 7—thermometer; 8—orifice plate flowmeter; 9—gas inlet; 10—U-shaped differential pressure gauge for total pressure drop; 11—head tank; 12—gas outlet; 13—liquid inlet; 14—five-hole probe; 15—U-shaped differential pressure gauge for rotor pressure drop; 16—rotating zigzag bed; 17—rotameter).

gas tangential velocities at different radial positions are measured using a five-hole probe.

When the liquid flow rate is zero, the gas tangential velocity as a function of the radial position under different conditions of gas flow rate and rotation speed is shown in Fig. 4.24. In Fig. 4.24, the solid line represents the variation of gas tangential velocity with radial position, and the dashed line represents the variation of rotor linear velocity with radial position. As the gas is introduced into the rotating zigzag bed in the tangential direction, the gas tangential velocity between the outer edge of the rotor and the internal wall of the cavity is equal to the gas flow velocity at the gas inlet, which is lower than the linear velocity of the rotor at the outer edge. According to the angular momentum theorem, the friction moment of the rotor outer edge causes the gas to rotate and leads to a rapid increase in the gas tangential velocity. As the gas enters the rotor, the gas tangential velocity increases continuously with the decrease of rotor radius until a maximum is reached. At this time, the gas tangential velocity is greater than the rotor linear velocity, and thus the friction moment will impede the rotation of the gas and the gas tangential velocity decreases with the decrease of rotor radius. In the rotor, the difference between gas tangential velocity and rotor linear velocity at the same radial position remains almost constant with the increase of rotation speed, but

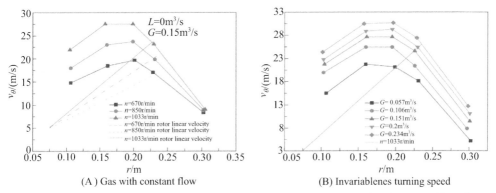

Figure 4.24 The gas tangential velocity as a function of radial position in the absence of liquid.

it increases with the increase of gas flow rate. This is because increasing the gas flow rate can increase the gas tangential velocity between the outer edge of the rotor and the internal wall of the cavity, which in turn increases the gas tangential velocity in the rotor and consequently leads to a greater difference between gas tangential velocity and rotor linear velocity.

The gas tangential velocity as a function of radial position in the presence of liquid is shown in Fig. 4.25. The overall gas tangential velocity is reduced, because the liquid is dispersed into small droplets by the small holes on the rotational baffles and then leaves the rotational baffles in the tangent direction. Thus, the tangential velocity of liquid droplets is equal to the linear velocity of rotational baffles (or the rotor linear velocity), and the gas tangential velocity is higher than the rotor linear velocity (or the tangential velocity of liquid droplets). Thus, the liquid droplets will impede the rotation of the gas and lead to a decrease in the gas's tangential velocity. It is also found the gas tangential velocity decreases slowly with the liquid flow rate.

According to the results in Figs. 4.24 and 4.25, the gas tangential velocity is expressed by the radial position, gas flow rate, and liquid flow rate, and then substituted into formula (4.1b) to obtain the centrifugal pressure drop Δp_c. The pressure drop caused by the contraction of the cross-sectional area (Δp_m) is obtained by formula (4.1a), and the rotor pressure drop Δp_R is obtained by experiments. Then, the coefficients α and β in formula (4.1c) are obtained to calculate the friction pressure drop Δp_f. The rotor pressure drop Δp_R and the pressure drops Δp_m, Δp_c, and Δp_f as a function of liquid flow rate, gas flow rate, and rotation speed are shown in Fig. 4.26. Because the pressure drop caused by the contraction of the cross-sectional area (Δp_m) is related only to the gas flow rate but independent of the liquid flow rate and rotation speed, the Δp_m curves for $L = 0 \text{ m}^3/\text{s}$ and $L = 0.0002778 \text{ m}^3/\text{s}$ are coincident in Fig. 4.26(B) and (C).

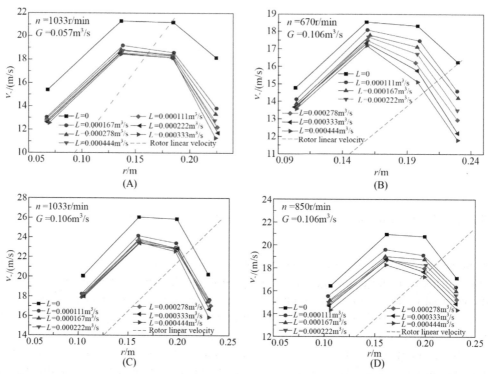

Figure 4.25 The gas tangential velocity as a function of radial position in the presence of liquid at different rotation speeds and gas flows.

As shown in Fig. 4.26, the pressure drop caused by the contraction of the cross-sectional area (Δp_m) is almost negligible, and the centrifugal pressure drop Δp_c and friction pressure drop Δp_f account for approximately 25%–38% and 62%–75% of the rotor pressure drop Δp_R, respectively. The friction pressure drop Δp_f increases with increasing gas flow rate but remains constant with increasing rotation speed because changes in rotation speed have no effect on the friction loss resulting from gas flow. The centrifugal pressure drop Δp_c decreases rapidly at first but then more slowly with increasing liquid flow rate, because the gas tangential velocity will decrease rapidly at first but then more slowly with increasing liquid flow rate. However, the friction pressure drop Δp_f first decreases and then increases slowly with the increase of liquid flow rate, because the introduction of the liquid into the rotor results in a decrease in the gas tangential velocity and consequently a smaller difference between gas tangential velocity and rotor linear velocity, and the friction loss of the gas passing through the small holes on the rotational baffles is reduced. However, as the liquid flow rate continues to increase, the liquid will occupy more space and thus lead to an increase in the friction loss of the gas rotation and eventually a slow increase in the friction pressure drop Δp_f.

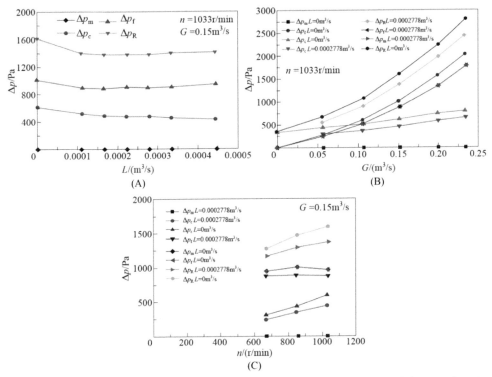

Figure 4.26 The rotor pressure drop Δp_R and the pressure drops Δp_m, Δp_c and Δp_f as a function of (A) liquid flow rate, (B) gas flow rate and (C) rotation speed.

The rotor pressure drop increases with increasing gas flow rate and rotation speed. As the liquid flow rate increases, the centrifugal pressure drop decreases more rapidly at first and then more slowly, the friction pressure drop first decreases and then increases slowly, and the pressure drop caused by the contraction of the cross-sectional area is negligible. Thus, the rotor pressure drop decreases continuously until a constant level is obtained. However, for packed and plate columns, the gas phase pressure drop increases with the increase of liquid flow rate; whereas for the rotating zigzag bed, the gas phase pressure drop first decreases and then remains almost constant with the increase of liquid flow rate because of the effect of the gas tangential velocity on the pressure drop.

The total pressure drop of the rotating zigzag bed can be obtained by the empirical formula. The total pressure drops under different conditions of gas flow rate, liquid flow rate, and rotation speed are determined by the experiments described in Fig. 4.23, and the results are shown in Fig. 4.27. It is found that the variation of the total pressure drop with gas flow rate, liquid flow rate, and rotation speed is similar to that of rotor pressure drop. More specifically, the total pressure drop increases with the increase of gas flow rate and rotation speed, but it first decreases and then remains

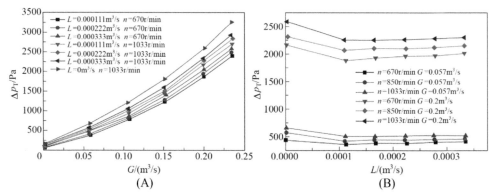

Figure 4.27 The total pressure drop of the gas phase as a function of liquid flow rate, gas flow rate, and rotation speed.

almost constant or increases slowly as the liquid flow rate increases. The empirical formula for the total pressure drop for wet and dry rotating zigzag bed is obtained from the experimental results in Fig. 4.27.

$$\Delta p_{Td} = \frac{1}{2}\rho_G \left(\frac{u_{Gr,avg}}{\varphi}\right)^2 \frac{f_{Td}}{d_h}(r_o - r_i) \tag{4.2a}$$

$$u_{Gr,avg} = \frac{G}{2\pi H r_m}$$

$$f_{Td} = (1.013 \times 10^7) Re_G^{-4.004} + 0.001942 Re_G^{-0.903} Re_\omega^{1.151}$$

$$\Delta p_{Tw} = \frac{1}{2}\rho_G \left(\frac{u_{Gr,avg}}{\varphi}\right)^2 \frac{f_{Tw}}{d_h}(r_o - r_i)$$

$$f_{Tw} = 0.1159 Re_G^{-1.308} Re_L^{0.0384} Re_\omega^{0.936} + 3.819 Re_L^{-0.0331} \tag{4.2b}$$

where Δp_{Td}—total pressure drop of dry bed, Pa;
Δp_{Tw}—total pressure drop of wet bed, Pa;
ρ_G—gas phase density, kg/m³;
$u_{Gr,avg}$—average superficial gas velocity, m/s;
G—gas flow rate, m³/s;
φ—aperture ratio of the gas flow cross-section;
d_h—hydrodynamic diameter of the rotational baffle, m;
r_i—rotor inner diameter, m;

r_o—rotor outer diameter, m;
r_m—rotor average radius, m;
f_{Td}—total pressure drop coefficient of dry bed;
f_{Tw}—total pressure drop coefficient of wet bed;
Re_G—gas phase Reynolds number;
Re_L—liquid phase Reynolds number;
Re_ω—gas phase rotate Reynolds number.

The flooding point is the upper limit for the operation of the gas-liquid mass transfer equipment. For a packed column, the flooding point is defined as the gas flow rate at which the pressure drop increases dramatically. For rotating zigzag bed, no substantial increase in gas phase pressure drop occurs with the increase of gas flow rate. Thus, the gas flow rate and rotation speed are kept constant, and the flooding point is defined as the liquid flow rate at which the total pressure drop of the gas phase increases dramatically. The experimental setup is shown in Fig. 4.23, where the casing diameter is 380 mm, the rotor outer diameter is 284 mm, the rotor inner diameter is 101 mm, the rotor height is 51 mm, and there are 9 rotational baffles, and 9 stationary baffles. Then, the gas flow rate, liquid flow rate, and rotation speed are integrated into two parameters $([u_{G,\Delta r}^2/(r_i\omega^2\Delta r)](\rho_G/\rho_L), (L_m/G_m)\sqrt{\rho_G/\rho_L})$. Then, $(L_m/G_m)\sqrt{\rho_G/\rho_L}$ is used as the x-coordinate and $[u_{G,\Delta r}^2/(r_i\omega^2\Delta r)](\rho_G/\rho_L)$ is used as the y-coordinate, and flooding points are denoted in the logarithmic coordinate system, as shown in Fig. 4.28. The following formula is obtained for the flooding point:

$$\lg\left[\frac{u_{G,\Delta r}^2}{r_i\omega^2\Delta r}\left(\frac{\rho_G}{\rho_L}\right)\right] = -2.281 - 0.9788\left[\lg\left(\frac{L_m}{G_m}\sqrt{\frac{\rho_G}{\rho_L}}\right)\right] - 0.1605\left[\lg\left(\frac{L_m}{G_m}\sqrt{\frac{\rho_G}{\rho_L}}\right)\right] \quad (4.3)$$

where $u_{G,\Delta r}$—flooding velocity calculated based on the annular area between the innermost rotational and stationary baffles, m/s;

r_i—rotor inner diameter, m;
ω—angular velocity, rad/s;
Δr—radial distance between the innermost rotational and stationary baffles, m;
ρ_G—gas phase density, kg/m³;
ρ_L—liquid phase density, kg/m³;
L_m—liquid mass flux based on the area of rotor inner edge, kg/(m² · s);
G_m—gas mass flux based on the area of rotor inner edge, kg/(m² · s).

The liquid flooding curve for the rotating zigzag bed is plotted in Fig. 4.28 based on formula (4.3). It is observed that the flooding curve of the rotating zigzag bed is close to but under the Sherwood curve. This is because the liquid flows outward in

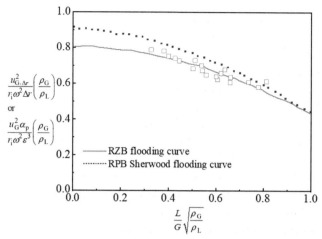

Figure 4.28 The flooding point curve in the rotating zigzag bed.

the radial direction under the centrifugal force in an RPB, but in the rotating zigzag bed, the liquid film that flows downward on the internal wall of the stationary baffle under gravity is more likely to be entrained by the upward flowing gas to produce liquid flooding.

4.4.1.4 Shaft power of rotating zigzag bed

The shaft power of the rotating zigzag bed consists of the power for the passage of the liquid through the rotor N_L, the friction between the rotating rotor and the gas in the rotor N_W, the mechanical friction of the shaft N_M, and the passage of the gas through the rotor N_G. As the liquid is repeatedly accelerated by rotational baffles and intercepted by stationary baffles, N_L accounts for the largest proportion of shaft power. It is determined by mathematical models and experiments. It is assumed that (1) the liquid flow is perfect with no friction loss in the rotor; (2) the tangential velocity of the liquid on the rotational baffle is equal to the linear velocity of the rotational baffle; and (3) the tangential velocity of liquid droplets leaving the rotational baffle in the tangential direction is equal to the linear velocity of the rotational baffle. A mathematical model is established, and the ideal power N_{LD} is obtained for the passage of the liquid through the rotor when the gas flow rate is zero.

$$N_{LD} = \rho_L L \omega^2 \left[r_{11}^2 + \sum_{i=1}^{m} \left(r_{11}^2 - \frac{r_{1,i-1}^2}{r_{1,i}^2} \right) \right] \tag{4.4}$$

where ρ_L—liquid density, kg/m³;
L—liquid flow rate, m³/s;
ω—angular velocity, rad/s;

r_{11}—radius of the innermost rotational baffle, m;
$r_{1,i}$—radius of rotational baffle i from the innermost rotational baffle, m;
m—number of rotational baffles.

According to Formula (4.4), N_{LD} is proportional to the square of the angular velocity and the first power of the liquid flow rate.

The experimental setup is shown in Fig. 4.23, where the casing diameter is 800 mm, the rotor outer diameter is 621 mm, the rotor inner diameter is 250 mm, the rotor height is 80 mm, and there are 6 rotational baffles and 6 stationary baffles. The motor input power under different conditions of gas flow rate, liquid flow rate, and rotation speed is measured by a three-phase kilowatt-hour meter equipped with a stopwatch. Fig. 4.29 reveals that the motor input power increases with the increase of liquid flow rate and rotation speed, but more slowly with the increase of gas flow rate. Increasing the liquid flow rate increases the N_{LD} and subsequently the shaft power and input power, whereas increasing the rotation speed increases the N_{LD}, N_W, and N_M and subsequently the shaft power and input power. The slow increase of motor input power with increasing gas flow rate suggests that N_G accounts for only a small proportion of shaft power.

The shaft power is the motor input power multiplied by the power factor and transmission efficiency. When both gas and liquid flow rates are zero, the shaft power under different rotation speeds is the sum of N_M and N_W. N_G is obtained by subtracting the sum of N_M and N_W from the shaft power under different gas flow rates and rotation speeds when the liquid flow rate is zero. N_{LR} is obtained by subtracting the sum of N_M and N_W from the shaft power under different liquid flow rates and rotation speeds when the gas flow rate is zero. N_L is obtained by subtracting the sum of N_M, N_W, and N_G from the shaft power under different gas flow rates, liquid flow rates, and rotation speeds.

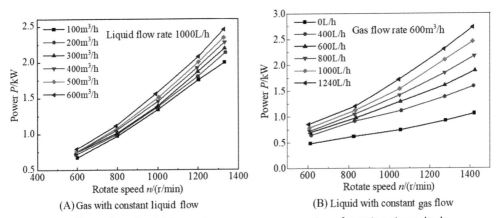

Figure 4.29 Effects of gas/liquid flow rate on power consumption of rotating zigzag bed.

The N_{LR}-N_{LD} and N_L-N_{LR} relationships are described as follows:

$$N_L = K_1 N_{LD} \tag{4.5a}$$

$$N_{LR} = K_2 N_{LD} \tag{4.5b}$$

where K_1, K_2—coefficients.

The coefficients K_1 and K_2 are obtained by experiments, and then N_L is calculated from formulas (4.4) and (4.5). The empirical formulas for N_M, N_W, and N_G are obtained from experimental data.

At rotation speeds >1000 r/min, N_L, N_W, N_M, and N_G account for 60.7%–66.1%, 3.7%–4.3%, 28.3%–35.0%, and 0.6%–1.3% of shaft power, respectively.

4.4.2 Characteristics of high-gravity distillation in rotating packed bed

4.4.2.1 Rotating packed bed structure

The packing used for high-gravity distillation in an RPB is shown Fig. 4.30. Either single-stage or multi-stage packing can be used, and the gas always flows countercurrent to the liquid in the RPB. The structure of the one-stage RPB is shown in Fig. 4.1, while that of multi-stage RPB is shown in Figs. 4.31 and 4.32.

4.4.2.2 Mass transfer performance of rotating packed bed

The experimental setup for testing the mass transfer performance of the two-stage RPB is shown in Fig. 4.33. One stage is used for distillation and the other stage is

(A) Screen corrugated packing

(B) Orthogonal mesh packing

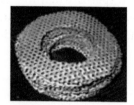
(C) Corrugated disc packing

(D) Finned guide plate packing

(E) Threaded bobbin packing

Figure 4.30 The packing used for high-gravity distillation in an RPB. *RPB*, Rotating packed bed.

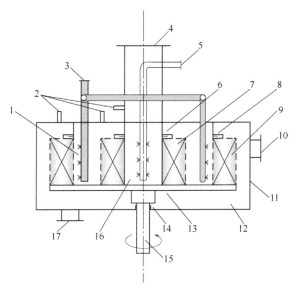

Figure 4.31 Two-stage high-gravity distillation equipment (1—liquid distributor; 2—pressure measurement port; 3—feed solution inlet; 4—gas outlet; 5—reflux inlet; 6, 8—seals; 7—first-stage packing; 9—second-stage packing; 10—gas inlet; 11—casing; 12—external cavity; 13—rotor; 14—bearing; 15—rotor shaft; 16—inner cavity; 17—liquid outlet).

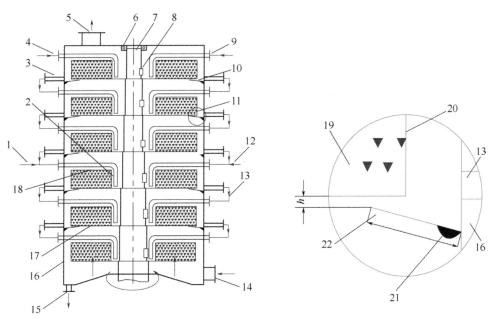

Figure 4.32 High-gravity distillation equipment with cross-flow in a single layer but overall countercurrent flow (1, 12—feed inlet; 2—liquid distributor; 3—sampling port; 4, 9—reflux inlet; 5—gas outlet; 6—bearing shaft; 7—rotor shaft; 8—setting wedge; 10—liquid receiver disk; 11—rotor; 13—downcomer; 14—gas inlet; 15—liquid outlet; 16—casing; 17—lower support disk of packing; 18—upper support disk of packing; 19—RPB specific packing; 20—rotor outer roller; 21—weld between liquid receiver disk and the casing; 22—inner edge of liquid receiver disk).

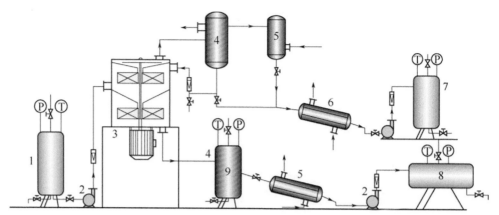

Figure 4.33 The experimental process for continuous high-gravity distillation (1—feed storage tank; 2—pump; 3—multi-stage rotating bed; 4—condenser; 5—total condenser; 6—heat exchanger; 7—product tank 1; 8—product tank 2; 9—reboiler).

used for stripping. The feed solution is pumped into the lower packing of the RPB and its flow rate is measured using a rotameter. As the liquid flows in the radial direction from the inner edge of the packing to the outer edge of the packing under the centrifugal force, it comes into contact with the steam introduced into the packing from the reboiler, leading to the vaporization of some feed solution and the condensation of some steam. The liquid flows into the reboiler through the liquid outlet, and uncondensed steam enters the upper packing, where it comes into contact with the reflux and causes the vaporization of a portion of the reflux and the condensation of a portion of the steam. Unvaporized liquid and feed enter the lower packing, and uncondensed steam is discharged at the top into the condenser for condensation. Some condensed liquid is returned to the upper packing as reflux, and some are stored in the product tank.

The mass transfer performance of RPB is closely related to the packing, rotor size, and operating parameters. The mass transfer performance of RPB for ethanol-water mixture is discussed, where the high-gravity factor is 1.99-48.88, the reflux ratio is 1-3.5, and the feed flow rate is 8-24 L/h.

(1) Effects of operating parameters on the mass transfer

The effects of various operating parameters on the mass transfer are investigated using rotor IV, where the feed flow rate is 15 L/h and the reflux ratio is 3 for packing I, II, and III, and the feed flow rate is 12 L/h and the reflux ratio is 2 for packing IV. The theoretical plate number as a function of high-gravity factor is shown in Figs. 4.34-4.37. The theoretical plate number increases with the increase of high-gravity factor irrespective of the type of packing. Increasing the high-gravity factor can reduce the liquid film thickness and thus increase the gas-liquid contact area.

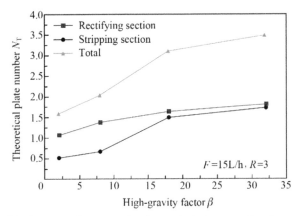

Figure 4.34 Effect of high-gravity factor on the theoretical plate number for packing I.

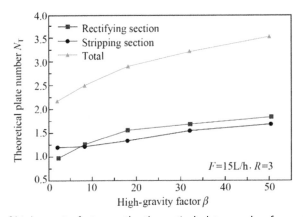

Figure 4.35 Effect of high-gravity factor on the theoretical plate number for packing II.

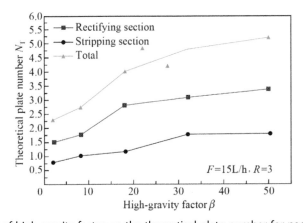

Figure 4.36 Effect of high-gravity factor on the theoretical plate number for packing III.

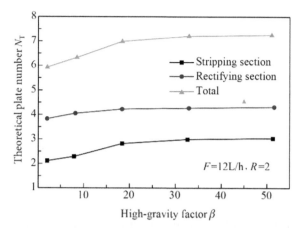

Figure 4.37 Effect of high-gravity factor on the theoretical plate number for packing IV.

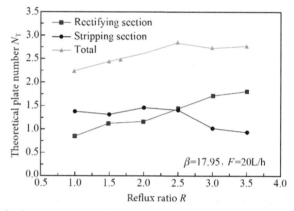

Figure 4.38 Effect of reflux ratio on the theoretical plate number for packing I.

These two factors contribute to improving the mass transfer process. The largest theoretical plate number is 4.3 (distillation section in Fig. 4.37) and the smallest is 0.5 (stripping section in Fig. 4.34), implying that the theoretical plate height of the RPB is 5.8–49.0 mm.

The theoretical plate numbers as a function of reflux ratio for packing I, II, III at a high-gravity factor of 17.95 and a feed flow rate of 20 L/h and for packing IV at a high-gravity factor of 18.54 and a feed flow rate of 12 L/h are shown in Figs. 4.38–4.41, respectively. It is found that the theoretical plate number in the distillation section and the whole system increases with the increase of reflux ratio; while in the stripping section, it decreases with the increase of reflux ratio for packing I, II, III but increases with the increase of reflux ratio for packing IV. Irrespective of the reflux ratio, the liquid flow rate in the distillation section is

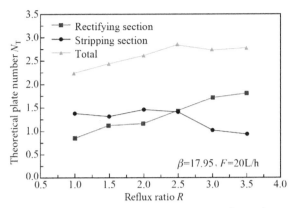

Figure 4.39 Effect of reflux ratio on the theoretical plate number for packing II.

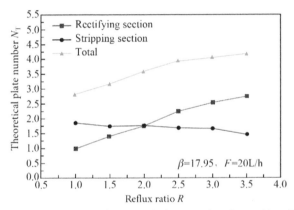

Figure 4.40 Effect of reflux ratio on the theoretical plate number for packing III.

always lower than that in the stripping section. At smaller reflux ratios, the liquid flow rate in the distillation section would be too low to completely wet the packing, which makes it difficult for the upward flow of the gas to contact with the reflux on the packing surface. As the reflux ratio increases, a larger area of the packing is wetted for mass transfer. In the stripping section, as the reflux ratio increases and the high-gravity factor is kept constant, the liquid film will become thicker which is not favorable for mass transfer. The largest theoretical plate number is 4.98 (distillation section in Fig. 4.41) and the smallest is 0.84 (stripping section in Fig. 4.38). The theoretical plate height is 4.93–29.76 mm.

The theoretical plate numbers as a function of feed flow rate for packing II at a high-gravity factor of 49.85 and a reflux ratio of 1 and for packing IV at a high-gravity factor of 18.54 and a reflux ratio of 1 are shown in Figs. 4.42 and 4.43, respectively. The theoretical plate number increases with increasing feed flow rate, and it is

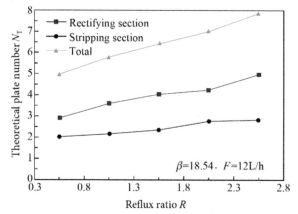

Figure 4.41 Effect of reflux ratio on the theoretical plate number for packing IV.

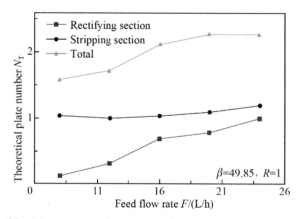

Figure 4.42 Effect of feed flow rate on the theoretical plate number for packing II.

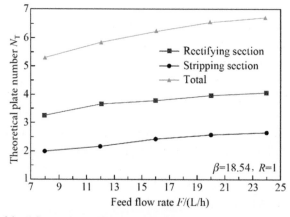

Figure 4.43 Effect of feed flow rate on the theoretical plate number for packing IV.

closely related to the wet area of the packing, liquid film thickness, and gas–liquid contact time.

(2) Mass transfer performance of different types of packing

The mass transfer performance of different types of packing is shown in Figs. 4.44 and 4.45, where rotor IV is used, the feed flow rate is 8 L/h, and the reflux ratio is 1. Irrespective of the packing type, the *HETP* increases with the increase of high-gravity factor, and it is in the range of 13.6–24.6 mm for packing I, 14.6–21.0 mm for packing II, 10.2–20.0 mm for packing III, and 6.9–8.4 mm for packing IV, respectively. Packing IV has the best mass transfer performance. The mass transfer in the high-gravity distillation process is closely associated with the

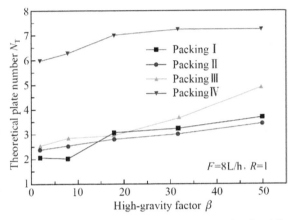

Figure 4.44 Effect of high-gravity factor on the theoretical plate number for different types of packing.

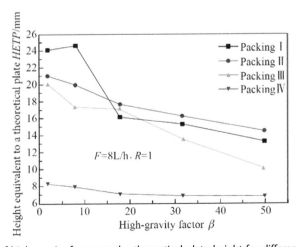

Figure 4.45 Effect of high-gravity factor on the theoretical plate height for different types of packing.

structure and characteristics of the packing. Of the four types of packing, packing I shows the smallest geometric specific surface area and consequently the smallest gas–liquid contact area. The disk packing has a simple structure and the liquid flows only on the plane disk, and the gas–liquid contact time is shorter than that of the wire mesh packing, which leads to poorer mass transfer performance compared to wire mesh packing. There is a difference in the effective specific surface area between packing II and III that are wire mesh packing with the same wire mesh diameter. Packing III shows lower bulk density and higher porosity, leading to a larger effective specific surface area compared to packing II. As packing III is irregular, more complex gas–liquid flow channels are formed, which leads to an increase in gas–liquid contact time and thus favors the mass transfer. Packing III shows better mass transfer performance than packing II. For packing I, II, and III, liquid flows almost in the radial direction and the residence time in the packing is short because of the effect of high gravity. For packing IV, the liquid flows in a 几乙 shaped path, which can greatly increase the gas–liquid contact time and consequently the separation efficiency.

The theoretical plate number and height as a function of reflux ratio for different types of packing at a feed flow rate of 12 L/h and a high-gravity factor of 17.95 is shown in Figs. 4.46 and 4.47, respectively. The theoretical plate number increases with the increase of reflux ratio irrespective of the type of packing. The theoretical plate height is in the range of 17.6-21.2 mm for packing I, 17.6-22.6 mm for packing II, 12.8-19.0 mm for packing III, and 6.59-8.84 mm for packing IV, respectively.

The theoretical plate number as a function of feed flow rate for different types of packing at a high-gravity factor of 17.95 and a reflux ratio of 1 is shown in Fig. 4.48. For each type of packing, the theoretical plate number increases with the increase in feed flow rate, which implies that the processing capacity of the

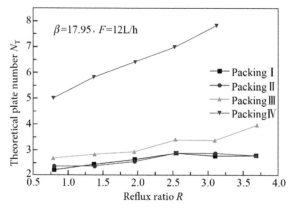

Figure 4.46 Effects of reflux ratio on the theoretical plate number for different types of packing.

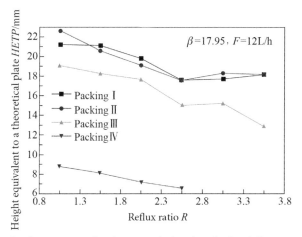

Figure 4.47 Effects of reflux ratio on the theoretical plate height for different types of packing.

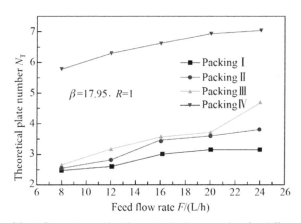

Figure 4.48 Effects of feed flow rate on the theoretical plate number for different types of packing.

equipment can be further enhanced. Fig. 4.49 shows the relationship between theoretical plate height and feed flow rate for different types of packing under the same operating conditions. It is found that the theoretical plate height is in the range of 18.7-25.4 mm for packing I, 11.8-23.6 mm for packing II, 15.2-25 mm for packing III, and 6.76-9.63 mm for packing IV, respectively, confirming again that packing IV has the best mass transfer performance.

(3) Mass transfer performance for different rotor structures

The gas and liquid flow patterns are very complex in the RPB. In the RPB with countercurrent gas-liquid contact, as the steam flows from the outer edge of the rotor to the inner edge of the rotor, the cross-sectional area of the channel is reduced and because of this, the gas flow velocity is increased. As the liquid flows

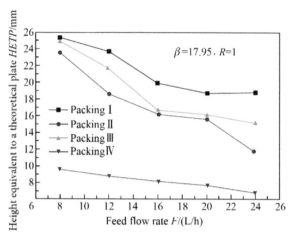

Figure 4.49 Effects of feed flow rate on the theoretical plate height for different types of packing.

from the inner edge of the rotor to the outer edge of the rotor, the cross-sectional area of the channel is increased, but the liquid flow velocity is increased because of the centrifugal force. Thus, the gas flow velocity reaches a maximum at the inner edge of the packing where the liquid flow velocity reaches a minimum. Attention should be paid to the design of rotor structure to ensure uniform distribution of the liquid in the packing in order to achieve better mass transfer performance.

Take packing III as an example. The mass transfer performances of three rotors are investigated under conditions of atmospheric pressure, total reflux, high-gravity factor ranging from 10.74 to 386.58, gas kinetic energy factor ranging from 0.22 to 0.91, and a reflux rate of 4.4 to 20 L/h. The four rotors differ in the inner and outer diameter, and the structural parameters are summarized in Table 4.1. The theoretical plate number and height as a function of high-gravity factor for the three rotors at a reflux rate of 10.4 L/h are shown in Figs. 4.50 and 4.51, respectively. For the three rotors, the theoretical plate number first increases and then decreases with the increase of high-gravity factor. The theoretical plate height is in the range of 10.9–14.0 mm for rotor I, 9.42–10.55 mm for rotor II, and 10.0–11.5 mm for rotor III, respectively. Rotor II shows the best mass transfer performance. At smaller inner diameters, the gas flow rate is high and liquid foam is likely to be entrained, which can impede mass transfer. The liquid is dispersed in a large area at the outer ring, which implies a large gas–liquid contact area and thus favors mass transfer. The gas–liquid contact time is long in packing II because of its large radial thickness (Table 4.2).

Figs. 4.52 and 4.53 show the theoretical plate number and height against gas phase kinetic energy factor curves for different rotors at a high-gravity factor of 171.8, respectively. As the gas phase kinetic energy factor increases, the maximum

Table 4.1 Characteristic parameters of the rotor.

Rotor	Outer diameter/mm	Inner diameter/mm	Height/mm	Thickness/mm	Porosity	Material
Rotor I	180	60	40	2	0.65	Stainless steel
Rotor II	180	80	40	2	0.65	
Rotor III	140	80	40	2	0.65	
Rotor IV	110	60	30	2	0.65	

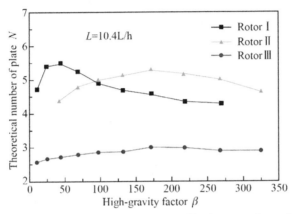

Figure 4.50 The theoretical plate number as a function of high-gravity factor for different rotors.

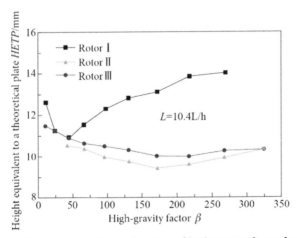

Figure 4.51 The theoretical plate height as a function of high-gravity factor for different rotors.

theoretical plate number appears for the three rotors. The gas flow velocity in the packing increases with increasing gas phase kinetic energy factor, increasing the gas–liquid turbulence and mass transfer. Under total reflux conditions, increasing the gas phase kinetic energy factor can also increase the reflux rate and the wet area, which favors the mass transfer. However, it also leads to a reduction of the residence time in the packing and a substantial increase in the gas flow rate at the inner edge of the packing. The violent gas flow can also carry the liquid out and leads to liquid foam entrainment. These two factors significantly reduce the mass transfer.

The theoretical plate height is 13.1-16.6 mm for rotor I, 9.4-10.0 mm for rotor II, and 10-11.4 mm for rotor III.

Figs. 4.54 and 4.55 show the theoretical plate number and height as a function of reflux rate for different rotors under total reflux conditions at a high-gravity factor of 171.8, respectively. The effect of the reflux rate on the mass transfer of different rotors is similar to that of gas-phase kinetic energy factor and rotor II shows the best mass transfer performance.

(4) Comparison of mass transfer characteristics with traditional distillation columns

The comparison of mass transfer characteristics between high-gravity distillation equipment and traditional distillation columns is shown in Table 4.3.

Table 4.2 Characteristic parameters of the packing.

Packing	Bulk density /(kg/m³)	Geometric specific surface area/(m²/m³)	Porosity	Material
Packing I (corrugated plate)	950	400	0.82	Stainless steel
Packing II (orthogonal mesh)	1100	1750	0.86	Stainless steel (0.285 mm in diameter)
Packing III (silk screen corrugated)	400	1100	0.95	Stainless steel (0.285 mm in diameter)
Packing IV (fin baffle)	735	450	0.5	Stainless steel

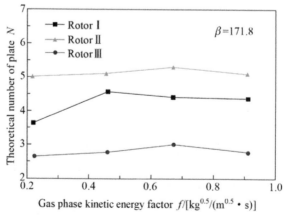

Figure 4.52 The theoretical plate number as a function of gas-phase kinetic energy factor for different rotors.

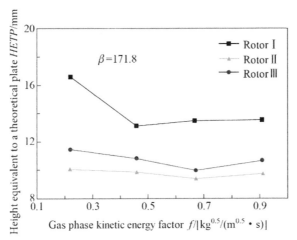

Figure 4.53 The theoretical plate height as a function of gas-phase kinetic energy factor for different rotors.

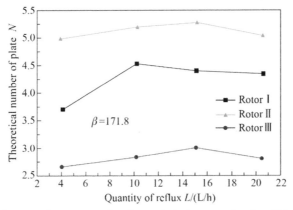

Figure 4.54 Effect of rotor reflux rate on the theoretical plate number for different rotors.

Table 4.3 reveals that the theoretical plate number per meter is 20.8–151.7 for high-gravity distillation equipment and 2.5–9 for traditional distillation column, indicating that the mass transfer efficiency of the high-gravity distillation equipment is 1–2 orders of magnitude higher than that of the traditional packed column.

4.4.2.3 Fluid mechanics of rotating packed bed

In the distillation process, the factors affecting the momentum transfer include operating conditions, equipment structure, and the physical properties of the packing. The distillation process in RPB under atmospheric pressure and total reflux conditions is schematically shown in Fig. 4.56, where the rotor inner diameter is 60 mm,

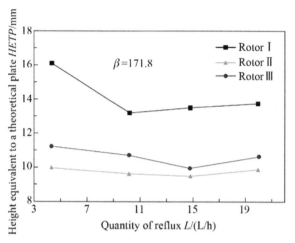

Figure 4.55 Effect of rotor reflux rate on the theoretical plate height for different rotors.

Table 4.3 Comparison of mass transfer characteristics between high-gravity distillation equipment and traditional distillation columns.

Distillation equipment	Packing	Specific surface area/(m²/m³)	Theoretical plate number per meter
Traditional packed column [53]	Orifice corrugated packing	250-450	2.5-3.5
	Mesh corrugated packing	650	6-9
	Ceramic corrugated packing	470	4-6
High-gravity distillation equipment	Corrugated plate packing	400	20.8-76.9
	Orthogonal mesh packing	1750	47.2-56.8
	Wire mesh corrugated packing	1100	56.8-91.7
	Fin baffle packing	450	101.6-151.7

the rotor outer diameter is 180 mm, the axial direction height is 40 mm, the diameter of the wire mesh corrugated packing is 0.285 mm, and the porosity of the packing is 0.95.

The liquid mixture to be separated is vaporized in the reboiler, and its flow rate, pressure, and temperature are measured using a gas flow meter, a pressure gauge, and a thermometer, respectively. It is introduced into the external cavity of the RPB through the gas inlet tube and then flows through the packing from the outer edge

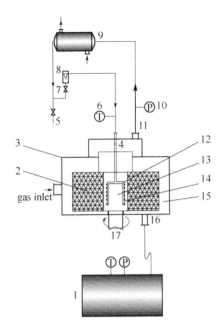

Figure 4.56 The distillation process in RPB (1—reboiler; 2—packing; 3—shell; 4—liquid inlet; 5—sampling port; 6—thermometer; 7—valve; 8—flowmeter; 9—condenser; 10—U-shaped differential pressure meter; 11—gas outlet; 12—liquid distributor; 13—center tube; 14—nozzle; 15—external cavity; 16—liquid outlet; 17—rotor shaft). *RPB*, Rotating packed bed.

to the inner edge driven by the pressure gradient. After that, it is collected in the central tube of the RPB and then flows into the condenser from the gas outlet. The pressure at the outlet is measured. The condensed liquid is introduced into the central distributor of RPB, and its flow rate and temperature are measured using a liquid rotameter and a thermometer, respectively. Then, it is sprayed onto the inner edge of the packing through the nozzle. The packing rotates at a high speed driven by the motor, creating a high gravity field that can force the liquid to flow outwards. Finally, the liquid flows into the reboiler again through the liquid outlet for subsequent reuse.

The relationship of the gas phase pressure drop (Δp) with the gas phase kinetic energy factor, high-gravity factor, and reflux rate in the RPB is shown in Figs. 4.57–4.59, respectively. All other conditions being equal, the gas phase pressure drop increases with the increase of gas phase kinetic energy factor, high-gravity factor, and reflux rate, and the total pressure drop is 54–411 Pa.

Table 4.4 shows the gas phase pressure drop per theoretical plate of the RPB under different conditions of high-gravity factor and gas phase kinetic energy factor. The gas phase pressure drop is in the range of 14.84–88.75 Pa, and the pressure drop at the highest mass transfer efficiency is 31.18–37.41 Pa.

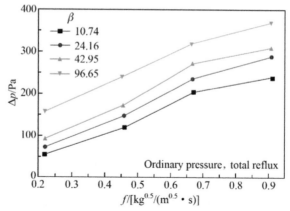

Figure 4.57 The relationship between gas phase pressure drop and gas phase kinetic energy factor.

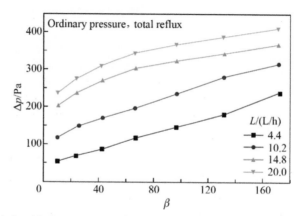

Figure 4.58 The relationship between gas phase pressure drop and high-gravity factor.

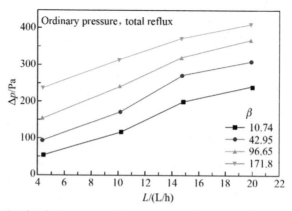

Figure 4.59 The relationship between gas phase pressure drop and reflux rate.

Table 4.4 Gas phase pressure drop per theoretical plate (Unit: Pa).

β	$f/[kg^{0.5}/(m^{0.5} \cdot s)]$			
	0.22	0.46	0.67	0.91
10.74	14.84	24.78	46.74	54.06
24.16	18.60	27.37	55.47	62.64
42.95	22.5	31.18	56.03	64.75
67.11	29.21	37.41	57.91	72.32
96.65	38.99	48.16	68.51	74.21
131.55	45.14	59.48	74.56	83.07
171.8	66.25	68.70	82.77	88.75

Table 4.5 Comparison between high-gravity distillation equipment and traditional packed column.

Distillation equipment	Packing	Specific surface area/(m²/m³)	Mass transfer unit height (range)/mm	Theoretical plate number per unit bed height	Minimum pressure drop per theoretical plate/Pa
Traditional packed column	Orifice corrugated packing	250-450	280-400	2.5-3.5	38.1
	Mesh corrugated packing	650	120-180	6-9	29.6
	Ceramic corrugated packing	470	180-250	4-6	100
High-gravity distillation equipment	Wire mesh corrugated packing	7100	10.9-17.6	56.8-91.7	14.8
	Fin baffle packing	450	4.82-11.88	101.6-151.7	13.5

4.4.3 Comparison between high-gravity distillation equipment and traditional distillation columns

Table 4.5 shows the comparison between high-gravity distillation equipment and traditional packed column reported in a previous study [71]. The mass transfer unit height of the high-gravity distillation equipment is 4.82-17.6 mm, which is about one order of magnitude higher than that of a traditional packed column (120-400 mm). The high-gravity equipment has better mass transfer efficiency than that of the traditional packed column. The equipment size could be reduced by up to several tens of times. The unique advantages of high-gravity equipment, such as small size, low space requirement, low investment, low power consumption, low liquid flooding, and high gas flow rate, make it a promising choice in the industry. It is shown in Table 4.5 that the packing with a large specific surface area can be used in high-gravity equipment, which is not possible for the packed column. In general, the packing with a high specific surface area has large bulk density but low porosity that may increase the gas resistance in the distillation column. In high-gravity distillation, the liquid passes through the packing in the forms of micro/nano liquid droplets, filaments, and films, and the liquid-gas ratio is much lower than that in a packed column. Thus, even though the specific surface area is 10 times higher than that of common distillation columns, the gas resistance (pressure drop) is not higher than half of the resistance of packed column. Despite its unique advantages, new problems may arise in the large-scale application and industrial scale-up.

4.5 Application examples

The rotating zigzag bed as a novel high-efficiency distillation equipment has many advantages over traditional distillation columns, such as small equipment height and size. Since the successful operation of the first rotating zigzag bed in 2004 in China, several hundreds of such beds have been used for ordinary or special distillation processes, such as methanol-water, ethanol-water, acetone-water, DMSO-water, DMF-water, ethyl acetate-water, methanol-methylal-water, methanol-tertiary butanol, ethyl acetate-methylbenzene-water, methanol-tertiary butanol, chlorobenzene-isohexane, triethylamine-methyl isopropylamine-water, methylene chloride-siloxane, morpholine-methanol-water, methylene chloride-water, THF-methanol-water, absolute alcohol, and anhydrous acetonitrile. Some beds have been operating stably for several decades.

4.5.1 Ordinary distillation

The fine chemical industry, such as medicine, dye, and pesticide, requires the use of a wide variety of solvents, but their amounts are generally low. In this circumstance, the rotating zigzag bed is suitable for the distillation of solvents because of low liquid holdup and short time needed for startup and shutdown. From the engineering perspective, the rotating zigzag bed is characterized by small volume, flexible installation, and easy operation and maintenance. The use of the rotating zigzag bed for a continuous distillation process is shown in Figs. 4.60 and 4.61. The rotating zigzag bed has been used for the distillation of ethanol-water mixture in a pharmaceutical company in Zhejiang Province, China, where the rotor is 630 mm in diameter and has two layers, the diameter of the casing is 800 mm, and the height is 550 mm. Continuous distillation is carried out. The ethanol volume fraction of the feed is 40%, and the reflux ratio is 2.5. The ethanol volume fraction is 95% in the product and 0.5% in the residual liquid in the reboiler, and the output is 4.5 t/d.

Figure 4.60 The rotating zigzag bed used for distillation of ethanol-water mixture.

Figure 4.61 The rotating zigzag bed used for distillation of methanol-water mixture.

The rotating zigzag bed is also used for distillation of methanol–water mixture in a chemical company in Jiaxing, Zhejiang Province, China, where the rotor is 750 mm in diameter and 80 mm in thickness and has three layers, the diameter of the casing is 830 mm, and the height is 800 mm. Continuous distillation is carried out. The methanol mass fraction is 70% in the feed, and the reflux ratio is 1.5. The methanol mass fraction is 99.7% in the final product and 0.5% in the residual solution in the reboiler, and the output is 12 t/d.

Table 4.6 reveals the comparison of height and volume between rotating zigzag bed and traditional packed column under the same conditions. The use of rotating

Table 4.6 Comparison between rotating zigzag bed and traditional packed column.

Materials	Equipment	Diameter/m	Height/m	Volume/m³	Height/volume ratio
Ethanol-water	Rotating zigzag bed	0.8	0.55	0.276	16.4/4.1
	Packed column	0.4	9.0	1.13	
Methanol-water	Rotating zigzag bed	0.83	0.8	0.433	13.8/7.2
	Packed column	0.6	11.0	3.11	

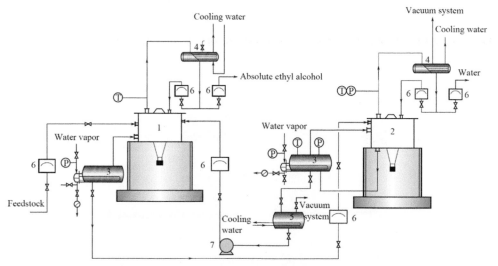

Figure 4.62 Extractive distillation for the production of absolute alcohol using a rotating zigzag bed (1—anhydrous ethanol rotating bed; 2—extractant recovery rotating bed; 3—reboiler; 4,5—condenser; 6—flowmeter; 7—pump).

zigzag bed can significantly reduce the equipment height and size, and it is a small-scale and resource-saving gas-liquid mass transfer equipment. These two projects provide a good example of continuous distillation in a high-gravity environment.

4.5.2 Special distillation

The rotating zigzag bed has been used for extractive distillation for the production of absolute alcohol in a pharmaceutical company. The flowchart is shown in Fig. 4.62, and the on-site photo is shown in Fig. 4.63. The ethanol mixture with a mass fraction of 90% is introduced to the rotating bed 1. The absolute alcohol with an ethanol content higher than 99.7% is obtained at the top of the bed, and the mixture of extractant and water at the bottom is transferred to the rotating bed 2 for recovery of extractant

Figure 4.63 The on-site photo of the extractive distillation for the production of anhydrous alcohol using a rotating zigzag bed.

Table 4.7 The size and major operating parameters for the two high-gravity beds.

No.	Rotor		Casing		Power/kW	Reflux ratio	Processing capacity/(t/d)	Moisture content in product/%
	Diameter/m	Number of layer	Diameter/m	Height/m				
1	0.75	3	0.83	0.8	11	1	9.6	<0.3
2	0.75	2	0.83	0.6	5.5	1	–	<0.2

that is operated at a negative pressure. The water is distilled from the top of the bed and the extractant on the bottom is cooled for reuse. Table 4.7 shows the size and major operating parameters for the two high-gravity beds.

4.5.3 Distillation of high-viscosity and heat-sensitive materials

In the fine chemical industry, such as medicine, pesticide, and dye, there is a need for distillation of high-viscosity and heat-sensitive materials. In these specific contexts, the use of a conventional distillation column may not be appropriate because the residence time is long and the decomposition or polymerization of heat-sensitive materials is likely to cause the blockage of the column. In comparison, the residence time is much short in a high-gravity bed and thus little or even no thermal decomposition occurs. The bed will be less likely to be blocked by high-viscosity materials because of the high gravity, and these materials can be supplied at the center of the rotor. In a company, the packed column is originally used for the removal of solvents in the production of vitamin B5. In order to avoid the blockage of the packing, they are fed at the bottom of the column and subjected to steam heating. The column is operated intermittently. The solvent content is high in the final product, and the residence time of these materials in the column is long which causes obvious decomposition. In order to prevent product decomposition, vacuum operation is used to reduce the temperature, and solvent vapor needs to be condensed by refrigerating fluid and thus a set of vacuum and refrigerator units are installed. The product quality could not meet the standard and energy consumption is very high. With the use of rotating zigzag bed, the materials are directly introduced into the center of the rotor without the use of reboiler. The temperature can be properly increased because of the short residence time. The operation is conducted under atmospheric pressure instead of vacuum, and the solvent vapor is condensed using circulating cooling water without the use of vacuum and refrigerator units. The equipment investment, volume, and energy consumption are also dramatically reduced and the product quality is greatly improved. The process is shown in Fig. 4.64, and the on-site photo is shown in Fig. 4.65. Table 4.8 shows the comparison in the technological and economic indexes between new and old technologies.

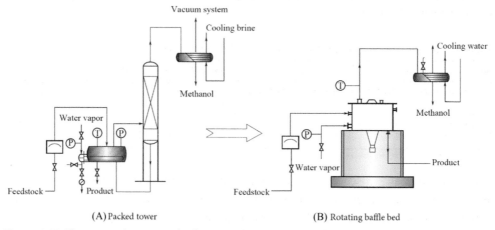

Figure 4.64 The removal process of solvents in the production of vitamin B5.

Figure 4.65 The removal process of solvents in the production of vitamin B5.

Table 4.8 Comparison of technological and economic indexes between rotating zigzag bed and packed column.

Index		Rotating zigzag bed	Racked column
Processing capacity/(t/h)		2	2
Product solvent content/%		0.1	0.3
Reboiler volume/m^3		No	8
Cooling liquid		Tap water	Brine chilling
Vacuum unit		No	Yes
Motor power/kW	Refrigerator unit	No	20
	Vacuum unit	No	15
	Rotating bed motor	18	No
	Total power consumption	18	35
Equipment floor requirement/m^2		4	16
Equipment space requirement/m^3		12	80

4.5.4 Distributed distillation

The rotating zigzag bed has several important advantages over traditional columns. From an economical perspective, the materials, space, and land required for manufacturing and installation of the rotating zigzag bed are substantially reduced because of its small volume and short height. From a social perspective, it avoids potential risks related to installation, testing, operation, and maintenance at heights and the damages caused by natural disasters such as earthquakes, typhoons, and lightning strokes, ensuring high operation safety and stability. The rotating zigzag bed can be installed flexibly at an appropriate position, which takes advantage of the available space and forms the distributed distillation mode. Many chemical plants in China have used rotating zigzag beds for recovery of solvents (Fig. 4.66) without the need for centralized treatment [Fig. 4.66(B)]. In this way, material transfer is dramatically reduced, which can substantially reduce the investment in construction of transfer systems and energy consumption, as well as avoid possible pollution resulting from solvent leakage. The in-situ recovery of solvents helps to save labor and improve production safety. Thus, the use of high-gravity distillation equipment brings substantial economic and social benefits. It also makes it easier for the purification of the intermediates used for multistep organic synthesis by equipping the reactor with a set of high-gravity distillation equipment. For instance, 100 rotating zigzag beds are used in a company for distributed recovery of solvents and separation of intermediates, as shown in Fig. 4.67.

4.5.5 Recovery of ammonia wastewater

The low-concentration ammonia wastewater can be used to prepare industrial ammonia with an ammonia content of over 25% for reuse in the industry. Twin columns are often

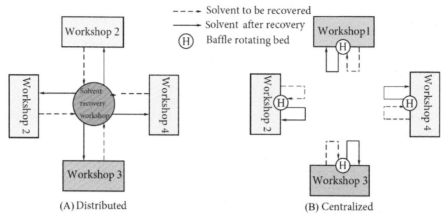

Figure 4.66 Centralized and distributed recovery of solvents.

Figure 4.67 Distributed recovery of solvents and separation of intermediates in a rotating zigzag bed.

used for this purpose, one for distillation of ammonia wastewater and the other for the absorption of ammonia gas. In the latter column, an auxiliary cooler is needed to remove the heat liberated in the absorption of ammonia in water. Thus, the process is very complicated and requires enormous investment. In the rotating zigzag bed, the stationary baffle can be designed to have a heat exchange function to remove the heat liberated in the absorption process. The equipment structure becomes more compact and the heat removal efficiency is improved. As shown in Fig. 4.68, the low-concentration ammonia wastewater is introduced from the inlet located between the lower two rotors. The bottom rotor is used for stripping, and the upper rotor is used for distillation. It is heated directly by the steam, and the ammonia content in the discharged liquid at the bottom meets the national standard. The ammonia vapor from the distillation section is condensed to obtain industrial ammonia with an ammonia content of 25% (mass fraction).

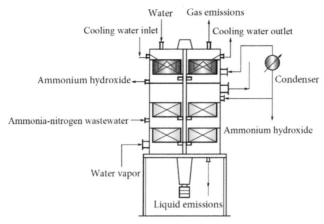

Figure 4.68 The use of rotating zigzag bed for removal of ammonia.

The tail gas containing ammonia is transferred to the top rotor and absorbed by clear water, and the final gas emissions meet the requirements. The stationary baffle of the top rotor is designed to have cooling capacity to remove the heat generated in the absorption of tail gas using cooling water. The technological parameters are summarized as follows: processing capacity >1 t/h, ammonia content in the mother liquid = 1.5%, ammonium hydroxide concentration >25%. The functions of traditional twin columns can be integrated into a single small-scale multi-layer rotating zigzag bed because of its advantages of high mass transfer efficiency, convenience in feed, and heat exchange capacity, which greatly simplify the process and save space and investment. The industrial use of the multi-layer rotating zigzag bed is shown in Fig. 4.69, and now several sets of such equipment have been successfully applied in the industry.

4.6 Prospects

Distillation is one of the most widely used and mature separation technologies and it plays a critical role in chemical production. However, it is still facing some problems that need to be addressed, such as huge equipment investment and energy consumption. The high-gravity packed bed and rotating zigzag bed we have proposed have solved some key technological problems, such as shaft seal, liquid distributor, and internal dynamic seals, and have been put into industrial use. The high-gravity technology has the potential to intensify the mass transfer process and makes the high-gravity distillation equipment miniaturized and the reflux ratio smaller. As a result, it has the advantages of low energy consumption, low investment, low space requirement, and convenient startup and shutdown. Now, high-gravity distillation is primarily used for separation of small amounts of mixtures in the fine chemical industrial

Figure 4.69 The multi-layer rotating zigzag bed for removal of ammonia.

such as medicine, dye, and pesticide. However, an ongoing effort should be undertaken to:

1. Establish more accurate heat and mass transfer models for high-gravity distillation in order to provide sound technological support for its industrial applications.
2. Develop novel high-efficiency structured packing in order to further improve the mass and heat transfer coefficient, gas-liquid contact efficiency, and consequently the separation capacity.
3. Solve the problem of bearing lubrication at operating temperatures in the distillation process.
4. Improve the design and manufacturing of high-gravity distillation equipment on the industrial scale.

The high-gravity distillation technology as an emerging technology has a long way to go before it gains widespread acceptance in the industry. We expect it will have widespread applications in the chemical, petroleum, food, and medicine fields.

References

[1] Todd DB, Maclean DC. Centrifugal vapor-liquid contacting. British Chemical Engineering, 1969,14(11):598-607.
[2] Ramshaw C. Higee distillation—an example of process intensification. The Chemical Engineer, 1983,2:13-14.
[3] Richard B. New mass-transfer find is a matter of gravity. Chemical Engineering, 1983:23-29.
[4] Chen WB, Jin GH, Liu CF. An investigation of new centrifugal mass transfer device. Journal of Chemical Industry and Engineering(China), 1989,5:635-639.

[5] Trevour K, James RF. Distillation studies in a high-gravity contactor. Industrial & Engineering Chemistry Research, 1996,35:4646-4655.
[6] Ji JB, Wang LH, Xu ZC. Study on the hydraulics of the rotating packed bed. Petro-chemical Equipment, 2001,30(50):20-23.
[7] Bao TH. Study on performance of fluid dynamics and mass transfer of rotating bed in high gravity. Hangzhou: Zhejiang University of Technology, 2002.
[8] Xu OG, Ji JB, Bao TH. Rotating zigzag bed for distillation. Zhejiang Chemical Industry, 2003,34(3):3-5.
[9] Xu ZC, Yu YL, Ji JB. Rotating zigzag high-gravity bed and its application in distillation. Petrochemical Technology, 2005,34(8):778-781.
[10] Ji JB, Yu YL, Xu ZC. Wetted wall cluster in high gravity field: zigzag rotating bed. Modern Chemical Industry, 2005,25(5):52-54.
[11] Yu YL, Ji JB, Xu ZC. Study on power consumption of the liquid phase. Journal of Zhejiang University of Technology, 2006,34(1):56-58.
[12] Lai SH, Zhu XJ. A new high efficient distillation device: zigzag rotating bed. Petro & Chemical Equipment, 2007,10(2):33-35.
[13] Chen ZD, Sui LT, Xu ZC. Research of acetonitrile recovery process using RZB. Zhejiang Chemical Industry, 2008,39(1):4-6.
[14] Zhang YH, Ruan Q, Li L. Study of distillation under total reflux in rotating packed bed packed with triangular spiral packing. Modern Chemical Industry, 2008,28(s1):29-32.
[15] Zhu XJ, Lai SH, Xu ZC. A new high efficient distillation device: zigzag rotating bed. Mechanical and Electrical Information, 2008,35:42-44.
[16] Yu YL, Xu ZC, Wang GQ. A new high efficient distillation device: zigzag rotating bed. Bulletin of Science and Technology, 2008,24(1):114-118.
[17] Sui LT, Xu ZC, Yu YL. Influence of the rotor structure of rotating zigzag bed on gas pressure drop. Journal of Chemical Engineering of Chinese Universities, 2008,22(1):28-33.
[18] Wang GQ, Xu ZC, Yu YL. Study on hydrodynamic and mass transfer performance of rotating zigzag bed. Modern Chemical Industry, 2008,28(s1):21-24.
[19] Sui LT. Study on gas pressure drop of rotating zigzag bed and its flow field simulation. Hangzhou: Zhejiang University of Technology, 2008.
[20] Deng DS, Wang RF, Zhang LZ, et al. Vapor-liquid equilibrium measurements and modeling for ternary system water plus ethanol + 1-butyl-3-methylimidazolium acetate. Chinese Journal of Chemical Engineering, 2011,19(4):703-708.
[21] Xie AY, Li YM, Xu ZC. Experiment study on liquid flooding and gas pressure drop of the rotating zigzag bed. Chemical Industry Times, 2009,23(2):14-16.
[22] Xie AY, Li YM, Xu ZC. Experimental study on gas pressure drop of two different rotating disc of the rotating zigzag high-gravity bed. Journal of Zhejiang University of Technology, 2010,38(1):23-25.
[23] Li YM, Ji JB, Yu YL. Mathematical model of liquid power consumption of rotating zigzag bed. Journal of Chemical Engineering of Chinese Universities, 2010,24(2):203-207.
[24] Wang GQ, Xu ZC, Yu YL. High gravity distillation technology and its industrial application. Modern Chemical Industry, 2010,30(s1):55-57.
[25] Wang Y, Li XH, Li YM. Mass transfer model and experiment for rotating jet high-gravity bed. Petrochemical Technology, 2011,40(4):392-396.
[26] Tong ZF, Li XH, Li YM. Hydrodynamics and mass transfer performance of rotating jet high-gravity bed. Petrochemical Technology, 2010,39(3):275-279.
[27] Li XH, Ji JB, Xu ZC. A new high efficient distillation device: zigzag rotating bed. Conference Proceedings of China Association of Traditional Chinese Medicine. China Association of Chinese Medicine, 2009:179-182.
[28] Xu ZC, Ji JB, Wang GQ. Rotating zigzag bed application in extractive distillation process of THF-methanol-water system. Modern Chemical Industry, 2012,32(6):94-96.
[29] Qian BZ. High-gravity rotating bed as a energy-saving equipment. Chemical Equipment Technology, 2012,1:9.

[30] Yao W, Li YM, Guo CF. Hydrodynamics and mass transfer performance of rotating compound bed with perforated sheet and packing. Chemical Industry Times, 2012,26(3):1-4.
[31] Guo CF, Wang GQ, Gao S. Mass transfer performance of a novel higee rotating zigzag bed. Petrochemical Technology, 2013,42(1):47-52.
[32] Yao W, Li YM, Guo CF. Mass transfer performance of rotating compound bed with perforated sheet and packing. Journal of Chemical Engineering of Chinese Universities, 2013,27(3):386-392.
[33] Yang S. Design and comparison on distributed layouts of solvent distillation recovery using HIGEE rotating zigzag bed. Hangzhou: Zhejiang University of Technology, 2014.
[34] Lu XH, Xu ZC, Ji JB. Effects of pulse ultrasound on adsorption of geniposide on resin 1300 in a fixed bed. Chinese Journal of Chemical Engineering, 2011,19(6):1060-1065.
[35] Li YM, Ji JB, Yu YL, et al. Hydrodynamic behavior in a rotating zigzag bed. Chinese Journal of Chemical Engineering, 2010,18(1):34-38.
[36] Li XP, Li N, Liu YZ. Process of mass transfer property of batch high gravity vacuum distillation. Chemical Industry and Engineering Progress, 2016,35(7):2001-2006.
[37] Li XP, Liu YZ, Zhang YH. Advances in the theoretical studies and applications of rotating packed bed. Journal of North University of China, 2003,24:140-143.
[38] Li XP, Liu YZ, Qi GS. Study on distillation in a rotating packed bed. Science & Technology in Chemical Industry, 2004,12(3):25-29.
[39] Li XP, Liu YZ, Liu LJ. Effect of different rotors on distillation mass transfer in rotating packed bed. Chemical Industry and Engineering Progress, 2005,24(3):303-306.
[40] Li XP, Liu YZ, Yang M. Mass transfer and hydrodynamics of rotating packed bed in distillation. The Chinese Journal of Process Engineering, 2005,5(4):375-378.
[41] Li JH, Li XP, Liu YZ. Mass-transfer studies in a rotating packed bed. Journal of North University of China(Natural Science Edition), 2005,26(1):42-45.
[42] Yu HB, Liu YZ, Li XP. Hydrodynamics characteristics of distillation in rotating packed bed. Energy Chemical Industry, 2006,27(2):24-25.
[43] Li XP, Liu YZ. Study on distillation in high gravity field. Modern Chemical Industry, 2006,26(z2):315-319.
[44] Li XP, Liu YZ, Li JH. Study on distillation in high gravity field. Modern Chemical Industry, 2008,28(s1):75-79.
[45] Li XP, Liu YZ, Li ZQ, et al. Continuous distillation experiment with rotating packed bed. Chinese Journal of Chemical Engineering, 2008,16(4):656-662.
[46] Chen J. Study of distillation experiment in rotating packed bed. Taiyuan: North University of China, 2009.
[47] Lei FB, Li XP, Liu YZ. Basic theory and application research for a novel distillation rotating bed. Journal of North University of China(Natural Science Edition), 2009,30(3):261-267.
[48] Chen J, Liu YZ, Li XP. Distillation process of methanol in rotating packed bed. Chemical Industry and Engineering Progress, 2009,28(8):1333-1336.
[49] Li XP, Liu YZ. Characteristics of fin baffle packing used in rotating packed bed. Chinese Journal of Chemical Engineering, 2010,18(1):55-60.
[50] Zhang ZC, Li XP, Liu YZ. Study on super two-phase separate gravity distillation. Modern Chemical Industry, 2010,30(4):79-81.
[51] Li XP, Liu YZ, Zhang ZC. Theoretical analysis on mass transfer process intensification of high gravity distillation. Chemical Engineering(China), 2010,38(12):8-11.
[52] Li X P, Li J H, Li J N. A type of liquid distributor for rotating packed bed. CN 102512913B, 2013-11-20.
[53] Liu Y Z, Li X P, Shen H Y. Liquid distributor and redistributor for multi-stage rotating distillation bed. CN 103272398B, 2015-03-25.
[54] Li X P, Liu Y Z, Li J H. Structured spiral packing. CN 101648129B, 2012-01-25.
[55] Liu Y Z, Li X P, Wang J W. Structured fin baffle packing for rotating packed bed. CN 101342477B, 2012-05-30.
[56] Li X P, Liu Y Z. High efficient rotating distillation bed. CN 101306258B, 2010-06-16.

[57] Li XP, Liu YZ, Wang XL. Study on heat and mass transfer mechanism in high gravity distillation process. Energy Chemical Industry, 2010,31(2):1-5.
[58] Zhang ZC, Li XP, Liu YZ. Application and development of super gravity distillation. Modern Chemical Research, 2010,7:14-17.
[59] Li XP, Liu YZ, Zhang ZC. Mass transfer property of high effective rotating distillation bed. Modern Chemical Industry, 2011,31(2):77-80.
[60] Li XP, Liu YZ, Zhang ZC. Distillation performance of multi-fin baffle packing rotating packed bed. Chemical Engineering (China), 2012,40(6):28-31.
[61] Li XP, Li JN, Liu YZ. Mass transfer property of multistage high gravity distillation process. Chemical Engineering (China), 2013,41(5):14-18.
[62] Li JN. Studies on the high gravity distillation for different types of packing. Taiyuan: North University of China, 2013.
[63] Wang XC. Higee distillation process simulation and optimization based on aspen plus. Taiyuan: North University of China, 2014.
[64] Song ZB, Li XP, Liu YZ. Application of higee distillation for pectin precipitation solvent recovery. Chemical Industry and Engineering Progress, 2015,34(4):1165-1170.
[65] Song ZB. Experimental studies of high gravity vacuum distillation for separating ethanol-water. Taiyuan: North University of China, 2015.
[66] Li DM, Li XP, Liu YZ. Study on the process of mass transfer property of high gravity distillation with extractive agent. Modern Chemical Industry, 2016,2:133-136.
[67] Li DM. Experimental studies of high gravity salt distillation for separating ethanol water. Taiyuan: North University of China, 2016.
[68] Lin CC, Tsungjen HO, Liu WZ. Distillation in a rotating packed bed. Journal of Chemical Engineering of Japan, 2002,35(12):1298-1304.
[69] Gao X, Chu GW, Zou HK. Studies of distillation in a novel multi-stage counter-current rotating packed bed. Journal of Beijing University of Chemical Technology(Natural Science Edition), 2010,37(04):1-5.
[70] Gao X. Distillation in a novel multi-stage counter-current rotating packed bed. Beijing: Beijing University of Chemical Technology, 2010.
[71] Wang ZA. Chemical Process Design Manual. Beijing: Chemical Industry Press, 2003.

CHAPTER 5

Liquid-liquid extraction

Contents

5.1 Overview	159
5.1.1 Definition of liquid-liquid extraction	160
5.1.2 Principle of liquid-liquid extraction	160
5.1.3 Equipment for intensification of liquid-liquid extraction	164
5.2 Mechanism of process intensification in the impinging stream-rotating packed bed	165
5.2.1 Intensification of mass transfer	166
5.2.2 Intensification of mixing	169
5.2.3 Coupling of mass transfer and mixing processes	175
5.3 Extraction operation in the impinging stream-rotating packed bed	180
5.3.1 Single-stage extraction	180
5.3.2 Multi-stage extraction	181
5.3.3 Calculation of the extraction operation in the impinging stream-rotating packed bed	182
5.3.4 Extraction characteristics of the impinging stream-rotating packed bed	185
5.4 Application examples	186
5.4.1 Extraction of phenol in wastewater	186
5.4.2 Concentration of acetic acid	190
5.4.3 Extraction of nitrobenzene	193
5.4.4 Extraction of dyes	196
5.4.5 Extraction of indium ions	198
5.4.6 Other applications	202
5.5 Prospects	204
References	204

5.1 Overview

Liquid-liquid extraction (or solvent extraction) is an important unit operation for separating useful components from a liquid mixture and it has a broad spectrum of applications in the chemical industry because of high separation efficiency, low energy consumption, low equipment investment, and rapid, continuous and safe operation. In recent years, there has been a growing interest in the use of liquid-liquid extraction for the separation of a wide variety of compounds and the extraction of high-purity substances. A better understanding of the liquid-liquid extraction process, especially the development of high-gravity extractors, such as impinging stream-rotating packed bed (IS-RPB), could provide a theoretical basis for the intensification of the extraction process and the development of novel extraction technologies.

5.1.1 Definition of liquid-liquid extraction

In liquid–liquid extraction, an immiscible or partially miscible extraction solvent is added into the homogeneous liquid mixture containing the target compounds, forming a two-phase system consisting of the feed solution and the solvent, and then the target components are extracted from the feed solution to the solvent phase based on the difference in partition coefficient of the components to alternating immiscible liquids. Liquid-liquid extraction exploits the difference in the solubility of the components in the two phases to extract the compounds into one solvent leaving the rest of the matrix in the other [1,2]. A typical liquid-liquid extraction process is shown in Fig. 5.1, where the component to be separated from the feed solution is called the solute and the solvent used to extract the solute is called the extraction solvent. In the extraction process, the extraction solvent is mixed with the feed solution and the solute will be continually diffused from the feed solution to the extraction solvent, forming the extraction phase (E) and the raffinate phase (R). Thorough mixing of the two phases is an important prerequisite for the success of liquid-liquid extraction. The extraction phase is still a mixture of solute and solvent, and further unit operations such as distillation or back extraction are needed to separate them and regenerate the extraction solvent for reuse.

5.1.2 Principle of liquid-liquid extraction

The basic principle of liquid-liquid extraction involves phase equilibrium and extraction kinetics. The equilibrium of the components in the two phases indicates the direction and intensity of mass transfer in the extraction process, while the extraction kinetics indicates the mass transfer rate of the components from one phase to the other [3].

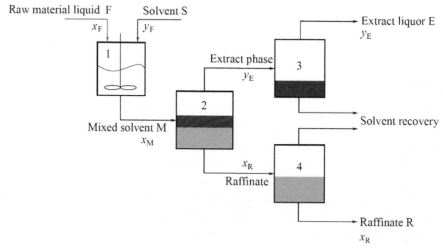

Figure 5.1 A typical liquid-liquid extraction process (1—mixing, 2—settling, 3—separation of the extraction phase, 4—separation of the raffinate phase).

5.1.2.1 Phase equilibrium

Liquid-liquid extraction can fall into two broad categories: physical extraction and chemical extraction. For the physical extraction, the solute exists mainly as single molecules in the two phases and no chemical reaction occurs between solvent and solute. For chemical extraction, the solute exists in different forms in two phases and a chemical reaction occurs between the solvent and solute. A variety of chemical extraction techniques have been proposed in recent years, including complexation, ion association, and synergistic extraction. The phase equilibrium in the physical and chemical extraction is described in the following paragraphs.

1. Phase equilibrium in the physical extraction

    Physical extraction is characterized by the existence of the solute in the form of neutral molecules in both feed and extraction phases and no chemical reaction between solvent and solute. The extraction solvent contains no complexing agent. A key determinant of extraction efficiency is the difference in the solubility of the solute between the two phases, assuming that the solute to be extracted is HA. As it exists in the same form in the two phases, the phase equilibrium can be expressed as:

    $$HA \rightleftharpoons \overline{HA} \qquad (5.1)$$

    where HA—component in the feed phase; $\overline{HA}$—component in the extraction phase.

    It is known from the thermodynamic principle that under constant temperature and pressure conditions, the change in Gibbs' free energy is zero at equilibrium. According to the Gibbs-Duhem equation:

    $$\alpha_{i(w)} = \alpha_{i(o)} \qquad (5.2)$$

    where $\alpha_{i(w)}$—the activity of component i in the feed phase; $\alpha_{i(o)}$—the activity of component i in the extraction phase.

    As the activity of component i is the same at phase equilibrium, the thermodynamic equilibrium constant can be expressed as:

    $$K_i = \alpha_{i(o)}/\alpha_{i(w)} \qquad (5.3)$$

    The activity coefficient is $\gamma_i = \alpha_i/x_i$. It is known from the definition of the activity coefficient γ_i that:

    $$x_{i(w)}\gamma_{i(w)} = x_{i(o)}\gamma_{i(o)} \qquad (5.4)$$

    As the partition coefficient, D_i is equal to the ratio of the equilibrium concentration of solute i in the two phases, then:

    $$D_i = \frac{x_{i(o)}}{x_{i(w)}} = \frac{\gamma_{i(w)}}{\gamma_{i(o)}} \qquad (5.5)$$

The partition coefficient can be calculated from the activity coefficient γ_i of solute i in the two phases. In thermodynamics, the mole fraction x is often used to indicate the concentration of a non-electrolyte solution, and γ is the activity coefficient of each solute in the pure state. For other concentration units and corresponding activity coefficients, transformation equations can be found in the literature.

2. Phase equilibrium in the chemical extraction

Chemical extraction involves a chemical reaction between solvent and solute. As the solute exists in different forms in feed and extraction phases, the equilibrium equation can be used to calculate phase equilibrium and partition coefficient. Take the extraction of acetic acid from dilute aqueous solutions with tributyl phosphate (TBP) as an example. The equilibrium equation is:

$$CH_3COOH + \overline{TBP} \rightleftharpoons \overline{CH_3COOH \cdot TBP} \quad (5.6)$$

The increment in each component is expressed as:

$$-dn_1 = -dn_2 = dn_3 \quad (5.7)$$

where n_1—amount of CH_3COOH; n_2—amount of TBP; n_3—amount of $CH_3COOH \cdot TBP$.

According to the thermodynamic principle at constant temperature and pressure, the change in Gibbs' free energy is zero at equilibrium. The thermodynamic equilibrium constant for the extraction of CH_3COOH with TBP is:

$$K = \frac{\alpha_3}{\alpha_1 \alpha_2} \quad (5.8)$$

where α_1—activity of CH_3COOH; α_2—activity of TBP; α_3—activity of $CH_3COOH \cdot TBP$.

The change in the standard free energy of formation in the extraction process is:

$$\Delta G^0 = \mu_3^0(T) - \mu_1^0(T) - \mu_2^0(T) = -RT\ln K \quad (5.9)$$

The thermodynamic equilibrium constant K will be kept constant under specific conditions, and it can be calculated from the equilibrium concentrations of components determined by experiments and their activity coefficients measured or calculated in order to predict the extraction equilibrium under other conditions. The partition coefficient D and the apparent extraction equilibrium constant expressed by the concentration may differ under different conditions. However, previous studies have often assumed that the concentration of each component in dilute organic solutions is proportional to the activity, which obviously exaggerates the application range of the apparent extraction equilibrium constant K. Such an assumption may lead to a small deviation for the separation of dilute solutions. For the extraction of dilute acetic acid (CH_3COOH) solution with TBP-kerosene, the apparent extraction equilibrium constant K can be expressed as:

$$K = \frac{[\overline{CH_3COOH \cdot TBP}]}{[CH_3COOH][\overline{TBP}]} \qquad (5.10)$$

Take the extraction of $UO_2(NO_3)_2$ by TBP-kerosene as an example. The equilibrium equation is:

$$UO_2^{2+} + 2NO_3^- + 2\overline{TBP} \rightarrow \overline{UO_2(UO_3)_2 \cdot 2TBP} \qquad (5.11)$$

where symbols without an overline represent a component in the feed phase, and symbols with an overline represent a component in the extraction phase.

The thermodynamic equilibrium constant for the extraction of $UO_2(NO_3)_2$ by TBP is:

$$K = \frac{a_4}{a_1 a_2^2 a_3^2} = \frac{x_4 \gamma_4}{x_1 x_2^2 \gamma_\pm^3 \cdot x_3^2 \gamma_3^2} \qquad (5.12)$$

where $\gamma_\pm$—average ionic activity coefficient, $\gamma_\pm^\nu = \gamma_+^{\nu+} \gamma_-^{\nu-}$; γ_+, γ_-—activity coefficient of negative and positive ions; ν^+, ν^-—the amount of negative and positive ions in water obtained per unit amount of electrolyte.

The standard free energy of formation in the extraction process is:

$$\Delta G^0 = \mu_4^0(T) - \mu_1^0(T) - 2\mu_2^0(T) - 2\mu_3^0(T) = -RT\ln K \qquad (5.13)$$

5.1.2.2 Extraction kinetics

The extraction process occurs in a heterogeneous system and the extraction rate is determined by the mass transfer rate of the solute between the two phases [4]. As the extraction solvent comes into contact with the feed liquid, there are two films on the interface, one is the water film for the feed phase and the other is the oil film for the extraction solvent. The mass transfer mechanism is shown in Fig. 5.2.

The relationship between the overall mass transfer coefficient and the double-film mass transfer coefficient is:

$$1/k_o + 1/k_w = 1/k \qquad (5.14)$$

where k_w—mass transfer coefficient across the water film; k_o—mass transfer coefficient across the oil film; k—total transfer coefficient.

Then, the total mass transfer equation is:

$$N = kA(c_w^* - c_w) \qquad (5.15)$$

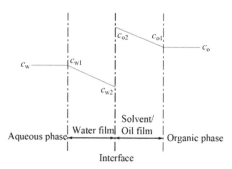

Figure 5.2 Schematic of mass transfer across water and oil films.

The mass transfer in liquid–liquid extraction is driven by concentration difference. It is proportional to the contact area between the two phases and related to the overall mass transfer coefficient. The main influencing factors of the mass transfer rate in liquid–liquid extraction include the interfacial area, driving force, and overall mass transfer coefficient. Accordingly, the mass transfer rate can be increased by increasing the driving force, interfacial area, and overall mass transfer coefficient.

5.1.3 Equipment for intensification of liquid-liquid extraction

The aim of process intensification is not only to reduce the experiment size, but also to minimize the energy consumption and the production of wastes and by-products, and finally to enhance the production efficiency and process safety, lower the production cost and reduce the environmental pollution [5]. Conventional extractors include mixer-settler extractor, spray column, and packed column. Williams et al. [6] developed a novel hollow fiber membrane contactor for the removal of ibuprofen and 4-isobutyl acetophenone by non-dispersive solvent extraction. Baier et al. [7] developed a taylor vortex extractor for liquid–liquid extraction. Bonam et al. [8] developed a rotating spray column for the removal of Cr(VI) by Aliquat 336. Impinging stream contactor has also been used for process intensification of liquid–liquid extraction. The rotating packing bed (RPB) devised based on high-gravity technology exhibits excellent performance in liquid–liquid extraction. In recent years, the high-gravity technique has been widely used in various fields such as wastewater treatment, polymerization, rubber, and biodiesel. Table 5.1 summarizes the mass transfer performance of common extractors used in the industry.

A higher overall volumetric mass transfer coefficient is obtained in the RPB and impinging stream extractor, implying that both of them are capable of intensifying the mass transfer process. Liu et al. have proposed an innovative extraction equipment called the IS-RPB [14] by integrating the advantages of RPB and impinging stream extractor, and its principle and application for intensification of liquid–liquid extraction process are discussed below.

Table 5.1 The overall volumetric mass transfer coefficients of different extractors.

Extractor	Chemical system	$Q_o; Q_a/[(m^3/s) \times 10^6]$	K_La/s^{-1}
Mixer-settler extractor [9]	Water-acetone-toluene	0.88; 0.78	0.0015-0.005
Spray column [10]	Water-acetone-toluene	16-130; 20-130	0.0005-0.008
Packed column [10]	Water-acetone-toluene	16-130; 8-65	0.0005-0.0055
Kuhin extraction column [11]	Water-acetone-toluene	3.9-8.9; 3.9-8.9	0.005-0.0125
Hollow fiber membrane contactor [6]	Water-ibuprofen-octanol	2.7-7.1; 3.1-7.9	0.0045-0.042
Impinging stream contactor [12]	Water-succinic acid-butanol	1.83-5.0; 1.83-5.0	0.077-0.25
Taylor vortex extractor [7]	Water-benzyl alcohol-paraffin oil	0.33; 0.33	0.002-0.0127
Rotating spray column [8]	Water-Cr(VI)-kerosene (Aliquat 336)	2.3-3.9; 2.5-6.8	0.06-0.12
RPB [13]	Water-methyl red-xylene	0.83-2.1; 4.16-20.83	0.015-0.205

Note: Q_o is the volumetric flow rate of the oil phase, Q_a is the volumetric flow rate of the water phase, K_La is the overall volumetric mass transfer coefficient.

5.2 Mechanism of process intensification in the impinging stream-rotating packed bed

The extraction process involves the mixing and phase separation of two immiscible liquid phases. IS-RPB can improve the mixing of the feed solution and the extraction solvent in the extraction process. In the IS-RPB, two sets of jet nozzles are coaxially and concentrically mounted at opposite ends in the central cavity of the rotor, and they are concentric with (or parallel to) the rotor shaft. The aperture and axial distance between the two nozzles are varied based on the actual situations. In general, the aperture of the nozzle is about several millimeters, and the axial distance between the two opposing nozzles is about 5-50 mm. The working principle of IS-RPB is shown in Fig. 5.3. Briefly, the feed solution and extraction solvent are pressurized by a centrifugal pump and their flow rates are measured. They are ejected from nozzles and then the opposing-jet streams impinge with each other, forming a round (fan-shaped) liquid mist that is perpendicular to the jet direction for micromixing and mass transfer between the two liquids. The edge of the mist enters the inner cavity of the RPB, and then the mixture flows towards the outer edge of the high-speed rotating packing, during which it is sheared, aggregated, and dispersed many

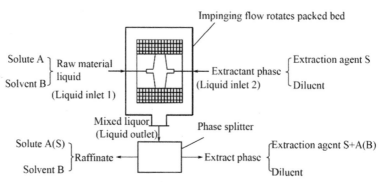

Figure 5.3 A schematic of the extraction process in the IS-RPB. *IS-RPB*, Impinging stream-rotating packed bed.

times and a thorough mixing is achieved. Finally, the mixture is thrown onto the interior wall of the casing and exits the IS-RPB through the liquid outlet.

IS-RPB allows the two liquid phases of approximately the same volume to flow coaxially but in opposite directions and impinge with each other. At the macroscopic level, the microliquid elements pass through the impact surface and penetrate into the counter-flow under the action of the inertia force. Because of the large kinetic energy, the kinetic energy is converted to static pressure energy that can change the flow direction of the two liquids and subsequently form an impact surface that is perpendicular to the original flow direction of the two liquids. At the microscopic level, the impinging of the two liquids flowing in opposite directions forms a narrow highly turbulent region, and the shear force resulting from the impinging of the two liquids can break the liquid into fine droplets and thus increase the surface area and the surface renewal of the liquid, and consequently lead to a higher mass transfer rate. After impinging, the liquid mist enters the rotating packing at a high speed and is then sheared into microliquid films, filaments, and droplets, which will repeatedly aggregate and disperse. As a result, rapid and uniform micromixing is obtained in the IS-RPB [15].

In the following sections, we will discuss the intensification of mass transfer and micromixing in the IS-RPB.

5.2.1 Intensification of mass transfer

In liquid–liquid extraction, the extraction rate is related to the chemical reaction rate, mixing degree, interfacial area, diffusion rate, and film thickness on both sides of the interface. The mass transfer rate of the solute from the feed solution to the extraction solvent is:

$$\mathrm{d}w/\mathrm{d}t = kA\Delta c \tag{5.16}$$

where dw/dt—mass transfer rate per unit time, g/s; k—overall volumetric mass transfer coefficient per unit volume, kg/[s(m²/m³)(g/m³)]; A—interfacial area per unit volume, m²/m³; Δc—difference between actual and equilibrium concentrations of the solute in the two phases, mg/L.

In typical extraction equipment, one liquid phase is used as the dispersed phase and dispersed in a continuous phase in order to increase the liquid–liquid contact area. The contact area per unit volume A is dependent on the liquid holdup of the dispersed phase and the liquid droplet size. The mass transfer process is rather complex in the extractor. The turbulent dispersion of the liquid in the extractor makes the mass transfer process fundamentally different from the simple molecular diffusion process. Although some empirical equations have been proposed based on experimental investigation and dimensional analysis, they could not be used in engineering calculation because of the substantial differences in the extraction system and the type and size of the extraction equipment used. Despite this, Eq. (5.16) is still useful in guiding the design of extraction equipment and analyzing how to intensify the mass transfer process [16]. It is known from Eq. (5.16) that the mass transfer rate can be increased by increasing the interfacial area, overall mass transfer coefficient, and mass transfer driving force.

5.2.1.1 Increasing the interfacial area A

One liquid phase can be dispersed into very tiny droplets in another liquid phase by applying external energy, such as stirring and pulse, in order to increase the contact area between the two phases. The smaller the liquid droplets are, the larger the mass transfer area will be. However, it is necessary to prevent emulsification resulting from the extremely high dispersion of the extraction solvent which may cause some unnecessary difficulties for subsequent separation of the solvent. For the system with low interfacial tension, adequate dispersion can be achieved once the density difference drives the solvent to pass through the sieve plate or the packing; while for the system with high interfacial tension, an external force is required to increase the dispersion. The mass transfer process consists of molecular diffusion and eddy diffusion, and the eddy diffusion coefficient may be several orders of magnitude higher than the molecular diffusion coefficient. Thus, the eddy diffusion induced by an external force can reduce the liquid droplet size and increase the contact area and the turbulence of the continuous phase.

5.2.1.2 Increasing the overall mass transfer coefficient k

The overall mass transfer coefficient is dependent on the physical and chemical properties of the system, equipment structure, and operating conditions. It can be increased by repeatedly breaking up and then agglomerating the liquid droplets of the dispersed phase or increasing the turbulence of the liquid phase. Because of the additive property of the overall mass transfer resistance, it is equal to the sum of the mass transfer resistances of the two phases in the physical extraction. In chemical extraction, if the chemical reaction occurs fast and the solute is at equilibrium between the two phases at the interface, there would be no mass transfer resistance at the interface. However, this assumption may cause large errors for extraction systems with a slow chemical reaction. The relationship between the overall mass transfer coefficient and the individual mass transfer coefficient considering the chemical reaction resistance at the interface is:

$$\frac{1}{k} = \frac{1}{k_w} + \frac{1}{k_r} + \frac{1}{k_o} \tag{5.17}$$

The physical meaning of this equation is that the overall mass transfer resistance is equal to the sum of the mass transfer resistance of the feed phase ($1/k_w$), the chemical reaction resistance at the interface ($1/k_r$), and the mass transfer resistance of the solvent ($1/k_o$). For a given system and equipment, the turbulence intensities of the two phases can be increased by increasing their flow velocities or by supplying an external energy in order to increase their individual mass transfer coefficients, especially the mass transfer coefficient of the phase that is controlled by mass transfer resistance.

5.2.1.3 Increasing the mass transfer driving force Δc

Under similar operating conditions, the countercurrent contact produces a higher driving force for mass transfer compared to other contact modes, and as a consequence, the solute concentration is high in the extraction phase but low in the raffinate phase. In a multi-stage extraction process, countercurrent contact should be used as much as possible in order to achieve the highest extraction efficiency. The use of a fresh solvent containing no solute can also increase the driving force for mass transfer.

The mass transfer rate can be increased by increasing the specific surface area or reducing the mass transfer resistance by creating high-intensity turbulence within and surrounding liquid droplets, both of which require the supply of an external force. It is difficult to achieve these two goals simultaneously with the use of traditional extraction equipment because tiny liquid droplets are difficult to flow at high velocities in the continuous phase. However, the two phases exist in the form of tiny liquid droplets, filaments, and films in a highly turbulent state in the IS-RPB. Therefore, the use of IS-RPB leads to a very large specific surface area and rapid surface renewal, and the mass transfer rates of the two phases are significantly increased. According to Eq. (5.16), IS-RPB has the potential to increase the mass transfer rate in the extraction process because of its ability to increase the mass transfer coefficient k and the mass transfer surface area A. However, the driving force for mass transfer Δc is determined by the extraction system itself and has little relevance to the extraction equipment.

Because of the unique structure and characteristics of IS-RPB, the two liquid phases are both dispersed phases, which can greatly improve the adaptation of the feed solution and the extraction solvent in the extraction process. In the IS-RPB, the flow of the liquid is driven by the external force, and mixing and separation are performed separately in the extraction, which can effectively prevent back mixing. The degree of back mixing in the large-scale traditional extraction equipment is expected to be higher than that in the laboratory-scale extraction equipment, and this effect should be taken into consideration in the design and selection of extraction equipment. IS-RPB can be used in industrial extraction processes due to its potential to minimize back mixing.

5.2.2 Intensification of mixing

Thorough mixing is a key requirement for any extraction process, and the mixing efficiency has a direct impact on the extraction efficiency. Liu et al. [17,18] investigated the micromixing performance of IS-RPB using the chemical coupling method. The competitive and consecutive reactions between α-naphthol (A) and diazotized sulphanilic acid (B) are shown in reactions (5.18) and (5.19), and the products are monoazo (R) and bisazo (S), respectively:

$$A + B \rightarrow R \quad (5.18)$$

$$R + B \rightarrow S \quad (5.19)$$

At 298K, reaction (5.18) is a fast reaction with $k_1 = 3800$ m^3/(mol·s), and reaction (5.19) is a slow reaction with $k_2 = 1.56$ m^3/(mol·s). The molar ratio of the two reactants is 1:2 when only bisazo is produced. At a molar ratio of 1:1, the reaction will be terminated because B is exhausted, and the effect of mixing on the reaction can be readily recorded. The concentrations of R and S (c_R and c_S) are determined by light absorption, and the mixing efficiency is indicated by the segregation index (X_S):

$$X_S = 2c_S/(2c_S + c_R) \quad (5.20)$$

It follows from Eq. (5.20) that X_S approaches 0 in the case of perfect or complete mixing; it is in the range of $0 < X_S < 1$ in the case of general mixing; and it approaches 1 in the case of complete segregation, which implies poorest or no mixing efficiency.

An orthogonal experimental design is used to evaluate the effects of high-gravity factor β, initial impact velocity u_O, impact distance L, nozzle diameter d, impact angle α (the angle between the central symmetry plane and the nozzle axis, which is 1/2 of the angle between the two nozzle axes), and packing type on the micromixing efficiency. The segregation index X_S decreases with the increase of high-gravity factor β

Table 5.2 Comparison of micromixing efficiency of IS-RPB, IS, and RPB.

Equipment	Structural parameters	Segregation index X_S
IS	$L = 5$ mm, $d = 1.5$ mm	$0.06 < X_S < 0.12$
RPB	$\beta = 150$	$0.05 < X_S < 0.1$
IS-RPB	$L = 5$ mm, $d = 1.5$ mm, $\beta = 150$	$X_S < 0.025$
Stirred tank	$D = 178$ mm	$X_S > 0.1$
Tubular reactor	—	$X_S > 0.15$
Tee mixer	—	$X_S > 0.1$

and stabilizes at $\beta > 100$, and at this time $X_S < 0.025$. It also decreases with the increase of initial impact velocity u_O, and $X_S < 0.025$ at $u_O > 10$ m/s. When the impact distance L is 2 times longer than the nozzle diameter, the segregation index is maintained constant. With the increase of the impact angle α from 30 to 90 degrees, the segregation index X_S shows a decreasing trend, and the best mixing efficiency is obtained at $\alpha = 90$ degrees. The comparison of the micromixing efficiency of IS-RPB, IS, and RPB is given in Table 5.2. IS-RPB shows the best micromixing efficiency, which is two times that of RPB and IS, and the micromixing efficiency is comparable between IS and RPB. In addition, the mixing efficiency of IS-RPB is 40 times that of the continuous stirred-tank reactor.

In recent years, the iodide/iodate method has been widely used to evaluate the micromixing process [19]. For instance, Jiao et al. [20,21] investigated the effects of impact angle α, high-gravity factor β, and impact distance L on the segregation index X_S using the iodide-iodate test reaction as the working system in an IS-RPB, and the micromixing time of the IS-RPB was calculated using the incorporation model. The main conclusions are summarized as follows.

5.2.2.1 Effect of impact angle α

The impact angle is an important factor affecting the mixing of impinging streams and the initial distribution of the liquid in the IS-RPB. Experiments are conducted under conditions of $[H^+] = 0.1$ mol/L, high-gravity factor $\beta = 106.2$, the flow rate of the buffer solution $Q_1 = 70$ L/h, the volume ratio of the two solutions $R = 7$, and the impact distance $L = 30$ mm. Fig. 5.4 shows the variation of the segregation index X_S with the impact angle α. The segregation index X_S decreases continuously with the increase of impact angle α, demonstrating that the larger the impact angle is, the better the micromixing efficiency of the IS-RPB will be. The impact of the two streams at larger angles leads to the formation of a fan-shaped impact surface that is perpendicular to the jet direction, and the surface becomes thinner farther away from the impact point, which causes an increase in the surface area and aerodynamics and the formation of expansion waves. The jetted liquid is broken into tiny droplets, and the high

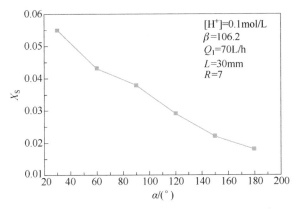

Figure 5.4 The variation of the segregation index X_S with the impact angle α.

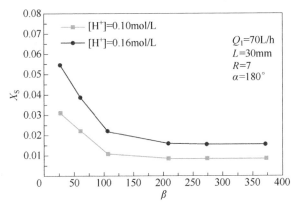

Figure 5.5 The relationship between the segregation index X_S and the high-gravity factor β.

turbulence of the jet provides additional energy for the break-up of the liquid droplets through hydrodynamic force or shock. As the impact angle α further increases, the impact becomes more violent, which can accelerate the break-up of the liquid droplets. At $\alpha = 180$ degrees, the segregation index X_S reaches a minimum of 0.018, implying that the two jets impact in opposite directions. The relative velocity of the fluid, as well as the turbulent kinetic energy, reaches the maximum, and the best micromixing efficiency is obtained. The liquid mist that is perpendicular to the jet direction is favorable for the initial distribution of the liquid.

5.2.2.2 Effect of high-gravity factor β

The high-gravity factor β is also an important influencing factor for the micromixing performance of the IS-RPB. Fig. 5.5 shows the variation of the segregation index X_S with the high-gravity factor β of the IS-RPB [21]. The segregation index X_S first decreases rapidly and then levels off with increasing high-gravity factor β. At higher β values, the liquid is broken into smaller droplets by the large shear force resulting from the high-speed rotation of the packing and the liquid film becomes thinner in the high-gravity field. The collision becomes more violent and the energy dissipation rate is increased. The residence time of the liquid in the packing is also reduced and the frequency of the aggregation and dispersion of the liquid is accelerated, leading to a reduction in the size of the dispersed liquid. These two factors lead to a decrease in the segregation index X_S with increasing high-gravity factor β and consequently the intensification of the molecule-level mixing in the IS-RPB.

5.2.2.3 Effect of impact distance L

Fig. 5.6 shows the effect of the impact distance L (the distance between two nozzles) on the segregation index X_S, where $[H^+] = 0.1$ mol/L, the high-gravity factor is $\beta = 106.2$, the flow rate of the buffer solution is $Q_1 = 70$ L/h, the volume ratio of the two solutions is $R = 7$, and the impact angle is $\alpha = 180$ degrees. The impact distance L has a significant effect on the segregation index X_S. Specifically, the segregation index X_S first decreases with the increase of the impact distance L until a minimum is reached at $L = 30$ mm, after which it increases with further increase of the impact distance L. Thus, it is necessary to select an appropriate impact distance in order to maximize the micromixing performance of the IS-RPB. For the IS-RPB with coaxial nozzles, the turbulence near the nozzle axis and the center of the impact surface is more pronounced than that in other regions (e.g., reflux region). In this case, the two liquids will flow to the region with low turbulence before they are well mixed in the region with high turbulence, leading to an increase in the segregation index X_S. At a small impact distance of L, the impact of the two jets with high turbulent kinetic energy leads to significant changes in fluid pressure and thus affects the coupling between the inner edge of the RPB and the impact mist. At an impact distance of

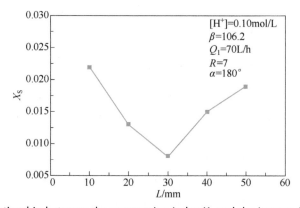

Figure 5.6 The relationship between the segregation index X_S and the impact distance L.

$L = 30$ mm, the two liquids will be well mixed in the region with high turbulent kinetic energy and then enter the RPB for further mixing. In this case, the optimal coupling between the inner edge of the RPB and the impact mist is obtained, and thus the segregation index X_S reaches the minimum. As the impact distance L further increases, no further improvement is observed for the micromixing performance of the IS-RPB, and the segregation index X_S is close to that in impinging stream equipment.

5.2.2.4 Effect of fluid viscosity μ

The variation of the segregation index X_S with the fluid viscosity μ is shown in Fig. 5.7. The segregation index X_S increases with increasing fluid viscosity μ, indicating that the micromixing efficiency is lower at higher μ values. The segregation index X_S increases with increasing flow rate, which is in good agreement with previous experimental results. According to the E-model [22], the micromixing characteristic time t_m is proportional to the square root of the kinematic viscosity and inversely proportional to the square root of the unit volume power. That is, $t_m \propto k\left(\frac{\nu}{\varepsilon}\right)^{1/2}$. Thus, t_m can be increased by increasing ν or decreasing ε, indicating that increasing the fluid viscosity can reduce the micromixing efficiency. When the rotation speed of the rotor (high-gravity factor) is kept constant in a high-gravity field, the unit volume power will remain unchanged and then the change in the fluid viscosity μ still has a significant effect on the segregation index X_S. Thus, change in the fluid viscosity μ has more pronounced effects on micromixing than unit volume power. The micromixing process is affected by both vortex entrainment and molecular diffusion [23].

5.2.2.5 Micromixing time t_m

The micromixing time t_m obtained by numerical simulation is one of the most straightforward indicators of the mixing performance of an extractor. The incorporation model

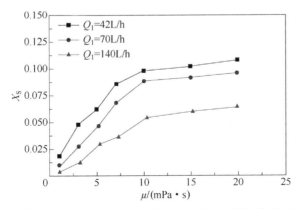

Figure 5.7 The relationship between the segregation index X_S and the fluid viscosity μ.

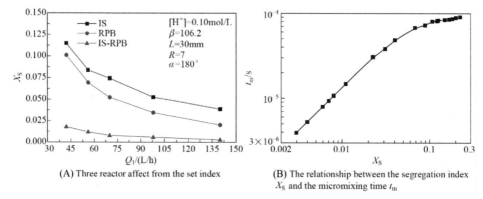

Figure 5.8 The relationship between the segregation index X_S and the micromixing time t_m.

proposed by Villermaux et al. [24] is initially used to calculate the micromixing time t_m of stirred tanks and then extended to continuous reactors such as Couette reactor and static mixer. The diffusion equations of ions in the reaction system are established, and then the differential equations are solved using Fortran or Matlab. The iteration is terminated when [H$^+$] approaches 0. The variation of the micromixing time t_m with the segregation index X_S is shown in Fig. 5.8. It is found that the micromixing time is 0.004–0.03 ms for IS-RPB, 0.05–1.6 ms for IS, and 0.02–1.4 ms for RPB, respectively.

5.2.2.6 Comparison of micromixing performance

Currently, there is a growing interest in the use of process intensification and process integration technologies for the intensification of micromixing in a reactor. In traditional batch reactors, energy is supplied by means of vigorous stirring or impact, and most energy is used for the overall mixing of the reactor. It is difficult to obtain a high local energy dissipation rate required for rapid micromixing. As the continuous reactor is intended to promote the dispersion or mixing of added materials, a high local energy dissipation rate is expected for micromixing. The micromixing efficiency of the continuous reactor can be increased by increasing the rotation rate of the rotor, and that of traditional batch reactors can be increased by introducing an external field or mixing elements. Previous studies have empirically investigated the micromixing performance of static mixer, pneumatic stirrer, tubular reactor, impinging stream reactor, T-shaped mixer, and rotor-stator mixer, as shown in Table 5.3. It is obvious that RPB is more effective than other reactors in intensifying the micromixing process.

Table 5.3 Comparison of micromixing performance of different reactors.

Reactor	Operation mode	Segregation index X_S	Micromixing time t_m/ms
Stirred tank [25]	Batch	0.18-0.28	5-200
Couette reactor [26]	Continuous	0.2-0.75	1-10
Sliding-surface mixer [27]	Batch	—	10-100
Taylor-Couette reactor [28]	Continuous	—	6-80
Submerged circulative impinging stream reactor [29]	Batch	0.02-0.12	87-192
Y-shaped micro-channel reactor [30]	Continuous	0.001-0.26	0.1-1
Rotor-stator reactor [31]	Continuous	0.004-0.037	0.01-0.05
RPB [32]	Continuous	0.008-0.024	0.01-0.1

Table 5.4 The relationship between the segregation index X_S and the partition coefficient under the same operating conditions.

Segregation index X_S	0.0239	0.0246	0.0257	0.0260	0.0296	0.0342	0.0468
Partition coefficient D'	48.0	47.6	44.7	44.2	39.9	37.5	38.2
$1/X_S$	41.8	40.65	38.91	38.46	33.78	29.24	21.37

5.2.3 Coupling of mass transfer and mixing processes

The mass transfer efficiency of the IS-RPB is closely associated with its micromixing efficiency. The higher the micromixing efficiency is, the higher the partition coefficient of the extractor will be.

5.2.3.1 Chemical extraction process

Under the same equipment and operating conditions, the micromixing (where the phase ratio is 1 and the concentration is 0.05 mmol/m^3) and mass transfer (where the phase ratio is 1 and 10% TBP is used as the solvent for extraction of phenol) performance are summarized in Table 5.4 [33,34].

The segregation index is a critical indicator of the degree of micromixing, and the smaller the segregation index is, the higher the degree of micromixing will be. The reciprocal of the segregation index $1/X_S$ is plotted against the partition coefficient D' to elucidate the relationship between micromixing and mass transfer efficiency, as shown in Fig. 5.9. It is found that at $1/X_S > 29.24$, the partition coefficient is linearly related to the micromixing efficiency, indicating that the mass transfer resistance for

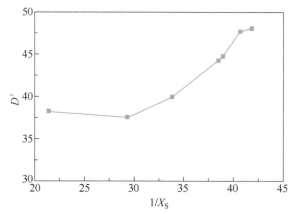

Figure 5.9 The relationship between the partition coefficient D' and the reciprocal of the segregation index $1/X_S$ for the extraction process with a chemical reaction.

the extraction of TBP and phenol is mainly derived from the diffusion process. The reaction rate between TBP and phenol is fast at the interface and thus the resistance is negligible. The extraction process is considered to be a diffusion process. The regression analysis yields the following equation:

$$D' = 11.68 + \frac{0.869}{X_S} \tag{5.21}$$

As the overall mass transfer resistance is equal to the sum of the mass transfer resistance of the feed phase, the chemical reaction resistance at the interface, and the mass transfer resistance of the solvent, increasing the mixing can significantly reduce the mass transfer resistance of the two phases and thus increase the interphase mass transfer rate and consequently the extraction efficiency. The chemical reaction resistance at the interface should not be neglected because of the short residence time of the two phases in the IS-RPB. At $1/X_S < 29.24$, increasing the $1/X_S$ could not increase the D' and the chemical reaction resistance at the interface dominates. In this case, there exists a limit for the degree of mixing, below which increasing the degree of mixing has little effect on the mass transfer rate, and above which increasing the degree of mixing has an effect on the mass transfer rate. It is necessary to understand the mass transfer mechanism of the extraction system in order to find the most effective approach to reduce the chemical reaction resistance and diffusion resistance and determine the degree of mixing. When the degree of mixing is determined, other operating parameters can be determined, such as the initial impact velocity and high-gravity factor. This will ensure high extraction efficiency without increasing the energy consumption to increase the degree of mixing.

5.2.3.2 Physical extraction process

Under the same equipment and operating conditions, the relationship between the extraction efficiency η and the segregation index X_S for the kerosene-benzoic acid-water system is shown in Table 5.5.

The extraction efficiency η is plotted against the reciprocal of the segregation index $1/X_S$ to elucidate the relationship between micromixing and mass transfer efficiency in the IS-RPB, as shown in Fig. 5.10.

Fig. 5.10 reveals that the extraction efficiency increases with increasing micromixing efficiency. The mass transfer resistance is equal to the diffusion resistance in the physical extraction process because there is no chemical reaction resistance at the interface, and the extraction equilibrium of the two phases can be achieved instantaneously at the interface. In physical extraction, the extraction efficiency is only related to the micromixing efficiency and thus it can be evaluated by the micromixing efficiency.

5.2.3.3 Equations for the mass transfer coefficient of the impinging stream-rotating packed bed

It is known from the mixing and mass transfer mechanism of the IS-RPB that:

$$k = f(D, d, u, \mu, \rho \cdots) \tag{5.22}$$

Table 5.5 The relationship between the extraction efficiency η and the segregation index X_S for the kerosene-benzoic acid-water system.

Segregation index X_S	0.0239	0.0246	0.0257	0.0260	0.0296	0.0342	0.0468
Extraction efficiency η/%	99.8	98.6	97.7	97.7	96.3	95.8	94.4
$1/X_S$	41.8	40.65	38.91	38.46	33.78	29.24	21.37

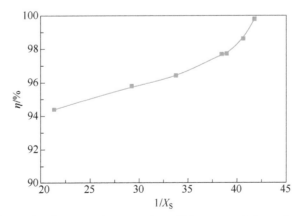

Figure 5.10 The relationship between the extraction efficiency η and the reciprocal of the segregation index.

where k—mass transfer coefficient; D—molecular diffusion coefficient of the two phases; u—flow velocity; μ—viscosity; ρ—density; d—linear dimension.

There are a large number of influencing factors, such as initial impact velocity, rotor size, and high-gravity factor. It would be time-consuming to explore the effect of each factor on the mass transfer. Dimensional analysis is often performed to determine the characteristic number equation [35]. The following characteristic numbers are often considered in the mass transfer of the liquid phase.

Sherwood number $Sh = kL/D = \frac{L}{D/k}$, where L is the characteristic dimension. It is a dimensionless number that contains the mass transfer coefficient k and is considered to be the ratio of the length L to the effective thickness of the retention layer D/k.

Schmidt number, $Sc = \frac{\mu}{\rho D} = \frac{\mu/\rho}{D}$, is the ratio of the dynamic viscosity μ/ρ to the molecular diffusion coefficient D.

Reynolds number, $Re = \frac{Lu\rho}{\mu}$, is a measure of fluid turbulence and is defined as the ratio of the inertial force to the viscous force.

A semi-empirical equation can be obtained based on dimensional analysis:

$$Sh = \phi(Re, Sc) \tag{5.23}$$

which can be expressed as follows in most circumstances:

$$Sh = Sh_0 + ARe^b Sc^c \tag{5.24}$$

where Sh_0—the lower limit of the mass transfer rate only considering the effect of molecular diffusion on the mass transfer coefficient; A, b, c—undetermined coefficients.

Eq. (5.24) can be transformed into:

$$\frac{(k-k_0)L}{D} = A\left(\frac{\mu}{\rho D}\right)^c \left(\frac{Lu\rho}{\mu}\right)^b \tag{5.25}$$

where $k - k_0$ represents an increment of mass transfer rate due to the increase of turbulence.

Then, Eq. (5.25) can be expressed as:

$$k - k_0 = AD^{(1-b)}\left(\frac{\mu}{\rho}\right)^{b-c} L^{(c-1)} u^c \tag{5.26}$$

According to Eq. (5.26), the increment of the mass transfer rate is also related to the representative linear dimension of the system and the flow velocity of the fluid.

In the IS-RPB, the effect of micromixing efficiency on the mass transfer coefficient can be divided into two parts. The first one concerns the impinging streams, where the characteristic size L is the nozzle diameter d_0 and u is the initial impact velocity u_0. The second one concerns the RPB, where the linear size L is the average radius of the RPB d_{Ave} and u is the average radial velocity of the liquid in the RPB u_{Ave}. The total increment of the mass transfer rate caused by the external energy-induced mixing is the sum of the increment of the mass transfer in the impinging stream and RPB. Then:

$$k - k_0 = AD^{(1-b)} \left(\frac{\mu}{\rho}\right)^{b-c} \left[d_0^{(c-1)} u_0^c + d_{Ave}^{(c-1)} u_{Ave}^c\right] \quad (5.27)$$

In the RPB, the relationship of the average radial velocity of the liquid in the packing with the liquid energy and angular velocity is:

$$u_{Ave} = BV^e (\omega^2 r)^f \quad (5.28)$$

where B, e, f—coefficients; V—liquid flow rate, m³/s; ω—angular velocity, rad/s; r—average radius of the packing, m.

In the IS-RPB, $V = \frac{1}{4}\pi d_0^2 u_0$ and $\omega^2 r = \beta g$. The effect of various equipment and operating parameters on the mass transfer rate is expressed as:

$$k - k_0 = AD^{(1-b)} \left(\frac{\mu}{\rho}\right)^{b-c} \left[u_0^c d_0^{(c-1)} + C d_{Ave}^{(c-1)} u_0^{ec} d_0^{2ec} \beta^{fc}\right]$$

$$C = B^e g^{fc} \left(\frac{1}{4}\pi\right)^{ec} \quad (5.29)$$

where A, b, c, e, f are undetermined coefficients.

Eq. (5.29) reveals that for a given system, the main factors that govern the mass transfer rate of the IS-RPB include the average radius of the packing d_{Ave}, nozzle diameter d_0, initial impact velocity u_0, and high-gravity factor β. Although the inclusion of a large number of parameters makes the equation less suitable for the calculation of the mass transfer rate, it provides important insights into how to improve the mass transfer rate in the IS-RPB. The mass transfer coefficient in Eq. (5.29) is the individual mass transfer coefficients of the two phases, and the overall mass transfer coefficient should be calculated from Eq. (5.16). In the liquid-liquid extraction, the overall mass transfer resistance is equal to the sum of the mass transfer resistance of the feed phase, the chemical reaction resistance at the interface, and the mass transfer resistance of the solvent. It is known from Eq. (5.17) that only the mass transfer coefficients of the two phases are considered. Because of the complex flow pattern of the fluid in the extractor and the diversity of extractors used, this method can only be used for qualitative analysis of the effects of various influencing factors on the mass transfer coefficient.

The mass transfer data obtained in previous pilot tests can be used as a reference for different systems and extractors under different operating conditions.

The ability of the IS-RPB to intensify the extraction process is related to its ability to improve the micromixing and reduce the mass transfer resistance at the interface. The unit operation of extraction in the IS-RPB consists of a continuous mixing process and a batch-phase separation process. For different separation purposes and requirements, multiple unit operations can be integrated to form a multi-stage extraction process, such as multi-stage cross-flow and countercurrent extraction.

5.3 Extraction operation in the impinging stream-rotating packed bed

5.3.1 Single-stage extraction

The experimental setup and procedure for single-stage extraction in the IS-RPB are schematically shown in Fig. 5.11. First, the feed phase and the extraction solvent stored in two storage tanks are pumped into IS-RPB using two centrifugal pumps, and their flow rates are measured using two rotameters. The flow rate and initial impact velocity can be regulated by control valves. The impact of the two phases leads to the formation of the impact mist, which in turn flows radially in the RPB (the rotation speed is adjustable) for further mixing and mass transfer. Then, the mixture flows into the liquid–liquid phase separator for phase separation, and the resulting extraction phase and raffinate phase

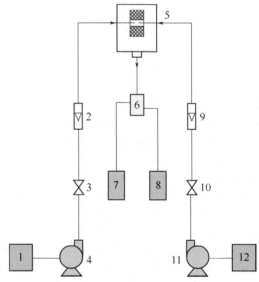

Figure 5.11 Experimental setup and procedure for single-stage extraction in the IS-RPB (1, 7, 8, 12—storage tank, 2, 9—rotameter, 3, 10—control valve, 4, 11—centrifugal pump, 5—IS-RPB, 6—liquid-liquid phase separator).

are transferred to storage tanks. The mixing process in the IS-RPB is a continuous process, but the phase separation process is a batch process. For different separation purposes, the raffinate phase resulting from the single-stage extraction can be recovered, discharged (provided that the discharge limit is met), or transferred to the next extraction process (when the discharge limit is not met). The solute in the extraction phase is separated from solvent by back extraction so that the solvent can be reused with minimum losses.

5.3.2 Multi-stage extraction

The multi-stage extraction processes in the IS-RPB are divided into cross-flow and countercurrent-flow extraction [36]. The multi-stage cross-flow extraction process in the IS-RPB is shown in Fig. 5.12. The feed solution is introduced into the first-stage extractor and fresh solvent is added at each stage. The raffinate phase obtained in the first-stage extractor is introduced into the second-stage extractor and comes into contact with the fresh solvent for further extraction. Then, the raffinate phase obtained in the second-stage extractor is introduced into the third-stage extractor for further extraction. This process is repeated until the solute concentration in the raffinate phase reaches the requirement. In theory, all solutes could be extracted from the feed solution if there are sufficient stages. The extraction phases obtained at all stages are mixed and the solute concentration is low in the mixture. The solute needs to be separated and the solvent is to be recovered. In the multi-stage cross-flow extraction, there is a large driving force for mass transfer because of the addition of fresh solvent at each stage, yielding a high extraction rate. However, the solute concentration is low in the extraction phase, and a high cost is incurred in the recovery of a large amount of solvent.

The flow chart of the three-stage countercurrent extraction in the IS-RPB is shown in Fig. 5.13. The feed solution and solvent are introduced into the first- and third-stage extractor, respectively. The extraction phase obtained at the next stage is refluxed to the previous stage for use as the solvent and comes into contact with the feed solution. The feed solution and solvent are brought into countercurrent contact

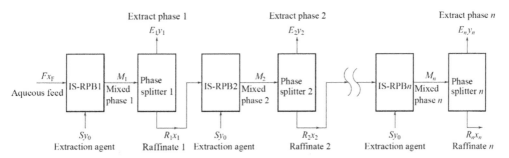

Figure 5.12 The flow chart of multi-stage cross-flow extraction.

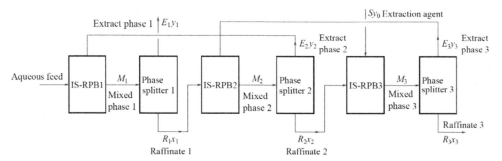

Figure 5.13 Flow chart of three-stage countercurrent extraction.

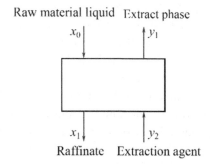

Figure 5.14 The material relationship in a single-stage extraction process in the IS-RPB.

and their compositions vary in a step-wise manner. In a countercurrent extraction process, the concentration of the solute extracted from the extraction phase is high and the total amount of solvent used is reduced, and the concentration of the solute in the extraction solvent is high.

5.3.3 Calculation of the extraction operation in the impinging stream-rotating packed bed

5.3.3.1 Operating line equation

The material relationship in a single-stage extraction process in the IS-RPB is shown in Fig. 5.14.

Let the initial concentration of the feed solution and the solvent phase be x_0 and y_0, the concentration of the raffinate phase and the extraction phase be x_1 and y_1 after the single-stage extraction, and the concentration of the raffinate phase and the extraction phase at complete equilibrium be x^* and y^*, respectively. The driving force for the mass transfer of the feed solution and the extraction phase is $x_0 - x^*$ and $y^* - y_0$, where x^* and y^* are the theoretical concentrations of extracted components in the raffinate phase and the extraction phase at complete equilibrium, respectively. An infinite time is required to reach the equilibrium concentration as the driving force is continuously reduced. In the IS-RPB, let the volumetric flow rate of the feed solution and the extraction solvent be L and Q, respectively. The volumes of the two phases are assumed to be constant throughout the extraction process. Then, the material balance equation in the single-stage extraction process can be obtained.

$$Lx_0 + Qy_0 = Lx_1 + Qy_1 \tag{5.30}$$

where x_0—initial concentration of the feed solution, mg/L; y_0—initial concentration of the extraction solvent, mg/L; x_1—concentration of the raffinate phase after the first-stage extraction, mg/L; y_1—concentration of the extraction phase after the first-stage extraction, mg/L; L—volumetric flow rate of the feed solution, L/h; Q—volumetric flow rate of the extraction solvent, L/h.

Then,

$$y_1 = -\frac{L}{Q}(x_1 - x_0) + y_0 \tag{5.31}$$

For the solvent containing no solute, Eq. (5.31) can be simplified into:

$$y_1 = -\frac{L}{Q}(x_1 - x_0) \tag{5.32}$$

Eq. (5.32) reveals the variation in the concentration of extracted components in the two phases of the single-stage extraction process, which is called the operating line equation. The operating line of the single-stage extraction is a straight line, where the slope is L/Q. In continuous operation, the ratio of the volumetric flow rate of the solvent and the phase to be treated is defined as the phase ratio (denoted as R).

In order to describe the single-stage extraction efficiency in the IS-RPB, the ratio of the concentration of the solute in the extraction phase to that in the raffinate solution after the first-stage extraction in the IS-RPB is defined as the partition coefficient D':

$$D' = \frac{y_1}{x_1} \tag{5.33}$$

For the extraction system with a constant partition coefficient, the operating line equation for the single-stage extraction can be simplified into:

$$y_1 = \frac{D'}{1 + RD'} x_0 \tag{5.34}$$

For the extraction system with a variable partition coefficient, it is difficult to obtain directly from the equation. In this case, it can be obtained using the graphical method to determine the intersection point of the operating line and the equilibrium line.

5.3.3.2 Characterization of the extraction efficiency

The extraction efficiency of the IS-RPB can be indicated by the extraction efficiency η in addition to the partition coefficient D. In reality, the two phases could not reach complete equilibrium after only one-time mixing and contact, and thus the theoretical extraction efficiency may not be achieved in the single-stage extraction process. In order to characterize the difference between actual and theoretical extraction efficiency, the extraction efficiency η is defined as follows:

$$\eta = \frac{x_0 - x_1}{x_0 - x^*} \times 100\% \text{(indicated by the feed solution)} \tag{5.35}$$

$$\eta = \frac{y_1 - y_0}{y^* - y_0} \times 100\% \text{(indicated by the extraction solvent)} \tag{5.36}$$

Substitution of equilibrium partition coefficient D^* and D' into η yields:

$$\eta = \frac{D'(1 + D^*)}{D^*(1 + D')} \times 100\% \tag{5.37}$$

in which $D^* = \frac{y^*}{x^*}$.

5.3.3.3 Calculation of the multi-stage extraction process

The calculation of the multi-stage extraction process in the IS-RPB is the same as that in other multi-stage extractors [37]. In multi-stage cross-flow extraction processes, if the partition coefficient D' and the phase ratio R are kept constant, the concentration of the solute in the raffinate phase after n-stage extraction can be calculated based on the principle of material balance:

$$x_n = \frac{x_0}{(1 + RD')^n} \tag{5.38}$$

When the initial concentration of the feed solution x_0, partition coefficient D', and extraction requirement x_n are known, the number of extraction stages can be calculated. When the equilibrium partition coefficient is not constant, the theoretical number of extraction stages can be determined by referring to previous results in the literature.

Unlike the multi-stage cross-flow extraction, the feed solution and the solvent flow counter are currently in each stage of the multi-stage countercurrent extraction. Assuming that the partition coefficient is the same at each stage and the solvent contains no solute, the concentration of components x_i in the feed solution at the outlet of extraction stage i can be obtained based on the principle of material balance.

$$x_i = x_0 \left[\frac{RD'^{(N+1-i)} - 1}{RD'^{(N+1)} - 1} \right] \tag{5.39}$$

5.3.4 Extraction characteristics of the impinging stream-rotating packed bed

5.3.4.1 Existence form of the liquid

The mixing and separation of two liquid phases are more difficult in liquid–liquid extraction compared to gas–liquid mass transfer because there is only a slight difference in the density between the two liquid phases but their viscosity and interfacial tension are high. In order to obtain a more complete extraction, it is necessary to increase the interfacial area between the two phases. In industrial extractors, one liquid phase is often dispersed in the other phase in the form of liquid droplets, and the selection of dispersed and continuous phases is based on the volume flow rates of the two phases, mass transfer direction, extraction system and equipment characteristics, and surface properties of the internal structure in order to ensure stable flux, extraction rate, and operation. It is clear from the structure and working principle of the IS-RPB that both of the two liquid phases are dispersed phases and they are uniformly mixed and dispersed. The diameters of liquid droplets are in the range of 0.5–5 mm in the extraction system but 3–55 μm in the RPB. IS-RPB exhibits better micromixing performance than RPB. The average diameter of liquid droplets in the IS-RPB is smaller than 55 μm, which leads to a larger contact area between the two phases and a shorter time needed to reach mass transfer equilibrium. In addition to liquid droplets, the liquid can also exist in the form of liquid films that have a large specific surface area and a high surface renewal rate in the IS-RPB, which can further increase the contact area between the two phases. All of these contribute to increasing the mass transfer in the extraction operation [38].

5.3.4.2 Residence time

The structure of IS-RPB suggests that the residence time of the liquid will be very short (mostly <0.2 s) in the IS-RPB, making it particularly suitable for fast reactions, exothermic systems, and treatment of special materials, such as degradable and radioactive materials.

5.3.4.3 Retention volume of the solvent

The size of IS-RPB is small and its packing volume is only 1/20 of that of the traditional packed column. Importantly, almost no solvent would be retained in the IS-RPB, which can reduce the amount of the extraction solvent and consequently the operating cost. Only simple washing and rinsing are required for IS-RPB in the replacement of the extraction system, but the same operation would be very difficult for extraction columns and tanks.

5.3.4.4 Extraction capacity

The extraction capacity is usually represented by the specific load, which is defined as the total flow rate of the two phases passing through the unit cross-sectional area of the equipment in unit time [$m^3/(m^2 \cdot h)$]. Table 5.6 summarizes the specific loads of IS-RPB and other extractors reported in the literature [39].

The specific load of IS-RPB is 178 $m^3/(m^2 \cdot h)$, which is approximately 1–2 orders of magnitude higher than those of other extractors with the same extraction capacity. Thus, the use of IS-RPB contributes to reducing the equipment size and raw material cost.

5.3.4.5 Adaptability

Most extraction processes are diffusion controlled. For extraction processes that involve a chemical reaction, diffusion control would still prevail if the reaction is fast. As the two phases are uniformly mixed and distributed in the packing of the IS-RPB, both of them are dispersed phases, which is favorable for diffusion-controlled extraction. Mass transfer and phase separation are performed separately in the IS-RPB, which makes IS-RPB have more potential applications. IS-RPB is applicable to a wide range of phase ratios because of its unique structure and working principle, and high mass transfer efficiency can be obtained for diffusion-controlled extraction.

5.4 Application examples
5.4.1 Extraction of phenol in wastewater

Phenol is a common contaminant in wastewater from various industrial processes and it has many negative environmental consequences if not properly disposed of. As the maximum allowable concentration of phenol in drinking water should be less than 0.002 mg/L, stringent requirements are imposed on the discharge of phenol wastewater into the environment [40]. Nevertheless, phenol is an important chemical used to manufacture various products, and thus recovery of phenol from wastewater is likely to offer both environmental and economic benefits. Solvent extraction is widely used for the treatment of phenol wastewater, especially at high concentrations. However, little is known about the equipment for treatment of phenol wastewater by solvent

Table 5.6 The specific load of IS-RPB and other extractors.

Extraction equipment	Specific load/[$m^3/(m^2 \cdot h)$]	Note
Mixer-settler extractor	0.2–1	—
Extraction column	2–20	—
Centrifugal extractor	40–80	Maximum extraction capacity <5 t/h
IS-RPB	178	No obvious amplification effect

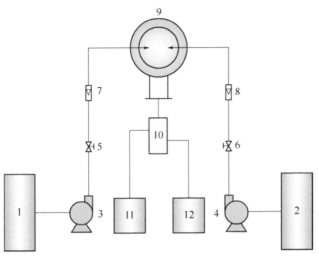

Figure 5.15 Single-stage extraction and back extraction (1,2—storage tank; 3,4—centrifugal pump; 5,6—control valve; 7,8—rotameter; 9—IS-RPB; 10—liquid-liquid phase separator; 11—light phase storage tank; 12—heavy phase storage tank).

extraction. Liu et al. [41] have investigated the extraction of tributyl phosphate (TBP) (kerosene)-phenol-water in the IS-RPB, where TBP was used as the extraction solvent, kerosene was used as the diluent, and phenol was used as the solute. The extraction process is shown in Fig. 5.15. Phenol is a typical Lewis acid and it can be readily extracted by complexation. TBP is a neutral organophosphorus extractant and the oxygen atom in the P=O bond has the potential to provide lone-pair electrons, and it is a moderate to strong Lewis base and shows a high equilibrium partition coefficient D^* for phenol. In the experiment, the aqueous phase phenol concentration was measured by 4-aminoantipyrine spectrophotometry, and the organic phase phenol concentration was calculated based on the principle of material balance. The D^* value of 100% TBP is over 400 for the dilute phenol solution, but because of the requirement on the solvent properties in the extraction process, TBP needs to be diluted with kerosene. The equilibrium partition coefficients obtained for different solvents with different volume fractions of TBP are given in Table 5.7.

The extraction performance is represented by the partition coefficient D' Eq. (5.33) and the extraction efficiency η Eq. (5.37). The effects of initial impact velocity (u_0), high-gravity factor (β), oil-water volume ratio (phase ratio, R), and solvent formulation on the partition coefficient and extraction efficiency are investigated.

5.4.1.1 Effect of u_0 on D' and η

Under conditions of constant TBP volume fraction and high-gravity factor, the variations of the partition coefficient D' and the extraction efficiency η with the initial

Table 5.7 The partition coefficients obtained at different volume fractions of TBP.

Extraction solvent	Water solubility/(kg/m³)	Equilibrium partition coefficient D^*
5% TBP (kerosene)	0.02	21.6
10% TBP (kerosene)	0.04	48.8
20% TBP (kerosene)	0.07	103.5
30% TBP (kerosene)	0.12	171.9

Note: Equilibrium partition coefficients are obtained by shaking for 30 min at a frequency of 150/min in a constant temperature water bath at 20°C in a conical flask, and the settling time is longer than 15 min.

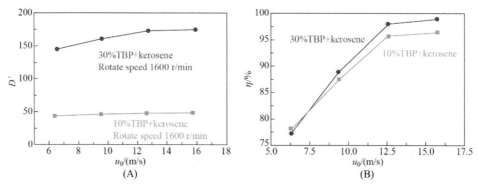

Figure 5.16 Effect of initial impact velocity u_0 on the partition coefficient D' (A) and extraction efficiency η (B).

impact velocity u_0 are shown in Fig. 5.16. Both D' and η increase with the increase of u_0 because increasing the initial impact velocity can increase the slip velocity between the two phases and consequently the turbulent kinetic energy in the impact region. As a result, the mass transfer area that is perpendicular to the jet direction is expanded. Increasing the initial impact velocity can also reduce the size of liquid droplets, as well as the diffusion path for mass transfer, which can further increase the mass transfer between the two liquid phases. However, as the initial impact velocity exceeds a certain limit, it will produce no further improvement to the extraction efficiency. Considering the energy consumption in the transfer process, the initial impact velocity should be set in the range of 10–12 m/s in practical applications.

5.4.1.2 Effect of β on D′ and η

All other operating parameters being fixed, the high-gravity factor β of the IS-RPB is varied to investigate its effect on D' and η, as shown in Fig. 5.17. It is seen that increasing the high-gravity factor β increases both D' and η. The liquid exists primarily in the form of liquid droplets, filaments, and films in the packing of the IS-RPB. As the high-gravity factor β increases, the liquid flows faster in the packing, the shear

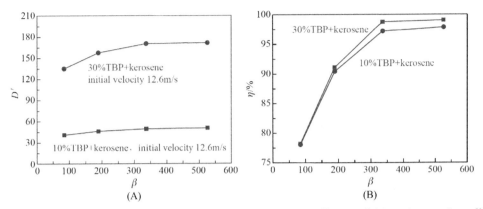

Figure 5.17 Effect of high-gravity factor β on the partition coefficient D' (A) and extraction efficiency η (B).

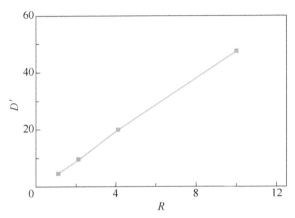

Figure 5.18 Effect of phase ratio R on the partition coefficient D'.

force of the packing on liquid droplets becomes larger, and the surface renewal of the liquid is accelerated. As a result, the liquid films on the packing are thinner and the liquid droplets are smaller, which can reduce the diffusion distance for mass transfer. There will be a higher frequency of aggregation and dispersion of liquid droplets. Thus, the mass transfer between the two liquid phases is greatly improved. Taking into consideration of the energy consumption, the β value should be controlled within 200–300.

5.4.1.3 Effect of R on D'

Under appropriate conditions, the effect of phase ratio R on the partition coefficient D' of the IS-RPB is shown in Fig. 5.18. It should be noted that the change in the phase ratio is accompanied by a change in the initial impact velocity u_0 because of the special structure of the IS-RPB. Fig. 5.18 reveals that the partition coefficient increases linearly with the increase of the phase ratio R. Increasing the phase ratio R can inevitably increase the volumetric flow rate of the oil phase (solvent) when the volumetric flow rate of the liquid phase is kept constant, and dramatically reduce the concentration of phenol in the aqueous (raffinate) phase. Also, increasing the volumetric flow rate of the oil phase can increase the turbulence in the mixing region, which is favorable for mass transfer. The partition coefficients of the IS-RPB under different phase ratios can be estimated for industrial applications.

5.4.1.4 Effect of solvent compositions on D'

The relationship between solvent composition and partition coefficient D' is shown in Table 5.8. It is seen that after the single-stage extraction in the IS-RPB, the partition coefficient almost reaches the equilibrium partition coefficient and the removal rate of phenol is higher than 95% under different solvent compositions, indicating that IS-RPB has high extraction efficiency. Based on environmental and economic considerations, it is appropriate to use 20% TBP/kerosene solution as the extraction solvent in the industry. In conclusion, IS-RPB has excellent extraction performance and thus has promising applications in the treatment of phenol wastewater.

5.4.2 Concentration of acetic acid

A large amount of dilute acetic acid solutions is generated in the production of acetic acid and other products using acetic acid as the raw material or solvent [42]. The concentration and recovery of acetic acid are not only economically profitable but also environmentally responsible. However, as the density and boiling point of acetic acid are close to that of water, complexation is more appropriate than distillation [43]. A large number of studies have been conducted on the equilibrium characteristics, kinetics, and mechanisms of the extraction of acetic acid by complexation. Qi et al. [44] investigated the extraction of acetic acid from dilute solution by complexation in the IS-RPB, where TBP diluted in kerosene was used as the extraction solvent. The

Table 5.8 Effect of solvent compositions on D'.

Extraction solvent	Water solubility/(kg/m³)	Equilibrium partition coefficient D^*	Partition coefficient D'	Removal rate of phenol/%
5% TBP (kerosene)	0.02	21.6	21.3	95.51
10% TBP (kerosene)	0.04	48.8	47.5	97.94
20% TBP (kerosene)	0.07	103.5	100.0	99.01
30% TBP (kerosene)	0.12	171.9	171.4	99.42

experimental procedure is shown in Fig. 5.19. The effects of initial impact velocity (u_0), high-gravity factor (β), and the number of extraction stages (n) on the extraction efficiency (η) were also investigated.

5.4.2.1 Effect of u_0 on η

The effect of initial impact velocity u_0 on the extraction efficiency η is shown in Fig. 5.20, where the high-gravity factor is kept at $\beta = 337$. It is found that increasing the initial impact velocity u_0 increases the extraction efficiency η, but it tends to level off at $u_0 > 12$ m/s. This is attributed to: (1) the higher turbulence in the impact

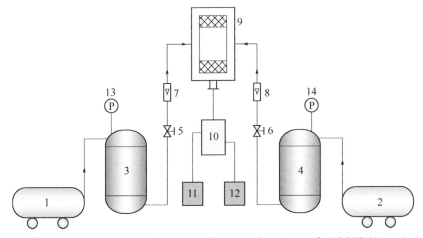

Figure 5.19 The extraction process of acetic acid by complexation in the IS-RPB (1,2—air compression pump, 3,4—pressure tank, 5,6—control valve, 7,8—rotameter, 9—IS-RPB, 10—liquid-liquid phase separator, 11,12—storage tank, 13,14—pressure gauge). *IS-RPB*, Impinging stream-rotating packed bed.

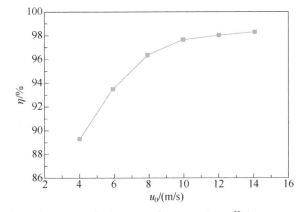

Figure 5.20 Effect of initial impact velocity u_0 on the extraction efficiency η.

region, which can enhance the diffusion of the two phases and reduce the mass transfer resistance between them; (2) the higher frequency of aggregation and dispersion of liquid droplets in the impact region that can enhance the mass transfer process; and (3) a larger impact area that can make the impact between the two phases at the inner edge of the packing more violent and thus enhance the mass transfer process. However, considering the transfer pressure of the liquid, the initial impact velocity is preferably controlled within 10–12 m/s.

5.4.2.2 Effect of β on η

At $u_0 = 12$ m/s, the effect of high-gravity factor β on the extraction efficiency η is shown in Fig. 5.21. It is seen that the extraction efficiency η increases with increasing high-gravity factor β, but the increasing rate is significantly lower at $\beta > 337$. The effect of high-gravity factor β on the extraction efficiency η is attributed to: (1) the higher shear force of the packing, which can reduce the liquid size and the mass transfer resistance; (2) the lower thickness of the liquid film and the larger interfacial area for mass transfer between the two phases; and (3) the higher frequency of aggregation and dispersion of liquid droplets in the impact region and the higher surface renewal rate, which can promote the exchange of acetic acid at the interface and thus intensify the mass transfer. In conclusion, the high-gravity factor should be controlled in the range of 200–300.

5.4.2.3 Effect of n on η

It is noted that although the extraction efficiency is up to 98% in the IS-RPB, the single-stage extraction rate of acetic acid is only 60% because the equilibrium partition coefficient of 60% TBP/kerosene is only 1.6 for acetic acid. Thus, three-stage cross-flow

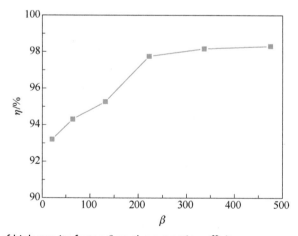

Figure 5.21 Effect of high-gravity factor β on the extraction efficiency η.

Table 5.9 The experimental results of three-stage cross-flow extraction.

Initial mass concentration of acetic acid/(g/L)	Mass concentration in the first-stage extraction raffinate/(g/L)	Mass concentration in the second-stage extraction raffinate/(g/L)	Mass concentration in the three-stage extraction raffinate/(g/L)	Total extraction rate of acetic acid/%
32.43	12.62	4.98	1.96	94.0

extraction experiments are performed, and the experimental results are given in Table 5.9. It is found that after the three-stage cross-flow extraction, the total extraction rate of acetic acid reaches 94%. IS-RPB can promote the complexation of TBP and acetic acid, and under conditions of $u_0 = 12$ m/s and $\beta = 337$, the extraction efficiency of acetic acid reaches 98%, and extraction equilibrium is obtained. The use of 60% TBP as the extraction solvent also produces a high extraction rate (94.0%) of acetic acid after the three-stage cross-flow extraction in the IS-RPB. This makes it clear that for some systems with low equilibrium partition coefficient, the extraction rate can be further increased by increasing the number of extraction stages. Therefore, IS-RPB has the potential to be used for systems with low equilibrium constants such as acetic acid.

5.4.3 Extraction of nitrobenzene

Nitrobenzene (NB) is an important chemical used in the production of nitrobenzene sulfonate, dinitrobenzene, dye intermediates, drugs, and pesticides. However, it is highly toxic and carcinogenic that can have many negative environmental consequences [45]. Liquid-liquid extraction allows the recovery of nitrobenzene from wastewater. Yang et al. [46] investigated the extraction of nitrobenzene from aqueous solutions in the IS-RPB using cyclohexane as the extractant, as shown in Fig. 5.22. The experimental results reveal that phase ratio (R), initial impact velocity (u_0), and high-gravity factor (β) have significant effects on the removal rate and extraction efficiency, and the most appropriate operating parameters are obtained for the treatment of industrial nitrobenzene wastewater.

5.4.3.1 Effect of phase ratio r on the removal rate and extraction efficiency of nitrobenzene

Under conditions of extraction temperature $T = 25°C$, pH $= 6.4$, liquid flow rate $Q = 50$ L/h (initial impact velocity $u_0 = 7.9$ m/s), and high-gravity factor $\beta = 10.5$, the effect of phase ratio R (V_{oil}: V_{water}) on the removal rate and extraction efficiency of nitrobenzene is shown in Fig. 5.23. It is found that as the phase ratio is increased from 1:5 to 1:1, the removal rate of nitrobenzene is increased from 61.4% to 94.9% and the extraction efficiency is increased from 64.7% to 100.0%. This is because the

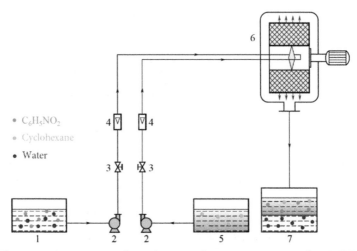

- $C_6H_5NO_2$
- Cyclohexane
- Water

Figure 5.22 The extraction process of nitrobenzene from wastewater in the IS-RPB (1—nitrobenzene wastewater storage tank, 2—pump, 3—valve, 4—liquid flowmeter, 5—cyclohexane storage tank, 6—IS-RPB, 7—liquid-liquid phase separator). *IS-RPB*, Impinging stream-rotating packed bed.

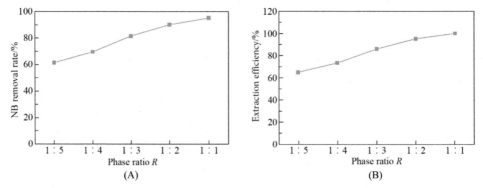

Figure 5.23 Effect of phase ratio R on the removal rate (A) and extraction efficiency (B) of nitrobenzene.

initial impact velocity and turbulent kinetic energy increase with increasing phase ratio, which creates a high turbulence region with a narrow distribution of energy and accelerates the diffusion of nitrobenzene between the two phases. At higher phase ratios, there is a higher frequency of collision between the two liquid phases in the packing, and the contact area is increased, which is favorable for mass transfer between the two phases. The removal rate reaches 94.9% at $R = 1:1$. The phase ratio is set at $R = 1:1$ for the IS-RPB in order to improve the extraction efficiency and reduce the cost and the amount of organic solvent used.

5.4.3.2 Effect of high-gravity factor β on the removal rate and extraction efficiency of nitrobenzene

Under conditions of extraction temperature $T = 25°C$, pH = 6.4, liquid flow rate = 50 L/h (initial impact velocity $u_0 = 7.9$ m/s), and phase ratio $R = 1:1$, the effect of high-gravity factor β on the removal rate and extraction efficiency of nitrobenzene is shown in Fig. 5.24. It shows that the removal rate and extraction efficiency of nitrobenzene first increase and then decrease with increasing high-gravity factor, and the maximum removal rate (94.9%) is obtained at $\beta = 10.5$. This is because as the high-gravity factor increases, the shear force of the packing on the liquid is increased and the interface is renewed rapidly in the aggregation and dispersion processes of liquid droplets so that better micromixing is achieved and the mass transfer coefficient is greatly increased [47]. However, at too high β values, the short residence time of the liquid in the packing leads to a decrease in the contact time between the two phases and consequently a decrease in the removal rate and extraction efficiency of nitrobenzene. The most suitable high-gravity factor is determined to be $\beta = 10.5$. However, the optimal rotation speed for IS-RPBs with different structures and sizes may differ slightly and should be determined based on the extraction system.

5.4.3.3 Comparison experiments

Under conditions of extraction temperature $T = 25°C$, pH = 6.4, liquid flow rate = 20-60 L/h ($u_0 = 7.9$ m/s), phase ratio $R = 1:1$, and $\beta = 10.5$, the removal rate and extraction efficiency of nitrobenzene in the IS-RPB, IS, and RPB are shown in Fig. 5.25. It is found that the removal rate of nitrobenzene in the IS is lower than that in the IS-RPB and RPB. At a liquid flow rate of 50 L/h, the removal rate of nitrobenzene reaches 94.9% in the IS-RPB. This is because the initial impact velocity is

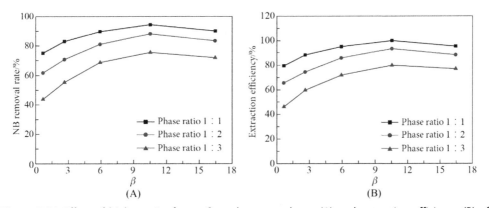

Figure 5.24 Effect of high-gravity factor β on the removal rate (A) and extraction efficiency (B) of nitrobenzene.

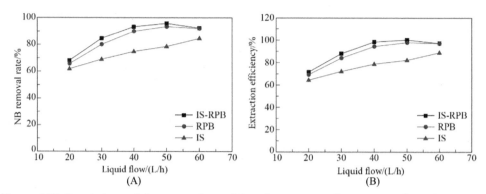

Figure 5.25 Comparison of the removal rate (A) and extraction efficiency (B) of nitrobenzene in the IS-RPB, IS, and RPB. *IS-RPB*, Impinging stream-rotating packed bed.

higher at higher liquid flow rates, which leads to the formation of smaller liquid droplets and a larger contact area between the two phases. Thus, as the liquid flow rate is increased from 20 to 60 L/h, the removal rate of nitrobenzene in the IS-RPB is increased by 35.3%. At low liquid flow rates, the packing could not be completely wetted and most part of the packing is not available for liquid-liquid mass transfer. In this case, impinging plays a dominant role. At a liquid flow rate of 20 L/h, the removal rate of nitrobenzene in the IS-RPB is only 10.3% higher than that in the IS. As the flow rate increases, the interface renewal rate is accelerated and the mass transfer efficiency is significantly improved, and then the shear effect of the packing plays a dominant role. Thus, at a liquid flow rate of 50 L/h, the removal rate of nitrobenzene in the IS-RPB is 22.0% and 2.8% higher than that in the IS and RPB, respectively. As the liquid flow rate further increases, the liquid film on the packing surface becomes thicker and the interface renewal rate is reduced because of the limited shear effect of the packing, which leads to a reduction in the mass transfer rate. Under similar conditions, the removal rate of nitrobenzene in the IS-RPB is 2.5%-21.9% higher than that in the RPB and IS, and the extraction efficiency is significantly improved. IS-RPB is capable of intensifying the mixing and mass transfer processes. Notably, the removal rate in the IS increases continuously with increasing liquid flow rate because the impinging at high velocities leads to the occurrence of emulsification between oil and water phases. However, this is less pronounced in the RPB. It is concluded that the use of IS-RPB can improve mass transfer and reduce the risk of emulsification.

5.4.4 Extraction of dyes

Dyes have been widely used to color products in the industries of textile, leather, paper making, plastics, food, and cosmetics, and most dyes and their derivatives are highly toxic and carcinogenic and can inhibit the photosynthesis of aquatic plants [48].

Jayant et al. [13] investigated the liquid–liquid extraction of methyl red in the RPB using xylene as the extraction solvent and the effects of high-gravity factor (β), aqueous phase flow rate (Q_a), and phase ratio (R) on the extraction efficiency and the overall volumetric mass transfer coefficient $K_L a$. Fig. 5.26 shows the effects of high-gravity factor β and aqueous phase volumetric flow rate Q_a on the extraction efficiency under different volumetric flow rate conditions. It is found that the higher the β and Q_a values are, the higher the extraction efficiency will be. At low Q_a/Q_o ratios, the stage efficiency is improved by 4.2%. The packing will have a stronger shear effect on the liquid as the high-gravity factor increases, which can accelerate the interface renewal and increase the mass transfer coefficient. However, increasing the Q_a leads to more violent turbulence in the impact region and a larger impact area. As a result, there is a higher frequency for the collision of liquid droplets and subsequently, the mass transfer between the two phases is increased. At low Q_a/Q_o ratios, increasing the volumetric flow rate of the solvent causes a substantial decrease in the concentration of methyl red in the aqueous phase (raffinate). When $\beta = 50.8$ and Q_o is fixed, the extraction efficiency increases with the increase of Q_a; while when Q_a is fixed, increasing the Q_o leads to a 0.55%–2.2% increase in the extraction efficiency, which is attributed to the combined effect of the high driving force and the strong turbulence of the organic phase.

The effects of the high-gravity factor and the flow rate of the aqueous phase on the overall mass transfer coefficient $K_L a$ are shown in Fig. 5.27. It is found that the higher the high-gravity factor β is, the higher the overall mass transfer coefficient $K_L a$ will be. The extraction efficiency is high at high Q_a/Q_o ratios. At $Q_a = 1.25 \times 10^{-6}$ m³/s, $Q_o = 2.08 \times 10^{-6}$ m³/s, and $\beta = 50.8$, the $K_L a$ value is close to 100%. This is because more holes are required for the distributor when the flow velocity is high, leading to an increase in the dispersion of the liquid and the surface renewal rate.

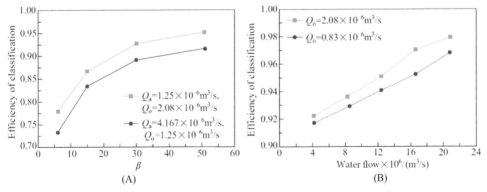

Figure 5.26 Effects of high-gravity factor β (A) and aqueous phase flow rate Q_a (B) on the extraction efficiency.

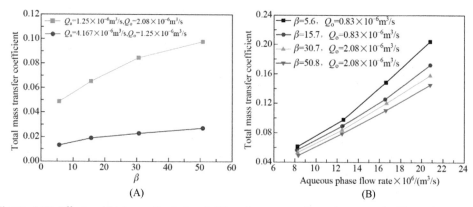

Figure 5.27 Effects of high-gravity factor β (A) and aqueous phase flow rate Q_a (B) on the overall mass transfer coefficient.

Increasing the Q_a can also increase the $K_L a$ because, at a given high-gravity factor, the number of liquid pathways in a given radial distance in the RPB will increase with increasing liquid flow rate, which in turn can increase the wet surface area of the packing and the interfacial contact area between the two liquid phases. A high overall mass transfer coefficient could be obtained at low β values because the residence times of the two liquid phases are sufficiently long in the packing and emulsification is effectively inhibited at lower rotation speeds.

In conclusion, the use of a high-gravity extractor greatly improves the mass transfer efficiency of dyes and provides an effective means for the purification of dye wastewater.

5.4.5 Extraction of indium ions

Indium (In) is widely used in the fields of aerospace, electronics, medicine, national defense, and energy because of its high ductility, plasticity, corrosion resistance, optical permeability, and conductivity [49]. In is mainly used as indium-tin-oxide (ITO) target for the production of liquid crystal displays and flat screens. The growing demand for In in recent years underlines the need for the recovery of In from the by-products resulting from the hydrometallurgy of zinc, and the zinc oxide dust containing In is an important source of In. The leachate of zinc oxide dust contains a variety of impurities such as zinc oxide and iron that can be separated by adsorption, ion exchange, and solvent extraction. Although solvent extraction is the most commonly used extraction method, non-equilibrium solvent extraction should be used for systems that are limited by thermodynamic equilibrium such as Fe and In [50]. Based on the non-equilibrium solvent extraction theory, Chang et al. [51] investigated the separation of In(III) and Fe(III) from the sulfate leaching solution in the IS-RPB using di-2-ethylhexylphosphoric acid

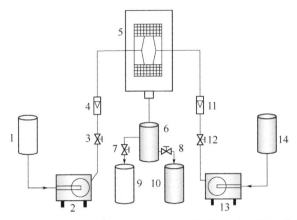

Figure 5.28 The experimental procedure for the separation of In(III) and Fe(III) in the IS-RPB (1,9—organic phase storage tank, 2,13—peristaltic pump, 3,7,8,12—valve, 4,11—rotameter, 5—IS-RPB, 6—separation storage tank, 10,14—aqueous phase storage tank). *IS-RPB*, Impinging stream-rotating packed bed.

(D2EHPA) as extraction solvent. The effects of the flow rate of the organic phase (Q_o), phase ratio ($R = Q_a/Q_o$), high-gravity factor (β), feed solution pH, D2EHPA concentration, and initial Fe(III) concentration on the extraction efficiency η of In(III) and Fe(III) and the separation factor $\xi_{In/Fe}$ (the ratio of the partition coefficient of In and Fe) were also investigated. The experimental procedure is shown in Fig. 5.28.

5.4.5.1 Effect of Q_o on η and $\xi_{In/Fe}$

The effect of the organic phase flow rate Q_o on the separation of In and Fe from the sulfate leaching solution was investigated under conditions of phase ratio $R = 2:1$, 25% D2EHPA/75% kerosene, temperature = 30°C, high-gravity factor = 83, and acid concentration of the feed solution = 0.3 mol/L. Fig. 5.29 reveals that the extraction rate of In is over 99.0%, but that of Fe is below 5.0%, and the separation factor is over 3000. However, the extraction rates of In and Fe are low at $Q_o = 10$ L/h. This is because at low flow velocities, no turbulence is formed in the impact region and there is a low frequency for the impact between the two liquid phases and the aggregation and dispersion of liquid droplets. The optimal flow rate of the organic phase is determined to be 30 L/h. The high flow velocity will produce a high-velocity jet in order to ensure the homogeneous mixing and mass transfer between the two liquid phases. As the flow rate exceeds 30 L/h, the extraction efficiency of In and Fe is reduced. In this case, more liquid droplets are formed, which is not good for the mixing of the liquid. Emulsification occurs at high flow velocities, which can reduce the phase separation process. To sum up, excellent separation efficiency can be obtained at $Q_o = 30$ L/h.

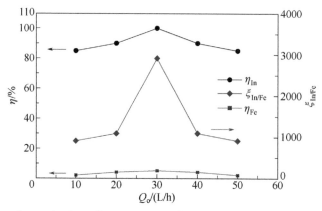

Figure 5.29 Effect of organic phase flow rate Q_o on the separation of Fe and In.

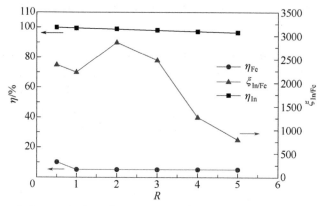

Figure 5.30 Effect of phase ratio R on the separation of In and Fe.

5.4.5.2 Effect of R on η and $\xi_{In/Fe}$

The effect of phase ratio R on the separation of In and Fe from the sulfate leaching solution was investigated under conditions of the flow rate of the organic phase = 40 L/h, 25% D2EHPA/75% kerosene, temperature = 30°C, high-gravity factor = 83, and the acid concentration of the feed solution = 0.3 mol/L. Fig. 5.30 shows that as the phase ratio is increased from 0.5 to 5, the extraction rate of In is reduced from 99.7% to 96.6%, and that of Fe is reduced from 10.3% to 3.3%. At a constant flow rate of the organic phase, the increase of the phase ratio R corresponds to the increase in the flow rate of the aqueous phase (continuous phase). Decreasing the phase ratio R can increase the residence time of the two phases in the IS-RPB, as well as the liquid holdup of the dispersed phase. Thus, the smaller the R-value is, the larger the contact area of the dispersed phase on the packing surface will be. As a result, the extraction efficiency of both In and Fe is obviously increased. In order to separate In and Fe, it is necessary to reduce the extraction efficiency of Fe and increase the extraction efficiency of In. The experimental results reveal that at $R=2$, the extraction efficiency of In is over 99.0% but that of Fe is below 5.0% with a separation factor of 2871. The phase ratio is an important parameter for the separation of heavy metal ions in the IS-RPB, and excellent selective separation can be obtained at $R=2$.

5.4.5.3 Effect of β on η and $\xi_{In/Fe}$

The effect of high-gravity factor β on the separation efficiency of In and Fe is shown in Fig. 5.31. At $\beta<83$, the extraction rate of In and Fe increases with increasing high-gravity factor, but the increasing rate is reducing; while at $\beta>83$, the extraction rate decreases with increasing high-gravity factor. At $\beta\leqslant 83$, increasing the high-gravity factor can reduce the size of the liquid in the voids of the packing and increase the interfacial area for mass transfer, reduce the thickness of liquid films on the packing surface, increase the contact area between the two phases, and increase the frequency of aggregation and dispersion of the liquid. At $\beta>83$, increasing the rotation speed leads to a reduction in the residence time of the two phases. The optimal high-gravity factor for the separation of In and Fe is determined to be 83, under which the extraction rate reaches 99% and the In/Fe separation factor reaches 3000. It is concluded that the high-gravity factor is a key parameter for extraction in the IS-RPB.

5.4.5.4 Comparison of different extractors

The separation efficiency of In(III) and Fe(III) obtained by different extractors is shown in Table 5.10.

IS-RPB shows high separation performance for Fe and In, and entrainment and emulsification rarely occur in the mixing process. Of the three extractors, separating

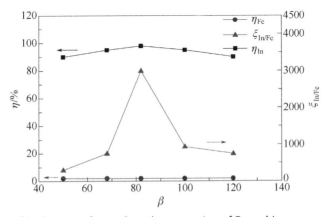

Figure 5.31 Effect of high-gravity factor β on the separation of Fe and In.

Table 5.10 The separation efficiency of In(III) and Fe(III) obtained by different extractors.

Extractor	IS-RPB (impinging stream-rotating packed bed)	SF (separating funnel)	ACC (annual centrifugal contactor)
Operating conditions	$R = 2$; organic phase flow rate = 30 L/h; high-gravity factor = 83	$R = 2$; vibrated for 5 min	Rotor diameter = 20 mm; rotation speed = 2500 r/min; total flow rate = 0.6 L/h; $R = 2$
Extraction efficiency of In	99.16	98.29	99.17
Extraction efficiency of Fe	3.68	16.32	6.85
Separation factor $\xi_{In/Fe}$	3090	287	1625
Regeneration rate of In	99.85	99.67	99.79

funnel (SF) shows the lowest extraction efficiency for In and the highest extraction efficiency for Fe. The extraction efficiency of ACC is in between IS-RPB and SF. In summary, IS-RPB has promising applications for the selective separation of Fe and In.

5.4.6 Other applications

5.4.6.1 Extraction of benzoic acid by physical methods

Li et al. [52] investigated the physical extraction of water/benzoic acid/kerosene in the IS-RPB using water as the extraction solvent and benzoic acid as the solute. The mass concentration of benzoic acid measured by acid-base titration is 0.15-0.2 g/L. The effects of impact initial velocity (u_0) and high-gravity factor (β) on the extraction efficiency were investigated under conditions of phase ratio $R = 1$ and impact angle $\alpha = 180$ degrees. As the partition coefficient of benzoic acid is not a constant in the two phases over the concentration range investigated, the extraction efficiency is measured by means of the stage partitioning coefficient, and the extraction efficiency η is evaluated to investigate the effect of u_0 and β, as shown in Fig. 5.32.

The extraction efficiency shows an increasing trend as the impact initial velocity and high-gravity factor increase in the IS-RPB, and the extraction efficiency reaches 99%. The high-gravity technology can intensify the mass transfer process, but cannot change the equilibrium partitioning coefficient. In view of this, chemical extraction is suggested for systems with low equilibrium partitioning coefficients. In conclusion, IS-RPB can greatly improve the mass transfer in both physical and chemical extraction processes. However, appropriate IS-RPB parameters should be used in order to avoid possible emulsification.

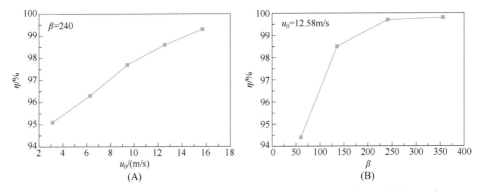

Figure 5.32 The effect of impact initial velocity u_0 (A) and high-gravity factor β (B) on the extraction efficiency η.

5.4.6.2 Continuous wet extraction of copper

Liu et al. [15] investigated the extraction of copper from the leaching solution resulting from hydrometallurgy of copper in a plant in China using LIX984N as the extraction solvent and kerosene as the diluent in the IS-RPB. The experimental results reveal that under conditions of the volume fraction of the solvent = 5%, $R = 1$, $\beta = 135$, and flow rate = 80 L/h, the extraction rate reaches 98.8%. After phase separation, IS-RPB was used for back extraction of the copper-rich oil phase (LIX984N/kerosene) using 180 g/L H_2SO_4 as the back extraction solvent. The first-stage back extraction efficiency is high with a back extraction rate of 95% under conditions of $R = 1$, $\beta = 135$, and flow rate = 80 L/h. After phase separation, the back extraction solvent is recovered for reuse. The use of IS-RPB can significantly improve the extraction rate of copper and reduce the operating cost in a simple and convenient manner.

5.4.6.3 Extraction crystallization of sodium carbonate

Extraction crystallization is used for the crystallization of salt by adding an organic solvent into a saturated salt aqueous solution because of mutual solubility of the solvent and water. Pan et al. [53] used extraction crystallization for recovery of sodium carbonate from saturated sodium carbonate solution/n-butanol in the IS-RPB, and investigated the effects of high-gravity factor, volumetric flow rate ratio of sodium carbonate solution and n-butanol, and impact initial velocity on the recovery rate of sodium carbonate. The results reveal that the recovery rate of sodium carbonate reaches a maximum of 72.1% under conditions of $\beta = 98.8$, $R = 1:1$, and $u_0 = 8.9$ m/s. However, it should be noted that mixing and phase separation are performed separately in the IS-RPB, which can effectively prevent back mixing that frequently occurs in the traditional extractors. In summary, IS-RPB can be used either as a continuous extractor or a continuous back extractor with a short residence time of the liquid, low occurrence of back mixing, and a limited amount of liquid retained in the IS-RPB. It is convenient for the treatment of special materials and exchange of extraction solvent. Impressively, the extraction efficiency almost reaches the equilibrium efficiency.

5.5 Prospects

Liquid-liquid extraction has widespread applications in hydrometallurgy, wastewater treatment, separation of gases and organic compounds, biological and medical separation, chemical sensors, and ion-selective electrodes. Although IS-RPB has excellent micromixing performance in liquid-liquid extraction, several challenges have to be addressed in the future:

1. It is essential to develop the fluid mechanics and micromixing theory for complex heterogeneous fluid in liquid-liquid extraction and advanced manufacturing platform for IS-RPBs. More pilot tests are needed for high-gravity separation in order to provide technological support for the continuous treatment of industrial organic wastewater.
2. It is crucial to better understand the extraction and separation of heavy metals in coking, printing and dyeing, and nuclear wastewater, and to develop thermodynamic and kinetic models for heavy metal ions in the high-gravity field. It is also important to recover valuable metals in wastewater in order to maximize the utilization of resources.
3. It is necessary to couple high-gravity technology with ultrasound, microwave, and supercritical fluid technologies for extraction of natural medical components such as anti-cancer drugs (e.g., taxol and its derivatives), antimalarial drugs (e.g., artemisinin), and cardiovascular drugs (e.g., ginkgolide).

References

[1] Shi J, Wang JD. Chemical Engineering Handbook. Beijing: Chemical Industrial Press, 1996.
[2] Zhang LX. Mass Transfer & Separation Technology. Beijing: Chemical Industrial Press, 2009.
[3] Dai YY. Principles of Liquid-Liquid Extraction. Beijing: Chemical Industrial Press, 2015.
[4] Guo YJ. Industrial Wastewater Treatment Engineering. Shanghai: East China University of Science and Technology Press, 2016.
[5] Stankiewicz A. Reactive separations for process intensification: an industrial perspective. Chemical Engineering and Process, 2003,42:137-144.
[6] Williams NS, Ray MB, Gomaa HG. Removal of ibuprofen and 4-isobutyl acetophenone by nondispersive solvent extraction using a hollow fiber membrane contactor. Separation and Purification Technology, 2012,88:61-69.
[7] Baier G, Graham MD, Lightfoot EN. Mass transport in a novel two-fluid Taylor vortex extractor. AIChE Journal, 2000,46:2395-2407.
[8] Bonam D, Bhattacharyya G, Bhowal A, et al. Liquid-liquid extraction in a rotating spray column: removal of Cr(VI) by Aliquat 336. Industrial & Engineering Chemistry Research, 2009,48:7687-7693.

[9] Hossein A, Ali MM, Reza RS. The effects of a surfactant concentration on the mass transfer in a mixer-settler extractor. Iranian Journal of Chemistry & Chemical Engineerin—International English Edition, 2006,25:9-15.

[10] Seibert AF, Fair JR. Hydrodynamics and mass transfer in spray and packed liquid-liquid extraction columns. Industrial & Engineering Chemistry Research, 1988,27:470-481.

[11] Alireza H, Meisam TM, Mehdi A. Mass transfer coefficients in a Kühni extraction column. Chemical Engineering Research & Design, 2015,93:747-754.

[12] Dehkordi AM. Novel type of impinging streams contactor for liquid-liquid extraction. Industrial & Engineering Chemistry Research, 2001,40:681-688.

[13] Jayant BM, Avijit B, Siddhartha D. Extraction of dye from aqueous solution in rotating packed bed. Journal of Hazardous Materials, 2016,304:337-342.

[14] Liu YZ. Chemical Process Intensification Methods and Technologies. Beijing: Chemical Industrial Press, 2017.

[15] Liu YZ. Research progress of high-gravity technology of IS-RPB to intensify liquid-liquid contact. Chemical Industry and Engineering Progress, 2009,28:1101-1108.

[16] Liu YZ. High-gravity chemical process and technology. Beijing: National Defence Industry Press, 2009.

[17] Jiao WZ, Liu YZ, Qi GS. Micromixing characteristics of IS-RPB reactor with chemical coupling method. Chemical Engineering(China), 2007,35:36-39.

[18] Zhao HH, Ouyang CB, Liu YZ. Contrasting of micromixing function of three reactors. Energy Chemical Industry, 2003,24:31-33.

[19] Guichardon P, Falk L, Andrieu M. Experimental comparison of the iodide-iodate and the diazo coupling micromixing test reactions in stirred reactors. Chemical Engineering Research & Design, 2001,79:906-914.

[20] Jiao WZ, Liu YZ, Qi GS. A new impinging stream-rotating packed bed reactor for improvement of micromixing iodide and iodate. Chemical Engineering Journal, 2010,157:168-173.

[21] Jiao WZ, Liu YZ, Qi GS. Micromixing efficiency of viscous media in novel impinging stream-rotating packed bed reactor. Industrial & Engineering Chemistry Research, 2012,51:7113-7118.

[22] Engelmann U, Schmidt NG. Influence of micromixing on the free radical polymerization in a discontinuous process. Macromolecular Theory and Simulations, 1994,3:855-883.

[23] Bourne JR. Mixing and selectivity of chemical reactions. Organic Process Research & Development, 2003,7:471-508.

[24] Fournier MC, Falk L, Villermaux J. A new parallel competing reaction system for assessing micromixing efficiency-determination of micromixing time by a simple mixing model. Chemical Engineering Science, 1996,51:5187-5192.

[25] Guichardon P, Falk L. Characterisation of micromixing efficiency by the iodide-iodate reaction system. Part I: experimental procedure. Chemical Engineering Science, 2000,55:4233-4243.

[26] Liu CI, Lee DJ. Micromixing effects in a couette flow reactor. Chemical Engineering Science, 1999,54:2883-2888.

[27] Rousseaux JM, Falk L, Muhr H, et al. Micromixing efficiency of a novel sliding-surface mixing device. AIChE Journal, 1999,45:2203-2213.

[28] Judat B, Racina A, Kind M. Macro- and micromixing in a taylor-couette reactor with axial flow and their influence on the precipitation of barium sulfate. Chemical Engineering Technology, 2004,27:287-292.

[29] Wu Y, Xiao Y, Zhou YX. Micromixing in the submerged circulative impinging stream reactor. Chinese Journal of Chemical Engineering, 2003,11:420-425.

[30] Yang K, Chu GW, Shao L, et al. Micromixing efficiency of viscous media in micro-channel reactor. Chinese Journal of Chemical Engineering, 2009,17:546-551.

[31] Chu GW, Song YH, Yang HJ, et al. Micromixing efficiency of a novel rotor—stator reactor. Chinese Engineering Journal, 2007,128:191-196.

[32] Yang HJ, Chu GW, Zhang JW, et al. Micromixing efficiency in a rotating packed bed: experiments and simulation. Industrial & Engineering Chemistry Research, 2005,44:7730-7737.

[33] Liu YZ, Qi GS, Yang LR. Study on the mass transfer characteristics in impinging stream-rotating packed bed extractor. Chemical Industry and Engineering Progress, 2003,22:1108-1111.

[34] Qi GS, Liu YZ, Yang LR. Single stage research for treating phenol-containing wastewater by impinging stream-rotating packed bed. Journal of Chemical Industry & Engineering, 2004,25:9-12.
[35] Qi GS. Extraction performance and application of IS-RPB. Taiyuan: North University of China, 2004.
[36] Si JF. Multistage Separation Process. Beijing: Chemical Industrial Press, 1991.
[37] Li YG, Li Z, Fei WY. Liquid-Liquid Extraction Process and Equipment. Beijing: China Atomic Energy Press, 1981.
[38] Chen JF. High-gravity Technology and Application. Beijing: Chemical Industrial Press, 2002.
[39] Shi JJ, Liu YZ, Yu HB. Technology and application of new type solvent extraction. Energy Chemical Industry, 2005,26:15-18.
[40] Zhang J, Li GB, Ma J. Detriment and treatment method of wastewater containing phenols. Chemical Engineering(China), 2001,2:36-37.
[41] Qi GS, Liu YZ, Yang LR. Complexation extraction of TBP and phenol in impinging stream-rotating packed bed. Chemical Production and Technology, 2004,11:13-16.
[42] Zhang CY, Guo WG, Liu YL. Recycling acetic acid in wastewater by the technology of extraction-back extraction. Environmental Protection in Petrochemical Industry, 2004,27:30-33.
[43] Dil B, Yang YY, Dai YY. Extraction of acetic acid from dilute solution by reversible chemical complexation. Journal of Chemical Engineering of Chinese Universities, 1993,7:174-179.
[44] Qi GS, Liu YZ, Jiao WZ. Experimental research on extraction of acetic acid from dilute solution by chemical complexation impinging stream-rotating packed bed. Modern Chemical Industry, 2008, (11):65-67.
[45] Xu JB, Jing TS, Yang L. Effects of nitrobenzenes on DNA damage in germ cells of rats. Chemical Research in Chinese University, 2006,22:29-32.
[46] Yang PF, Luo S, Zhang DS, et al. Extraction of nitrobenzene from aqueous solution in impinging stream-rotating packed bed. Chemical Engineering Process: Process Intensification, 2018,124:255-260.
[47] Guo K, Guo F, Feng YD, et al. Synchronous visual and RTD study on liquid flow in rotating packed-bed contactor. Chemical Engineering Science, 2000,55:1699-1706.
[48] Akgül M. Enhancement of the anionic dye adsorption capacity of clinoptilolite by Fe^{3+} grafting. Journal of Hazardous Materials, 2014,267:1-8.
[49] Li CX, Wei C, Xu HS, et al. Kinetics of indium dissolution from sphalerite concentrate in pressure acid leaching. Hydrometallurgy, 2010,105:172-175.
[50] Alfantazi AM, Moskalyk RR. Processing of indium: a review. Miner Engineering, 2003,16:687-694.
[51] Chang J, Zhang LB, Du Y, et al. Separation of indium from iron in a rotating packed bed contactor using di-2-ethylhexylphosphoric acid. Separation and Purification Technology, 2016,164:12-18.
[52] Li TC, Qi GS. Impinging stream-rotating packed bed extractor. Petro-Chemical Equipment, 2004,33:26-28.
[53] Pan HX, Liu YZ, Qi GS. Study on extraction crystallization for recovery of sodium carbonate by high-gravity technology. Modern Chemical Industry, 2010,11:76-78.

CHAPTER 6

Liquid membrane separation

Contents

6.1 Overview	207
6.2 Mechanism of intensification of liquid membrane preparation and separation by impinging stream-rotating packed bed	209
6.2.1 Mechanism of emulsion liquid membrane separation	209
6.2.2 Mechanism of intensification of liquid membrane preparation by impinging stream-rotating packed bed	211
6.2.3 Mechanism of intensification of liquid membrane separation by impinging stream-rotating packed bed	212
6.3 Key technologies of emulsion liquid membrane separation in the impinging stream-rotating packed bed	216
6.3.1 Emulsion liquid membrane preparation	216
6.3.2 Emulsion liquid membrane separation	219
6.3.3 Demulsification	221
6.4 Application examples	221
6.4.1 Treatment of phenol wastewater	221
6.4.2 Treatment of aniline wastewater	230
6.5 Prospects	238
References	239

6.1 Overview

Liquid membrane separation has been an important branch of membrane separation processes since its first application to reverse osmosis desalination in the 1960s. In the mid-1960s, Li and Somernet[1] observed a stable membrane with no mechanical support at the interface between the aqueous solution of a surfactant and the oil solution in the investigation of their interfacial tension using the Du Nouy ring method, and they were granted a patent on liquid membranes for separating hydrocarbons with similar physical and/or chemical properties in 1968. Since then, there has been a growing interest in the study and use of liquid surfactant membranes. In the early 1970s, Cussler incorporated active carriers into the liquid membrane in order to enhance its selectivity. Bloch et al. [2] studied the separation of metal ions by dialysis through solvent membranes. Ward and Robb [3] developed an immobilized aqueous bicarbonate-carbonate solution membrane for CO_2-O_2 separation. In the late 1980s, Marr et al. [4] recovered Zn from wastewater by liquid membrane-permeation on an industrial scale. Liquid membrane separation has been shown to be effective in separating, enriching and recovering solutes, and this is attributed to its ability to combine the advantages of both solid membrane separation and extraction technologies, such as high efficiency, rapid process, high selectivity, and energy efficiency [5]. It is widely used in hydrometallurgy, wastewater treatment, gas separation, drug extraction, petrochemical industry, chemical sensors, and ion-selective electrodes [6].

A liquid membrane is a very thin layer of liquid (either aqueous or organic solution) that separates two mutually miscible solutions of different compositions, and it is immiscible with these two phases and allows the selective transfer of one or more components from one phase to another [7]. A liquid membrane is composed of solvent ($>90\%$), surfactant (1%-5%), and mobile carrier (1%-5%). The solvent is used as the substrate material of the liquid membrane because of its viscosity and mechanical strength, the surfactant—containing hydrophilic and/or hydrophobic groups is used to enhance the stability of the liquid membrane because of directional alignment, and the mobile carrier is responsible for the selective transfer of desired solutes or ions, which plays a decisive role in the selectivity and flux of the liquid membrane. In general, according to the shape, liquid membrane is divided into two categories: diaphragm membrane and spherical membrane. Spherical membrane can be further divided into single-drop liquid membrane and emulsion liquid membrane (ELM). ELM is also referred to as surfactant membrane and it is a double emulsion membrane with either water in oil in water (W/O/W) or oil in water in oil (O/W/O) emulsion [8]. The typical structure of ELM is schematically shown in Fig. 6.1. The internal phase of the emulsion is miscible with the continuous external phase, both of which are immiscible with the membrane phase. The emulsion can be either oil-in-water (O/W) or water-in-oil (W/O). In the former case, an oil membrane is formed, which is applicable to the separation of solutes from aqueous solutions, while in the latter case, a water membrane is formed, which is applicable to the separation of solutes from oil solutions. This chapter discusses only ELM and readers interested in other liquid membrane separation processes are referred to other sources.

Two immiscible phases are subjected to high shear forces to produce an emulsion. The resulting emulsion is dispersed in a third continuous phase, and a liquid membrane is formed between the internal phase and the continuous external phase. It allows for selective transfer of one or more solutes from the external phase to the internal phase because of their differences in selective permeability. The internal droplets are several microns in diameter, the W/O emulsion is about 0.1-1 mm in diameter, and the effective thickness of the membrane is 1-10 μm, which is much thinner than common solid membranes. Therefore, ELM is characterized by large specific surface area, rapid permeation and high separation efficiency [9]. Although the internal droplets are merged with various reagent droplets, the concentrations of these reagents are assumed to be zero because the droplets are very small. The solute diffuses to the outside of droplets through membrane phase diffusion. Therefore, ELM is considered to be equivalent to a vacuole with a constant thickness [10].

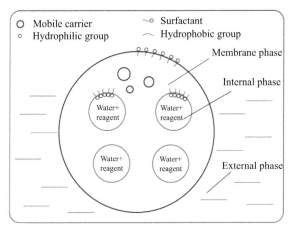

Figure 6.1 A schematic diagram of the structure of ELM. *ELM*, emulsion liquid membrane.

6.2 Mechanism of intensification of liquid membrane preparation and separation by impinging stream-rotating packed bed

6.2.1 Mechanism of emulsion liquid membrane separation

The mechanism of ELM separation involves selective permeation and facilitated transport based on the mass transfer process [11].

6.2.1.1 Selective permeation

Selective permeation is essentially a physical mass transfer process that is dependent primarily on the dissolution of components to be separated. The components of a mixture have different permeation rates through the membrane because of their differences in solubility and diffusion coefficient. It is shown in Fig. 6.2(A) that for a mixture consisting of two components (A and B), the permeation rate of A through the membrane is higher than that of B, which allows selective transfer of A from the external phase to the internal phase but retention of B in the external phase. The separation process stops once the concentrations of A on both sides of the membrane reach an equilibrium. Therefore, selective permeation could not be used to concentrate the components to be separated.

6.2.1.2 Facilitated transport

Facilitated transport is a process in which the permeation of a solute through the liquid membrane is chemically augmented. Specifically, the solute can react with the internal phase or the membrane phase to increase the driving force for mass transfer and eventually the mass transfer efficiency. Facilitated transport can be divided into type I and type II facilitated transport according to the difference in chemical reaction.

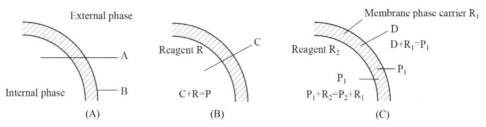

Figure 6.2 Mechanism of ELM separation. (A) Selective permeation; (B) type I promotes transmission; and (C) type II promotes transmission. *ELM*, emulsion liquid membrane.

In type I facilitated transport, the solute to be separated from the feed solution first dissolves into the membrane phase and then reacts with the reagent in the internal phase at the interface between the membrane phase and the internal phase to form a new species that is insoluble in the membrane phase and could not diffuse back to the external phase through the membrane. The solute concentration in the internal phase is reduced, and the resulting concentration gradient between internal and external phases facilitates the continuous transfer of the solute from the external phase to the internal phase and eventually concentrates the solute in the internal phase. The chemical reaction is shown in Fig. 6.2(B). Briefly, solute C diffuses from the continuous external phase to the membrane phase and then reacts with reagent R in the internal phase to form a new species P which is insoluble in the membrane phase. As a result, the concentration of C in the internal phase is zero, and the process continues until R is exhausted. Therefore, the concentration gradient of C between internal and external phases is maximized to facilitate the separation of C.

In type II facilitated transport (or carrier-facilitated transport), a carrier (extractant or complexing agent) is incorporated in the membrane phase to augment the transfer rate. A complexation reaction occurs between carrier and solute at the interface between the membrane phase and the feed solution, and the resulting complex dissolves into the membrane phase and diffuses to the interface between the membrane phase and the internal phase, where it reacts with the decomplexing agent in the internal phase. The solute is decomplexed into the internal phase, and the free carrier continues to react with the solute in the external phase. As shown in Fig. 6.2(C), solute D in the feed solution reacts with carrier R_1 in the membrane phase at the interface to form complex P_1, which diffuses into the internal phase and reacts with reagent R_2 to form complex P_2 that is insoluble in the membrane phase. At this time, R_1 is released. Thus, R_1 acts as the carrier to facilitate the transfer of the solute.

The mechanism of ELM separation can be divided into immobile carrier-based ELM separation [Fig. 6.2(A) and (B)] and mobile carrier-based ELM separation [Fig. 6.2(C)] based on whether there is a mobile carrier to facilitate the transfer of the solute through the liquid membrane [12].

6.2.2 Mechanism of intensification of liquid membrane preparation by impinging stream-rotating packed bed

A major challenge for the industrial application of liquid membrane separation is to develop high-efficiency emulsification devices. The preparation process of ELM in the IS-RPB (impinging stream–rotating packed bed) is controllable and the resulting ELM is small, uniform and highly stable. The preparation conditions can be optimized by adjusting the parameters of the high-gravity field in order to better meet the requirements of subsequent phase separation (demulsification). The mechanism of how IS-RPB improves emulsion preparation is described as follows. The membrane and internal aqueous phases of approximately the same volume flow along coaxial feed pipes and are injected into the IS-RPB through nozzles which are arranged coaxially and concentrically at opposite sides of the IS-RPB. The two phases impinge at the middle of opposing nozzles, and the conversion of the kinetic energy to the static pressure energy causes changes in the flow direction of the two phases and subsequently the formation of a fan-shaped impingement surface that is perpendicular to the jet direction for mixing and mass transfer between the two phases. This surface becomes thinner further away from the impingement point, resulting in an increase in the contact area and flow velocity and the formation of expansion waves. Eventually, the surface is broken into liquid ligaments and droplets. At high initial impact velocities, the two phases will be more uniformly dispersed and mixed within a very short period of time for preliminary emulsification [13]. The impingement of the two opposing jets leads to strong radial and axial components of the turbulent velocity, and the liquid mixture is uniformly dispersed and flows at a high velocity in the RPB, where it is sheared into small liquid films, filaments, and droplets by the high-speed rotating packing [14]. The high centrifugal force resulting from the high gravity field in the RPB leads to high turbulence and rapid interfacial renewal, so that the mixture can be repeatedly broken, aggregated, and dispersed in a short period of time for thorough emulsification. Ouyang et al. [15] have shown that under microgravity conditions ($g \rightarrow 0$), there is no interphase flow for the liquid and the intermolecular force plays a leading role, making it difficult for sufficient contact between the two phases and thus resulting in low interphase mass transfer and micromixing. However, the higher the buoyancy factor $\Delta(\rho g)$ is, the higher the slip velocity between the two fluids will be. The bed that rotates at a high speed can generate large shear force on the liquid, which enhances the collision, aggregation, and dispersion of droplets and consequently the micromixing process. It becomes clear that increasing the high-gravity factor can increase the buoyancy factor $\Delta(\rho g)$, which in turn can increase the mass transfer and the impinging frequency and mixing of droplets.

The preparation process of liquid membrane using the IS-RPB is shown in Fig. 6.3. Oil (O) and water (W) are injected into the IS-RPB for impingement, mixing, and dispersion in the inner cavity of the IS-RPB, and the mixture flows radially from the inner edge to the outer edge of the packing driven by the centrifugal force.

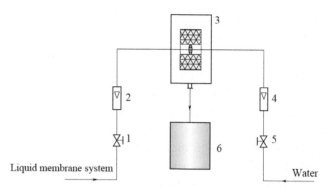

Figure 6.3 The preparation process of liquid membrane using the IS-RPB(1,5—control valve;

After that, the mixture is thrown onto the wall of the casing and then discharged into the storage tank from the liquid outlet at the bottom of the IS-RPB. This results in obtaining desired emulsion (W/O). The two-step emulsification process in the IS-RPB leads to better micromixing of the two phases in a shorter period of time in comparison with a stirred tank. Liu et al. [16,17] investigated the preparation and properties of methanol-diesel-emulsified fuel in the IS-RPB. The computational fluid dynamics (CFD) simulation reveals that the emulsification of methanol-diesel is attributed to vigorous impingement, high-gravity, baffle, and atomization. Therefore, The unique advantages of IS-RPB, such as small emulsion particle size, narrow size distribution, high stability, high swelling rate, short residence time, small equipment size, low space requirement, low energy consumption, easy scaling up, and continuous operation, make it suitable for emulsion preparation on an industrial scale.

6.2.3 Mechanism of intensification of liquid membrane separation by impinging stream-rotating packed bed

The mechanism of intensification of ELM separation by IS-RPB is discussed taking phenol and aniline as examples.

6.2.3.1 Mechanism of emulsion liquid membrane separation
6.2.3.1.1 Mass transfer mechanism

The driving force for mass transfer in wastewater treatment using ELM is derived from the difference in chemical potential of the solute between internal and external phases. The diffusion of the solute through the membrane is controlled by concentration gradient. The ELM separation of phenol or aniline from wastewater mostly follows the facilitated transport mechanism without a carrier. Because of the chemical reaction, there is a large concentration gradient of the solute between the two sides of the membrane to facilitate the transfer of the solute and subsequent enrichment in the internal phase.

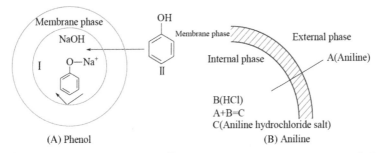

Figure 6.4 Schematic of phenol (A) and aniline (B) extraction by emulsion membrane.

Fig. 6.4(A) shows the removal process of phenol by ELM. The phenol that exists in the form of molecules in the external aqueous phase (feed solution) dissolves in the membrane phase and diffuses to the membrane-internal phase interface. Then, it reacts with NaOH in the internal aqueous phase to form sodium phenolate which is insoluble in the membrane phase and thus retained in the internal aqueous phase [18]. Because of the chemical reaction, the concentration of phenol in the form of molecules is maintained to be 0 in the internal aqueous phase, which maintains a continuous high concentration gradient of phenol between internal and external phases to drive the transfer of phenol from the external phase to the internal phase. As long as there is sufficient NaOH in the internal phase, the phenol in the external phase will be continuously transferred to the internal phase. As shown in Fig. 6.4(B), the driving force for the mass transfer of aniline using ELM is the difference in the chemical potential of the solute between external and internal phases. Similar to sodium phenate, ionic aniline hydrochloride is also insoluble in the membrane phase and could not diffuse back to the external phase [19]. The chemical reaction leads to a zero effective concentration of aniline in the form of molecules in the internal phase so that a high concentration gradient of aniline is maintained between internal and external phases to continuously drive the transfer of aniline from the external phase to the internal phase. Most aniline diffuses into the internal phase and is removed from the wastewater.

6.2.3.1.2 Kinetics of mass transfer

ELM is featured by large specific surface area, high permeability, high selectivity, and high separation efficiency. Both phenol and aniline are highly soluble in the membrane phase and diffuse rapidly through the membrane, yielding a high flux. An irreversible chemical reaction occurs in the internal phase and the products (sodium phenate and aniline hydrochloride) could not diffuse back to the external phase [20].

$$C_6H_5OH + NaOH \longrightarrow C_6H_5ONa + H_2O \qquad (6.1)$$

$$C_6H_5NH_2 + HCl \longrightarrow C_6H_5NH_3^+Cl^- \qquad (6.2)$$

The diffusion of the solute in the liquid membrane and the facilitated transport in the internal phase are the two major steps. The controlling step is the diffusion of phenol and aniline in the liquid membrane, which is expressed as follows:

$$\frac{dc}{dt} = -DA\frac{\Delta c}{\delta} \qquad (6.3)$$

where c—permeate concentration, mol/L;
t—time, s;
D—permeability coefficient, $1/(m \cdot s)$;
A—effective permeation area, m^2;
δ—membrane thickness, m.

According to formula (6.3), the mass transfer rate can be increased by either increasing the driving force for mass transfer or reducing the mass transfer resistance. This is achieved by increasing the permeability coefficient, mass transfer area, or permeate concentration, or reducing the membrane thickness. For a given system and equipment, the turbulence of the two phases can be increased by appropriately increasing their flow velocity or supplying external energy to increase the permeability coefficient. A higher degree of turbulence can also accelerate the surface renewal of the liquid and increase the relative concentration of the permeate. The supply of external energy contributes to reducing the droplet size and consequently increasing the mass transfer area, and reducing the oil/water ratio contributes to reducing the membrane thickness.

6.2.3.2 Mechanism of intensification of liquid membrane separation by impinging stream-rotating packed bed

In the following section, we will discuss the mechanism of intensification of liquid membrane separation by IS-RPB from the perspectives of micromixing and liquid distribution in the packing.

6.2.3.2.1 Micromixing

Li [21] compared the micromixing efficiency (represented by the segregation index X_S) of IS-RPB, IS and RPB with traditional mixers, as shown in Table 6.1.

It is seen from Table 6.1 that IS-RPB has the best micromixing efficiency, followed by RPB and IS. Specifically, the micromixing efficiency of IS-RPB is about

Table 6.1 Comparison of the micromixing efficiency of different mixers.

Novel mixers	Micromixing efficiency	Traditional mixers	Micromixing efficiency
IS-RPB	$X_S < 0.025$	CSTR	$X_S > 0.1$
IS	$0.06 < X_S < 0.12$	Tee mixer	$X_S > 0.1$
RPB	$0.05 < X_S < 0.1$	Tubular reactor	$X_S > 0.15$

two times that of IS and RPB and four times that of traditional mixers (CSTR, Tee mixer, and tubular reactor). Therefore, the micromixing process is greatly intensified in the IS-RPB.

Li et al. [22] found that the micromixing time t_m of IS was approximately 1 ms at an impingement velocity of 6 m/s. As the micromixing time t_m is 0.04–0.4 ms for RPB and 5–50 ms for traditional stirred tank reactor, respectively [23], RPB and IS are expected to have better micromixing efficiency than traditional stirred tanks. Zhou [24] found that the average residence time in the RPB was less than 0.5 s at rotation speeds higher than 1000 r/min, and the higher the rotation speed and the liquid flow rate are, the shorter the average residence time will be. Therefore, IS-RPB could be a promising mixer with higher micromixing efficiency than traditional stirred tank and it is more suitable for the rapid micromixing process.

6.2.3.2.2 Liquid distribution in the packing

Zhang [25] observed how liquid flowed in the RPB by using a stroboscope. The results have demonstrated that the liquid in the packing mainly exists in the form of droplets, films, and ligaments. Specifically, the liquid mainly exists in the form of droplets at the inner edge of the packing but in the form of films in the packing. At rotation speeds of 300–600 r/min, the liquid mainly flows in the form of films covering the surface and voids of the packing, while at rotation speeds of 800–1000 r/min, the liquid mainly flows in the form of films on the packing surface and droplets in the voids of the packing. The image analytic method was used to measure the diameters of droplets in the RS wire mesh packing, and the results showed that the diameters ranged from 0.1 to 0.3 mm. When the force acting on the liquid in the wire mesh packing in the high-gravity field is simplified, the average diameter of liquid droplets can be calculated as follows:

$$d = 0.7284 \left(\frac{\sigma}{w^2 R \rho} \right)^{0.5} \tag{6.4}$$

where σ—surface tension, N/m;
w—rotation speed, r/s;
R—rotor radius, m;
ρ—liquid density, kg/m^3.

Guo [26] measured the thickness of the liquid film on the packing surface using the image analytic method and found that the liquid film thickness was 10-30 and 20-80 μm on the wire mesh packing and foam metal packing, respectively. Guo et al. [27] established a mathematical model for liquid film thickness in an RPB:

$$\delta = 4.20 \times 10^8 \frac{vL}{a_f w^2 R} \qquad (6.5)$$

where v—kinematic viscosity, m²/s;
  L—liquid flux, m³/(m² · s);
  a_f—specific surface area of packing, m²/m³;
  w—rotation speed of packing, r/min;
  R—packing radius, m.

According to formulas (6.4) and (6.5), the liquid films in the packing become thinner and the droplets on the wire mesh packing become smaller with increasing rotation speed. Also, the liquid is highly dispersed into micro-elements (e.g., ligaments, droplets, and films) in the high-speed rotating packing, which greatly increases the contact area between the two liquid phases and consequently the liquid-liquid mixing.

In summary, the use of RPB permits rapid and uniform liquid-liquid micromixing. The two liquid phases are brought into intimate contact in the packing in a high-gravity field that is hundreds and even thousands of times higher than the gravity, and the mixture is broken into nano-sized ligaments, droplets, and films by the high shear force resulting from the high-speed rotation of the packing. In the IS-RPB, emulsification is significantly improved, the liquid droplets in the internal phase are smaller with a narrow particle size distribution, and the emulsion is more uniformly distributed in the wastewater. As a result, the mass transfer area of the emulsion is increased and the thickness of the liquid membrane is reduced, which leads to not only an increase in the driving force for mass transfer but also a decrease in the mass transfer resistance.

6.3 Key technologies of emulsion liquid membrane separation in the impinging stream-rotating packed bed

The key technologies of ELM separation in the IS-RPB mainly involve ELM preparation, ELM separation, and demulsification.

6.3.1 Emulsion liquid membrane preparation

Selection of an appropriate mobile carrier, ELM composition, and ELM preparation is critical to obtain desired ELM in the IS-RPB [28].

6.3.1.1 Selection of mobile carrier

Liquid membrane separation is essentially an extraction process, and a mobile carrier (corresponding to the extractant) is incorporated in the membrane phase in order to increase the flux of metal ions across the membrane and facilitate molecular diffusion. In general, any extractant used for solvent extraction can be used as mobile carrier for liquid membrane separation. The mobile carrier should be soluble in the membrane phase but insoluble in internal and external phases, and its complex with the component to be separated should be stable to some extent. The carrier forms a complex with the component to be transported at the feed-membrane interface and transfers the complex to the membrane-receiving phase interface, where the complex is dissociated. The carrier should not react with the surfactant otherwise, it will reduce the stability of the membrane.

6.3.1.2 Emulsion liquid membrane composition

The membrane solvent is the major constituent of the liquid membrane. The stability of the liquid membrane and the solubility of the solute should be considered in selecting the solvent. The criteria for the selection of membrane solvent include the following: (1) it is insoluble in internal and external phases, (2) it is not volatile with a low boiling point, (3) no liquid-to-solid phase transition occurs during use, (4) it has an adequate level of viscosity to maintain the mechanical strength, (5) water is the preferred solvent for the separation of hydrocarbons, (6) for the liquid membrane without a carrier, the solvent must preferentially dissolve the solute to be separated, while for the liquid membrane with a carrier, the solvent must dissolve only the carrier but not the solute, and (7) a non-polar organic solvent is often used in the oil membrane in order to ensure the carrier to have the maximum solubility and prevent the miscibility with water.

The stability of the liquid membrane is proportional to the surfactant concentration. The higher the surfactant concentration is, the more stable the liquid membrane will be. However, an extremely high concentration of surfactant has an impact on the permeability of the membrane because of the increased thickness and viscosity of the membrane. The criteria for the selection of surfactant include the following: (1) it has low solubility in internal and external phases to preferentially facilitate the transfer of the desired solute through the membrane, (2) it will not affect the selectivity of the liquid membrane with a mobile carrier, and (3) for the separation of ions from the aqueous solution, a nonionic surfactant contributes to increasing the stability of the liquid membrane. Since 1998, Wan et al. [29] have developed a series of lubricating oil additives polyisobutylene succinimide with a similar structure to ENJ-3029, such as Shang-205, Lan-113A, and Lan-113B, for use in liquid membrane.

6.3.1.3 Preparation of liquid membrane

In order to obtain a stable ELM, it is necessary to maintain the average diameter of internal phase droplets in the range of 1-3 μm. This is often achieved by vigorous stirring. A commercially available high-speed stirrer is often used on a laboratory scale, such as Tekmar homogenizer and Waring mixer with a stirring speed of up to 20,000 r/min [30]. Another option is the ultrasonic emulsifier. Colloid mill is often used for the large-scale preparation of emulsion. However, dynamic emulsifiers are susceptible to corrosion caused by stripping acid on pilot and industrial scales, and the use of corrosion-resistant materials will substantially increase the cost. Static emulsifiers with a production capacity of up to 600 L/h has been developed to address these challenges, and they are widely used for preparing emulsion used for the removal of Zn from wastewater in the viscose fiber industry. There is a practical need to develop high-efficiency emulsifiers in order to increase the industrial applicability of liquid membrane separation. Currently, the most common emulsifiers include mechanical stirrers, high-shear emulsifiers (e.g., IS-RPB), high-pressure homogenizers, colloid mills, and ultrasonic emulsifiers [31].

Stirrer is the earliest and most widely used equipment for emulsion preparation, and it is characterized by simple operation and low space occupancy. However, the resulting emulsion often shows low dispersion, large internal phase particles, low stability, long emulsification time, and batch operation. In their study of the extraction of acetic acid with ELM prepared using LMS2 or Span80 as surfactant, kerosene as solvent, and NaOH as the internal aqueous reagent in a stirrer, Liu et al. [32] found that a stable emulsion was obtained at a stirring speed of 3000 r/min for 20 min, and the removal rate of acetic acid reached 96% and the breakage rate of the membrane was only about 5%.

The high-shear emulsifier exhibits much better emulsification efficiency but shorter emulsification time than the stirrer, but it is more applicable to laboratory situations because of the low processing capacity. Qiao et al. [33] prepared ELM with petroleum ether as the membrane phase using a high-shear emulsifier and found that the most stable emulsion was obtained at a stirring speed of 4000 r/min for 15 min. The breakage rate was less than 10% after standing for 24 h and was about 5% after extraction for 20 min.

The high-pressure homogenizer shows low thermal effect and the resulting emulsion droplets are small and uniform. However, more energy would be consumed and, more importantly, it is prone to damage that requires considerable labor for the maintenance and is not applicable to materials with high viscosity [34]. Shen et al. [35] prepared emulsion using a high-pressure homogenizer at a stirring speed of 5000 r/min for 3 min. Under appropriate conditions, the removal rate of alanine reached 80% with a breakage rate of about 5% and a swelling ratio of about 40%.

Colloid mill is featured by simple structure, easy maintenance, and wide applicability. However, premixing is required. In many industrial situations, the operation is too complicated to allow automatic control. The properties of the material may change due to the generation of substantial heat from the friction between the rotor and the material and stator. Also, long-term continuous production is not possible. Wan et al. [36] used liquid membrane for treatment of wastewater containing high concentrations of phenol. A stable emulsion was obtained after pre-mixing for 1 min using a stirrer and treatment with a colloid mill for 3 min. After secondary treatment, the phenol concentration was reduced from 1000 to 0.5 mg/L.

Ultrasonic emulsifier is featured high emulsification efficiency and high emulsion stability, but it is only applicable to laboratory settings because of the low processing capacity and long emulsification time.

Stirrer and ultrasonic emulsifier are suitable for batch preparation or long-time small-scale preparation of emulsion, while high-pressure homogenizer and colloid mill have the disadvantages of high consumption of energy and emulsifying agent and low emulsification efficiency, which may reduce the mass transfer rate and increase the breakage rate of the membrane in the extraction process. IS-RPB as a representative of high-shear homogenizer allows continuous operation and has excellent mass transfer and micromixing performance, and it is an efficient device for emulsion preparation.

6.3.2 Emulsion liquid membrane separation

The ELM separation process is schematically shown in Fig. 6.5. ELM separation can be divided into batch and continuous processes. In batch processes, batch mixing is adopted in both emulsion preparation and ELM extraction, which are only suitable for laboratory settings or small-scale emulsion preparation and wastewater treatment, whereas in continuous processes, ultrasonic emulsifier, colloid mill and IS-RPB are used for emulsion preparation, and extraction columns (e.g., rotating disk column) are used for ELM extraction, which are suitable for large-scale applications. A typical continuous ELM separation process is shown in Fig. 6.6.

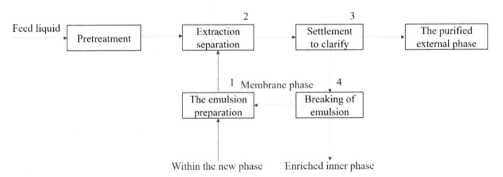

Figure 6.5 A schematic diagram of ELM separation process. *ELM*, emulsion liquid membrane. Note: the operation sequence is 1→2→3→4.

220 HiGee Chemical Separation Engineering

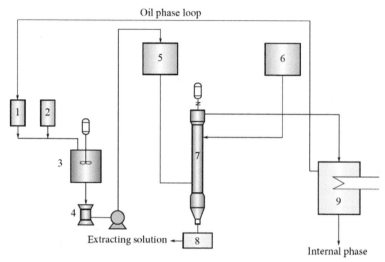

Figure 6.6 A schematic diagram of the typical continuous ELM separation process (1—oil phase storage tank, 2—internal phase storage tank, 3—pre-stirrer, 4—colloid mill, 5—emulsion head tank, 6—wastewater head tank, 7—rotating disk column, 8—oil eliminator, 9—emulsion breaker). *ELM*, emulsion liquid membrane.

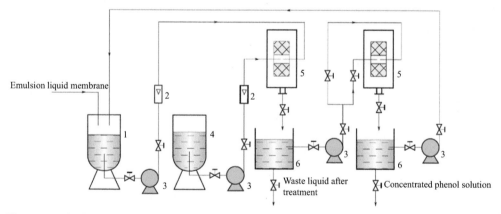

Figure 6.7 ELM separation process (1—emulsion tank, 2—flowmeter, 3—pump, 4—wastewater tank, 5—IS-RPB, 6—demixer). *ELM*, Emulsion liquid membrane.

A continuous ELM separation process is adopted in the IS-RPB. For such a process, a necessary prerequisite is the thorough mixing of the emulsion and the feed phase. IS-RPB is expected to be capable of intensifying ELM separation because of its potential to substantially increase the mass transfer and mixing between two liquid phases. As shown in Fig. 6.7, the emulsion and the feed phase (e.g., wastewater containing phenol or aniline) are impinged and dispersed in the inner cavity of the IS-RPB. As the mixture flows in the pores of the packing from the inner edge to the outer edge, it is broken into ligaments, droplets, and films by the high shear force from the high-speed rotation of the packing and then thrown onto the inner wall of the housing. Finally, it is transferred to the phase separator for stratification. The lower layer of the liquid is the residual external phase (extracted wastewater), and the upper layer of the liquid is the emulsion, which is pumped into a second IS-RPB for demulsification and then into the demixer to obtain concentrated solution and the membrane phase, the latter of which is recycled for reuse.

Swelling is a common phenomenon in liquid membrane separation attributable to the transfer of water from the external phase into the internal phase by permeation or entrainment. Thus, it can be classified into entrainment and osmotic swelling, but entrainment swelling is usually much greater than osmotic swelling. Swelling makes it more difficult to separate emulsion and water and causes incomplete demulsification. Thus, it is necessary to prevent swelling in the IS-RPB. Wang et al. [37] found that increasing the viscosity of the emulsion and membrane phase could significantly reduce entrainment and osmotic swelling of the liquid membrane. Osmotic swelling is caused by the difference in the activity between internal and external phases, but it is also related to the characteristics and concentration of the surfactant. Thus, osmotic swelling can be reduced with the use of a suitable surfactant [38]. Nakashio [39] found that the swelling ratio was very small in the early stage of liquid membrane separation, and thus the treatment time should be reduced as much as possible in order to avoid swelling.

6.3.3 Demulsification

Demulsification refers to the breaking up of the emulsion by means of heating or electrocoalescence. The solute in the internal phase is retrieved by distillation or crystallization, and the recovered membrane phase can be used for the preparation of a new emulsion. An attempt has to be made to minimize the loss of the membrane phase and the consumption of energy and reagent so that high demulsification efficiency can be achieved at low cost. Demulsification is classified into chemical demulsification and physical demulsification [40]. For chemical demulsification, demulsifiers are used to break the water/oil emulsion, while for physical demulsification, the external forces are used such as centrifugation, heating, grinding, ultrasonic wave, and electrostatic field. Each technology has its own advantages and disadvantages depending on the application scenarios. In recent years, much effort has been devoted to developing new demulsification technologies or combining existing ones, such as high-pressure pulse, microfiltration membrane, microwave, IS-RPB, and combination of demulsifier and ultrasonic wave [41].

6.4 Application examples

6.4.1 Treatment of phenol wastewater

Phenol is an important chemical with a wide range of applications. Water pollution occurs when the concentration of phenol and its derivatives exceeds the self-purification capacity of water bodies. The concentration of phenol in industrial wastewater ranges from several hundred mg/L to tens of thousands mg/L, which far exceeds the self-purification capacity of water bodies and the national discharge standard (0.5 mg/L) [42]. Phenol wastewater has many negative environmental consequences if not properly disposed of, such as low dissolved oxygen concentration, low self-purification rate, poor water quality, ecological unbalance, and toxicity to aquatic animals and plants. The treatment of phenol wastewater is important for environmental protection and sustained ecological balance. The experimental setup for emulsion preparation and removal of phenol from wastewater in the IS-RPB is shown in Fig. 6.8. The characteristic parameters of the IS-RPB are: impingement angle = 90 degrees, nozzle diameter = 1.5 mm, housing outer diameter = 180 mm, rotor outer diameter = 140 mm, rotor inner diameter = 60 mm, and high-gravity factor β = 0-470. The emulsion yield, membrane stability, and membrane swelling of IS-RPB and traditional stirred tank reactor are compared, and the effects of water/oil ratio, water/emulsion ratio, high-gravity factor, extraction flow rate, and packing characteristics on the removal rate of phenol are also investigated [43]. We first describe emulsion preparation and extraction of phenol from wastewater.

Emulsion preparation: The membrane phase consisting of Span80, liquid paraffin and kerosene, and the internal phase (NaOH solution) stored in pressure tanks are pumped into the IS-RPB using air compressors. Their flow rates are measured using flowmeters. The two phases are impinged and mixed, and the mixture is dispersed in the inner cavity of the IS-RPB. As the mixture flows in the pores of the packing from the inner edge to the outer edge, it is broken into ligaments, droplets, and films by the high shear force resulting from the high-speed rotation of the packing and then thrown onto the inner wall of the housing. Finally, it is transferred to the phase separator for stratification to obtain the emulsion.

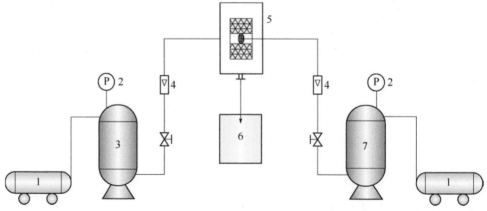

Figure 6.8 Emulsion preparation and extraction of phenol from wastewater in the IS-RPB (1—air compressor, 2—pressure gauge, 3,7—pressure tank, 4—rotameter, 5—IS-RPB, 6—phase separator). *IS-RPB*, impinging stream-rotating packed bed.

The emulsion yield (α) is used to characterize the emulsification efficiency. NaOH is taken as the indicator and its concentrations in the internal and external phases are measured. The emulsion yield is defined as follows:

$$\alpha = \left(1 - \frac{\text{NaOH amount outside the emulsion}}{\text{NaOH amount in the initial internal phase}}\right) \times 100\% \quad (6.6)$$

Also, NaOH is used as the indicator to calculate the breakage rate ε, which is defined as the amount of NaOH released into the external phase due to the breakage of the emulsion and is used to characterize the stability of the membrane:

$$\varepsilon = \frac{m_{wt} - m_{w0}}{m_0 - m_{w0}} \times 100\% \quad (6.7)$$

where m_{wt}—NaOH amount in the external phase at stirring time t, g;
m_{w0}—NaOH amount in the external phase at $t = 0$, g;
m_0—initial NaOH amount, g.

The emulsion containing phenol and water is allowed to stand for phase separation. Then, the volume of swollen emulsion is accurately measured and the swelling ratio η of the membrane is calculated as follows:

$$\eta = \frac{V_t - V_0}{V_0} \times 100\% \quad (6.8)$$

where V_t—emulsion volume at $t = t_s$, mL;
V_0—emulsion volume at $t = 0$, mL.

Extraction of phenol from wastewater: The as-prepared emulsion and wastewater containing 1 g/L phenol are stored in pressure tanks. The extraction process is similar to the emulsion preparation process. After extraction, the mixture is allowed to stand for phase separation. The upper layer of the liquid is the extracted emulsion, and the lower layer of the liquid is the dephenolized wastewater. The concentration of phenol in the aqueous phase after dephenolization is determined by 4-aminoantipyrine spectrophotometry using a 721model spectrophotometer, and the removal rate of phenol is calculated.

The emulsion preparation and separation of phenol from wastewater in the IS-RPB are affected by a wide variety of factors, including packing type, high-gravity factor, as well as amounts of NaOH, Span80, and paraffin. The effects of high-gravity factor β, phase ratio R_x and material properties are discussed [44,45] in the following section.

6.4.1.1 Effects of packing type on the emulsion yield and the removal rate of phenol

Emulsion is prepared under conditions of $\varphi_{Span80} = 4\%$ (where φ is the volume fraction), $\varphi_{paraffin} = 3\%$, $\varphi_{kerosene} = 93\%$, $w_{NaOH} = 5\%$ (where w is the mass fraction), $R_{oi} = 1: 1.5$

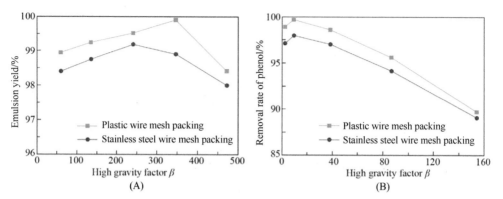

Figure 6.9 The relationships of packing type with the emulsion yield (A) and the removal rate of phenol (B).

(where R_{oi} is the oil/water ratio), and $Q_o = 60$ L/h (where Q_o is the oil phase flow rate). In the experiment, two types of packing (plastic and stainless-steel wire mesh packing) are used, and the variation of the emulsion yield with the high-gravity factor is shown in Fig. 6.9(A). The plastic wire mesh packing is slightly better than the stainless-steel wire mesh packing in terms of emulsion preparation. Despite their similar specific surface area, porosity, and equivalent diameter, the stainless-steel wire mesh packing has a regular shape with a smooth surface, while the plastic wire mesh packing has an irregular shape with a rougher surface and larger pores, which leads to a longer residence time of the feed liquid on the packing and eventually better mixing and mass transfer efficiency.

The extraction of phenol from wastewater is performed under conditions of $Q_e = 40$ L/h (where Q_e is the emulsion flow rate in the membrane phase) and $R_{ew} = 1:1$ (where R_{ew} is the ratio of the volume flow rates of emulsion and aqueous phases). The effect of packing type (plastic and stainless-steel wire mesh) on the removal rate of phenol is shown in Fig. 6.9(B). The plastic wire mesh packing is found to be slightly better than the stainless-steel wire mesh packing in terms of the extraction efficiency, which can also be attributed to the longer residence time of the feed liquid on the packing because of its rougher surface. The use of the plastic wire mesh packing with a rough surface also makes the IS-RPB more lightweight. However, an ongoing effort is needed to better understand how packing surface roughness affects the removal rate of phenol.

6.4.1.2 Effects of high-gravity factor β on the emulsion yield and the removal rate of phenol

Emulsion is prepared under conditions of $\varphi_{Span80} = 4\%$, $\varphi_{paraffin} = 3\%$, $\varphi_{kerosene} = 93\%$, $w_{NaOH} = 5\%$, $R_{oi} = 1:1.5$, and $Q_o = 60$ L/h. For the stainless-steel wire mesh packing, the variation of emulsion yield with high-gravity factor β is shown in Fig.6.9(A). The emulsion yield increases with increasing high-gravity factor until a maximum is reached, after which a decrease in the emulsion yield is observed. At first, increasing the high-gravity factor β increases the rotation speed of the packing, implying that the mixture is subjected to large shear forces and the resulting droplets are repeatedly broken, aggregated, and dispersed to increase the emulsion yield. However, at extremely high levels of high-gravity factor, the residence time of the two liquids in the IS-RPB is substantially reduced and they will leave the IS-RPB before thorough mixing occurs. It is clear that the high-gravity factor should be controlled within a reasonable range. In this study, $\beta = 300$ is selected considering the emulsion yield and energy consumption.

Under conditions of $Q_e = 40$ L/h and $R_{ew} = 1:1$, the high-gravity factor β is varied to explore its effect on the removal rate of phenol. Fig. 6.9(B) reveals that the maximum removal rate of phenol is reached at $\beta = 9.6$. At $\beta < 9.6$, increasing the high-gravity factor reduces the liquid membrane thickness and mass transfer resistance and increases the surface renewal of the liquid membrane, while at $\beta > 9.6$, the residence time of the liquid in the IS-RPB is substantially reduced, and the high mixing intensity facilitates the breakage of the emulsion. It is noteworthy that too high mixing intensity will lead to excessive emulsification and make subsequent phase separation more difficult, which is not good for the extraction process. Therefore, a high level of high-gravity factor is not always helpful for the separation of phenol. Taking into consideration of the removal rate of phenol and subsequent separation, $\beta = 10$ is considered to be a proper choice.

6.4.1.3 Effects of phase ratio R_x on the emulsion yield and the removal rate of phenol

Emulsion is prepared under conditions of $\varphi_{Span80} = 4\%$, $\varphi_{paraffin} = 3\%$, $\varphi_{kerosene} = 93\%$, $w_{NaOH} = 5\%$, and $\beta = 346.7$. The volume flow rate of the oil phase and the oil/water ratio (R_{oi}) is varied to explore their effects on the emulsion yield. Fig. 6.10(A) shows that the emulsion yield decreases with the increase of the volume flow rate of the oil phase at a

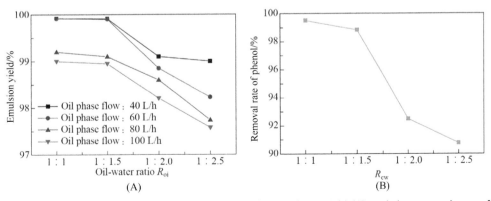

Figure 6.10 The relationships of phase ratio with the emulsion yield (A) and the removal rate of phenol (B).

given R_{oi}, but it decreases with the increase of the oil/water ratio at a given volume flow rate of the oil phase. However, the emulsion yields obtained at $R_{oi} = 1{:}1$ and $R_{oi} = 1{:}1.5$ are approximately the same ($>99.9\%$) when the volume flow rate of the oil phase is kept in the range of 40-60 L/h. Under optimal conditions, the two phases are brought into intimate contact in the IS-RPB for sufficient mixing. At $R_{oi} > 1{:}1.5$, the internal aqueous phase accounts for a large proportion and it can no longer be completely encapsulated in the oil phase. The liquid membrane formed is thin and prone to break. Considering the operational flexibility and cost, $Q_o = 40$ L/h and $R_{oi} = 1{:}1.5$ are recommended.

Under conditions of $Q_e = 40$ L/h and $\beta = 9.6$, the ratio of the volume flow rates of emulsion and aqueous phases R_{ew} is varied to explore its effect on the removal rate of phenol. Fig. 6.10(B) shows that the removal rate of phenol decreases continuously with increasing R_{ew} value. A higher R_{ew} value implies a lower relative amount of emulsion. The processing capacity per unit area is increased and the mass transfer capacity is decreased, resulting in a gradual decrease in the removal rate of phenol. Also, the impingement surface will deviate at all R_{ew} values except at $R_{ew} = 1{:}1$, which will affect the mixing and mass transfer.

6.4.1.4 Effects of material properties on the removal rate of phenol
6.4.1.4.1 Effects of membrane reinforcing agents on the stability of the liquid membrane

A number of reinforcing agents are currently available, such as n-butanol, isooctanol, tributyl phosphate, and liquid paraffin. The emulsion obtained in the IS-RPB is added to the external phase (phenol wastewater) and mixed for some time using a high-speed stirrer. Extraction experiments are conducted under conditions of $\varphi_{Span80} = 4\%$, $\varphi_{paraffin} = 3\%$, $\varphi_{kerosene} = 93\%$, $w_{NaOH} = 5\%$, $\beta = 346.7$, and stirring speed = 400-600 r/min. Fig. 6.11 reveals that the breakage rate of the liquid membrane increases over

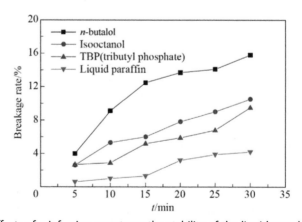

Figure 6.11 The effects of reinforcing agents on the stability of the liquid membrane.

time for all reinforcing agents, and the most stable liquid membrane is obtained with the use of paraffin, followed by isooctanol, tributyl phosphate, and n-butanol. Alkane is more effective than alcohol and esters in enhancing the stability of the liquid membrane because the hydrophilic groups of alcohol and esters can increase the swelling and thus decrease the stability of the emulsion. Also, the organic matter with a higher carbon content is more effective than that with a lower carbon content in enhancing the stability of the liquid membrane. Therefore, the use of liquid paraffin can significantly improve the stability of the liquid membrane.

6.4.1.4.2 Effects of additive amounts on the emulsion yield

Emulsion is prepared using Span80 as the surfactant, kerosene as the solvent, and liquid paraffin as the reinforcing agent under conditions of $R_{oi} = 1:1.5$, $\beta = 346.7$, and $w_{NaOH} = 5\%$. The effects of the amounts of Span80 and liquid paraffin on the emulsion yield are shown in Fig. 6.12. The emulsion yield increases nonlinearly as the amount of Span80 increases. The emulsion yield is low when 2% of Span80 is used, and a much higher emulsion yield is obtained at $\varphi_{Span80} = 4\%\text{-}6\%$. Emulsion is obtained by dispersing one liquid phase in the other and it is a thermally metastable system. The addition of surfactant contributes to reducing the interfacial tension of the membrane and increasing its stability. According to Gibbs' adsorption theory, surfactant is adsorbed at the interface, forming a membrane that prevents the aggregation of dispersed droplets when impinged and protects the liquid membrane against breaking. When the amount of Span80 is low in the membrane phase, fewer Span80 molecules will be adsorbed at the interface and the interface strength will be weak. As a result, the emulsion shows low stability and high breakage rate, resulting in a low emulsion yield. At higher amounts of Span80, the interface membrane is composed of adsorbed molecules that are tightly arranged and highly oriented and the

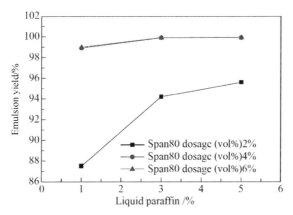

Figure 6.12 Effects of additive amounts on the emulsion yield.

interface strength is strong. This makes it difficult for emulsion droplets to aggregate and the resulting liquid membrane will be more stable. However, as the amount of Span80 further increases, almost all adsorbed molecules are aligned in an oriented fashion, which could not further increase the stability of the liquid membrane. Instead, an excessive amount of surfactant will reduce the removal rate of phenol by increasing the viscosity of the emulsion and the mass transfer resistance, and it will also increase the operating cost. Thus, 4% of Span80 is appropriate.

The emulsion yield increases continuously with increasing amount of liquid paraffin. Increasing the amount of liquid paraffin leads to an increase in the strength and viscosity of the liquid membrane, and the liquid membrane becomes more stable and the emulsion yield is increased. However, when the surfactant amount is higher than 3%, no further increase is observed for the emulsion yield with an increasing amount of liquid paraffin. The emulsion yield reaches 99.9% under the optimal conditions of $\varphi_{Span80} = 4\%$, $\varphi_{paraffin} = 3\%$, and $\varphi_{kerosene} = 93\%$.

6.4.1.4.3 Effects of NaOH solution concentration on the removal rate of phenol

Under conditions of $\beta = 9.6$, $Q_e = 40$ L/h, and $R_{ew} = 1:1$, the concentration (mass fraction) of NaOH solution in the internal phase is varied to explore its effect on the removal rate of phenol. Fig. 6.13 shows that the removal rate of phenol is high at a NaOH concentration of 2%–5% and decreases significantly at NaOH concentrations higher than 5%. The hydrolysis of ester bonds of Span80 in a strongly alkaline solution causes a drastic reduction in the stability of the membrane, which leads to severe damage to the membrane and ultimately a significant reduction in the removal rate of phenol. Considering the stability of the membrane and the allowable concentration range of phenol in wastewater, 5% NaOH solution is appropriate in practical applications.

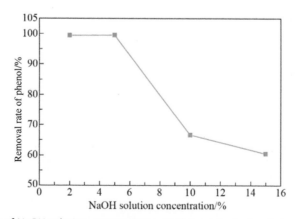

Figure 6.13 Effects of NaOH solution concentration on the removal rate of phenol.

6.4.1.5 Comparison of impinging stream-rotating packed bed with high-speed stirrer

In order to demonstrate the advantages of IS-RPB in emulsion preparation, the emulsion yields of IS-RPB and high-speed stirrer are compared. The experimental conditions for IS-RPB are: $\varphi_{Span80} = 4\%$, $\varphi_{paraffin} = 3\%$, $\varphi_{kerosene} = 93\%$, $w_{NaOH} = 5\%$, $R_{oi} = 1:1.5$, $\beta = 346.7$, and $Q_{o} = 60$ L/h, and the experimental conditions for high-speed stirrer (JB90-D electric stirrer) are: $\varphi_{Span80} = 4\%$, $\varphi_{paraffin} = 3\%$, $\varphi_{kerosene} = 93\%$, $w_{NaOH} = 5\%$, $R_{oi} = 1:1.5$, rotation speed $= 2800$ r/min, and 40 L of emulsion is prepared at a single time. The resulting emulsion yields are summarized in Table 6.2.

For the IS-RPB, the emulsion yield reaches a maximum of 99.9% almost instantaneously (<1 s), while for the high-speed stirrer, at least 25 min is needed to reach the maximum emulsion yield of 99.8%. More importantly, IS-RPB can also increase the stability of the liquid membrane. Fig. 6.14(A) shows that the breakage rate of the emulsion prepared by IS-RPB is 10 times lower than that prepared by a high-speed stirrer. Therefore, the use of IS-RPB can significantly improve the preparation efficiency of emulsion and the stability of the liquid membrane.

The swelling ratio is also an important indicator of the performance of the emulsion. Swelling can reduce the concentration of substances enriched in the membrane and the selectivity, change the viscosity and dispersion of the emulsion, and affect the

Table 6.2 Emulsion yields of IS-RPB (impinging stream-rotating packed bed) and high-speed stirrer.

Equipment	Emulsion yield/%						
	<1 s	5 min	10 min	15 min	20 min	25 min	30 min
IS-RPB	99.90						
High-speed stirrer		60.11	72.77	89.38	97.60	99.77	98.21

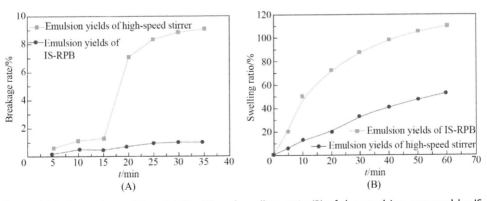

Figure 6.14 Comparison of the stability (A) and swelling ratio (B) of the emulsion prepared by IS-RPB and high-speed stirrer. *IS-RPB*, impinging stream-rotating packed bed.

Table 6.3 Removal rate of phenol by IS-RPB (impinging stream-rotating packed bed) and high-speed stirrer.

Equipment	Removal rate of phenol/%						
	<1 s	5 min	10 min	15 min	20 min	25 min	30 min
IS-RPB	99.72						
High-speed stirrer		20.25	42.16	65.35	75.80	84.21	89.57

separation of emulsion and the external phase as well as subsequent demulsification. The emulsion prepared by IS-RPB and high-speed stirrer is mixed with phenol wastewater at a rotation speed of 400 r/min. After phase separation and settling, the volume of swollen emulsion is accurately measured to calculate the swelling ratio of the liquid membrane. Fig. 6.14(B) reveals that the swelling ratio of the emulsion prepared by the IS-RPB is about three times higher than that prepared by the high-speed stirrer within 60 min of mixing. Because of its potential to enhance the mixing and mass transfer, IS-RPB will produce smaller and more uniform emulsion droplets under the same mixing conditions. There is a larger contact area with the external aqueous phase and more water is penetrated into the emulsion. The extraction process is completed within a short period of time in the IS-RPB because of the large mass transfer area, which minimizes the negative impact of swelling on the whole process.

The removal rates of phenol obtained by IS-RPB and high-speed stirrer are also compared, as shown in Table 6.3. Phenol can be removed instantaneously (<1 s) with a removal rate of 99.7% with the use of IS-RPB, but the removal rate of phenol by the high-speed stirrer is much lower (the maximum removal rate is only 89.6%) and it takes about 30 min to reach the maximum removal rate. Considering the energy consumption, the preparation conditions of emulsion using IS-RPB are suggested to be R_{oi} = 1 : 1-1 : 1.5 and β = 300, and the extraction conditions are R_{ew} = 1 : 1 and β = 10. It is important to develop lightweight, rough-surfaced wire mesh packing. In conclusion, IS-RPB is an efficient equipment for emulsion preparation and extraction because of the short operation time, high efficiency, easy scaling-up, and continuity. It is applicable to the preparation and extraction process of W/O and O/W emulsions and has promising industrial applications in liquid membrane separation.

6.4.2 Treatment of aniline wastewater

Aniline is an important chemical in the manufacture of dyes, pesticides, drugs, and military products. A large amount of aniline wastewater is released in the production and use of aniline, which has severe ecological and environmental consequences due to its toxicity and difficulty in biodegradation [46,47]. The main treatment methods of aniline wastewater include physical, chemical, and biological methods, all of which

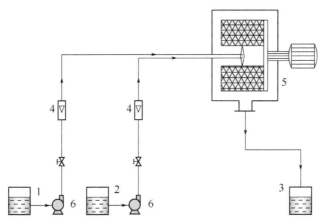

Figure 6.15 A schematic diagram of the emulsion preparation process in the IS-RPB (1—oil phase tank, 2—aqueous phase tank, 3—emulsion tank, 4—flowmeter, 5—IS-RPB, 6—diaphragm pump). IS-RPB, impinging stream-rotating packed bed.

have several common disadvantages, such as low efficiency and difficult operation. Therefore, it is of practical significance to develop novel and efficient technologies for removal of aniline in wastewater [48–50]. IS-RPB can be used for emulsion preparation and extraction of aniline in wastewater. The emulsion preparation process is shown in Fig. 6.15, where the diameter of the liquid inlet tube is 10 mm, the nozzle is 1.5 mm in diameter and spaced 3 mm apart. The inner diameter, outer diameter, and thickness of the packing are 60, 160, and 60 mm, respectively; moreover, the outer diameter of the housing is 200 mm, the rotation speed is 0–1600 r/min, and the stainless-steel wire mesh packing is used.

The as-prepared oil phase and the internal aqueous phase stored in storage tanks are pumped into the IS-RPB at a volume ratio of 1:1. The two streams are injected through nozzles at the same speed (5–15 m/s) and impinged to form a fan-shaped impingement surface that is perpendicular to the jet direction for preliminary emulsification. After that, the mixture flows in the pores of the high-speed rotating packing from the inner edge to the outer edge. The large slip velocity of the mixture with the rotating packing produces a strong shear force on the mixture and enhances the mixing and mass transfer between the two liquid phases for further emulsification. The mixture is thrown from the outer edge of the rotor to the housing under the action of centrifugal force and then flows to the emulsion tank through the outlet under the action of gravity to obtain W/O emulsion.

The mechanism of liquid membrane separation of aniline is described in Section 6.3. A traditional stirrer is often used for this purpose. The as-prepared W/O emulsion and the simulated aniline wastewater containing 1000 mg/L aniline are mixed at an emulsion/water ratio of 1: 20 in a stirrer at a rotation speed of 200 r/min.

A stable W/O/W ELM is obtained, and water samples are collected at regular intervals for the measurement of aniline concentrations. In this study, IS-RPB is used to prepare emulsion and the effects of different operating conditions (e.g., high-gravity factor β and initial impact velocity u_0) on the stability of the emulsion are investigated. A stirrer is used for liquid membrane separation and the effects of different operating conditions (e.g., surfactant amount m, high-gravity factor β, and initial impact velocity u_0) on the removal rate of aniline are investigated [51,52].

6.4.2.1 Effects of surfactant amount on the stability of the liquid membrane and the removal rate of aniline

Under preparation conditions of $\beta = 65$, $u_0 = 9.5$ m/s, and $R_{ow} = 1:1$, the amount of Span80 in the membrane phase is varied (2%, 4%, 6%, and 8%) to investigate its effect on the stability of the liquid membrane. Fig. 6.16(A) shows that the liquid membrane becomes more stable with increasing amount of Span80. At lower amounts of Span80, few molecules are adsorbed at the interface and loosely arranged, resulting in low strength of the membrane interface and unstable emulsion. At higher amounts of Span80, the interface membrane is composed of adsorbed molecules that are tightly arranged and highly oriented and the interface strength is increased. This makes it difficult for emulsion droplets to aggregate and the resulting emulsion is more stable. However, as the amount of Span80 further increases, the membrane becomes thicker and the stability and strength of the emulsion are increased. An increasing tendency is also observed for the breakage rate of the liquid membrane with the increase of the extraction time, which is primarily attributed to the swelling. For cost considerations, the concentration of surfactant should be 4%-6%. At this time, the liquid membrane is stable with a breakage rate of about 5%.

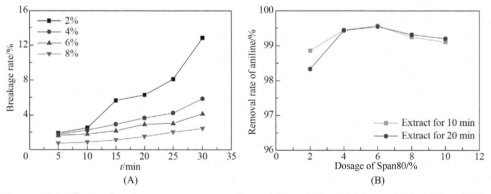

Figure 6.16 Effects of surfactant amount on the stability of the liquid membrane (A) and the removal rate of aniline (B).

The as-prepared liquid membrane was used for liquid membrane separation of aniline using 1 mol/L hydrochloric acid. Fig. 6.16(B) shows the removal rate of aniline as a function of the Span80 amount. The removal rate of aniline increases with an increasing amount of Span80 and then levels off, but an extremely high amount of Span80 is not good for the extraction of aniline. The interfacial tension is low at high amounts of Span80, which makes the liquid membrane more stable and thus is favorable for the transfer of aniline. However, the viscosity and thickness of the liquid membrane are increased with the use of excessive Span80, which causes a decrease in the diffusion rate of aniline. The swelling of the system can reduce the removal rate of aniline. When the amount of Span80 is too low, the extraction time is increased from 10 to 20 min and the removal rate is decreased significantly, which is attributed to the low strength of the liquid membrane. Therefore, the appropriate amount of Span80 ranges from 4% to 6%, at which the removal rate of aniline is greater than 99.0%.

6.4.2.2 Effects of high-gravity factor on the stability of the liquid membrane and the removal rate of aniline

Under condition of $u_0 = 9.5$ m/s, the high-gravity factor β is varied (β = 32, 65, and 110, corresponding to a rotation speed of 700, 1000, and 1300 r/min, respectively) to explore its effect on the breakage rate of the emulsion over time. Fig. 6.17(A) reveals that at $\beta = 32\text{-}110$, the breakage rate increases over time and the stability increases with increasing rotation speed of the packing. At $\beta = 32$, the breakage rate of the emulsion increases more rapidly over time, which is higher than that at $\beta = 65$ and $\beta = 110$. At $\beta > 65$, the breakage rate of the liquid membrane changes little over time and the maximum breakage rate is only 5%, implying that the liquid membrane is highly stable because the higher the rotation speed of the packing, the higher the frequency of the aggregation and dispersion of the liquid mixture in the IS-RPB, and

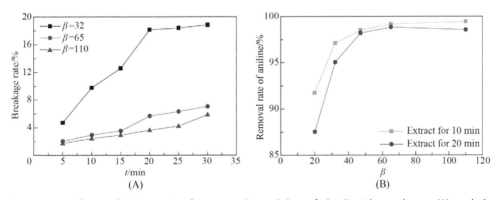

Figure 6.17 Effects of high-gravity factor on the stability of the liquid membrane (A) and the removal rate of aniline (B).

the resulting emulsion droplets become smaller and more stable. For energy consumption considerations, $\beta = 65$ is used for emulsion preparation in the IS-RPB.

Under conditions of $\varphi_{Span80} = 4\%$, $\varphi_{kerosene} = 96\%$, the concentration of hydrochloric acid in the internal phase = 1 mol/L, and $u_0 = 9.5$ m/s, the high-gravity factor is varied ($\beta = 20, 32, 47, 65$ and 110, corresponding to a rotation speed of 550, 700, 850, 1000, and 1300 r/min, respectively) to explore its effect on the removal rate of aniline. Fig. 6.17(B) reveals that the higher the high-gravity factor is, the higher the removal rate of aniline will be. At $\beta > 65$, the removal rate of aniline reaches 99.5% within 10 min and its concentration is reduced to 5 mg/L, which meets the third-level discharge standard. The high-speed rotation of the packing also increases the shear frequency and intensity, which makes the liquid droplets smaller and more uniform after repeated aggregation and dispersion processes and finally improves the stability of the emulsion. The residence time of the mixture in the packing is reduced because of the high centrifugal forces at high rotation speeds, which can reduce the effective emulsification time and increase the damage of the liquid membrane over time. Therefore, the removal rate of aniline could not be further increased. Given the impact of high-gravity factor on the stability of the emulsion and energy considerations, $\beta = 60\text{-}70$ is selected for extraction of aniline in the IS-RPB.

6.4.2.3 Effects of initial impact velocity on the stability of the liquid membrane and the removal rate of aniline

Under conditions of constant ELM formula and $\beta = 65$, the initial impact velocity is varied ($u_0 = 6.3, 7.9$, and 9.5 m/s, corresponding to a liquid flow rate of 40, 50, and 60 L/h, respectively) to explore its effect on the breakage rate of the liquid membrane over time. It is seen in Fig. 6.18(A) that at the same extraction time, the higher the initial impact velocity is, the lower the breakage rate of the emulsion will be. At higher initial impact

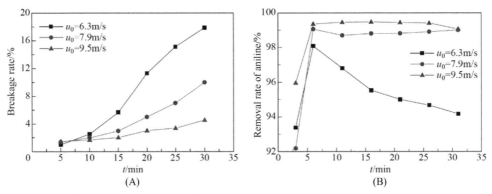

Figure 6.18 Effects of initial impact velocity on the stability of the liquid membrane (A) and the removal rate of aniline (B).

velocities, the impinging energy is large and the turbulent intensity in the impingement zone is increased. The radial velocity of the droplets flowing into the packing is increased. The packing has a stronger shear effect on the mixture, which can reduce the thickness of the liquid film and the particle size distribution range of liquid droplets. The high dispersion of emulsion droplets in wastewater is also favorable for the extraction of aniline.

Under conditions of $\varphi_{Span80} = 4\%$, $\varphi_{kerosene} = 96\%$, the concentration of hydrochloric acid in the internal phase $= 1$ mol/L, and $\beta = 65$, the initial impact velocity is varied ($u_0 = 6.3, 7.9, 9.5$ m/s, corresponding to a liquid flow rate of 40, 50, and 60 L/h, respectively) to explore its effect on the removal rate of aniline. Fig. 6.18(B) shows that the removal rate of aniline increases with increasing initial impact velocity. The greater the initial impact velocity is, the higher the impinging momentum and the energy dissipation rate in the impingement zone will be. The two liquids are more uniformly pre-mixed and the velocity of the mixture flowing into the packing is increased. The packing has a strong shear effect on the mixture, resulting in more uniform dispersion of the emulsion in wastewater and an increase in the surface area for mass transfer of aniline. At $u_0 = 9.5$ m/s, the removal rate of aniline reaches 99.5% and the concentration of residual aniline is about 5 mg/L, which meets the third-level discharge standard. In conclusion, $u_0 = 8\text{-}9$ m/s is a reasonable choice.

6.4.2.4 Comparison of impinging stream-rotating packed bed with high-speed stirrer

The emulsion preparation efficiency of IS-RPB and high-speed stirrer are compared in terms of particle size and distribution, settling stability, and liquid membrane stability.

6.4.2.4.1 Particle size and particle size distribution of emulsion

IS-RPB is capable of preparing emulsion in a continuous manner. The residence time of the liquid in the IS-RPB measured by the tracer method is about 0.5 s. The emulsion preparation process in a high-speed stirrer is a batch process. The particle size gradually decreases over time but remains largely unchanged after 20 min. The emulsion obtained by the IS-RPB has a smaller particle size (3-20 μm) with a narrow particle size distribution [Fig. 6.19(A)], while the emulsion obtained by stirring for 20 min in the high-speed stirrer has a larger particle size (7-38μm) with a wider particle size distribution [Fig. 6.19(B)]. In the IS-RPB, the two jets are impinged to either reduce the pressure pulse or produce high radial and axial components of the turbulent velocity, leading to excellent preliminary emulsification in the impingement zone. After that, the mixture flows into the high-speed rotating packing and is subjected to high shear force of the packing, which further enhances the micromixing and emulsification. Thus, the two liquids are thoroughly mixed in a very short time in the IS-RPB, but in the stirrer, micromixing occurs only near the agitating valve and a long time is needed.

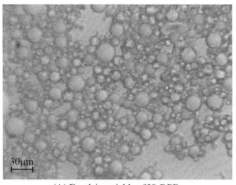 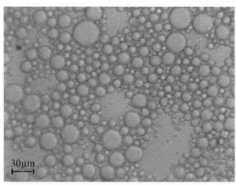

(A) Emulsion yields of IS-RPB   (B) Emulsion yields of high-speed stirrer for 20 min

Figure 6.19 Particle size distribution of emulsion particles.

Table 6.4 The breakage rates of emulsions obtained by IS-RPB (impinging stream-rotating packed bed) and high-speed stirrer.

Equipment	Breakage rate/%		
	Settling for 2 h	Settling for 6 h	Settling for 24 h
IS-RPB	13.0	18.3	29.6
High-speed stirrer	15.2	21.9	36.5

6.4.2.4.2 Stability at different settling times

The breakage rates of emulsions prepared by IS-RPB and high-speed stirrer at different settling times are compared. Table 6.4 shows that the emulsion prepared by IS-RPB is much more stable than that prepared by high-speed stirrer because the emulsion particles prepared by IS-RPB are smaller and more uniform, and according to Stokes equation, smaller particles have a lower settling velocity and thus better stability. The higher stability of the emulsion prepared by IS-RPB makes it easier for storage, transfer, and extraction process of the emulsion.

6.4.2.4.3 Liquid membrane stability and the removal rate of aniline

Under the same extraction conditions (where the emulsion/water ratio is 1:20, and the stirring speed is about 200 r/min), the stability of the liquid membrane prepared by IS-RPB and high-speed stirrer over time is shown in Fig. 6.20(A). The breakage rate of the emulsion prepared by IS-RPB is much lower than that prepared by high-speed stirrer. Especially, at 30 min of extraction, the breakage rate of the liquid membrane prepared by IS-RPB is about 6%, while that prepared by high-speed stirrer is 10%. In the IS-RPB, the two jets are mixed in the impingement zone and the radial and axial components of the turbulent velocity are high, leading to good preliminary

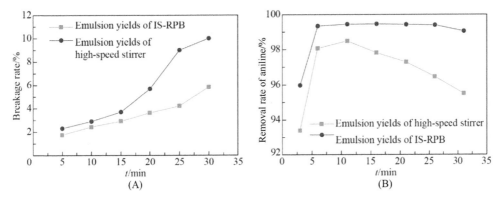

Figure 6.20 Effects of emulsification methods on the stability of the liquid membrane (A) and the removal rate of aniline (B).

emulsification. Subsequently, the emulsification efficiency is further enhanced because of the high shear force resulting from the high-speed rotation of the packing. Therefore, IS-RPB is advantageous over high-speed stirrer to improve the stability of the liquid membrane.

Under the same emulsion formula conditions, the emulsion prepared by IS-RPB and high-speed stirrer and the simulated wastewater containing 1 g/L aniline are mixed at an emulsion/water ratio of 1: 20 to investigate the removal rate of aniline over time. Fig. 6.20(B) shows that the removal rate of aniline reaches 99.5% within 5-10 min with the use of IS-RPB, which changes little over time and reaches 99% at 30 min. For the emulsion prepared by high-speed stirrer, the removal rate of aniline first increases and then decreases, and the maximum removal rate (98.5%) is reached at 10 min. However, the breaking of the liquid membrane leads to the transfer of aniline and hydrochloric acid from the internal phase to the external phase. At 30 min, the removal rate of aniline is reduced to 95.5%. Therefore, IS-RPB is advantageous over high-speed stirrer to improve the removal rate of aniline.

6.4.2.4.4 Comprehensive comparison

The comprehensive comparison of IS-RPB and high-speed stirrer in emulsion preparation and separation processes is summarized in Table 6.5.

The time needed by the IS-RPB for emulsion preparation is only 1/2400 of that by high-speed stirrer. More importantly, the emulsion preparation process in the IS-RPB is a continuous process with a processing capacity of 120 L/h, which is 40 times that in the high-speed stirrer. The emulsion prepared by IS-RPB is characterized by small and uniform particles, high stability at settling, and high stability of the liquid membrane. The use of IS-RPB allows for rapid and continuous preparation of emulsion, and the obtained ELM has good stability and high removal rate of aniline. In

Table 6.5 Comprehensive comparison of IS-RPB (impinging stream-rotating packed bed) and high-speed stirrer.

Conditions	IS-RPB	High-speed stirrer	Conditions	IS-RPB	High-speed stirrer
Rotation speed/(r/min)	1000	1000	Breakage rate after settling for 2 h/%	13.1	20.4
Time/s	0.5	1200	Breakage rate after extraction for 30 min/%	5	11
Processing capacity/(L/h)	120	3	Removal rate of aniline at 10 min/%	99.5	98.0
Emulsion particle size distribution /μm	3-20	7-38	Removal rate of aniline at 30 min/%	99.0	95.5

practical applications, the optimal amount of surfactant is 4%-6% and the optimal emulsification conditions are $\beta = 60\text{-}70$ and $u_0 = 8\text{-}9$ m/s.

In summary, IS-RPB can be used for emulsion preparation and separation with short preparation time, high yield, high emulsion stability, high removal rate of aniline, and continuous operation. Therefore, the development of IS-RPB can expand the applications of liquid membrane extraction on an industrial scale.

6.5 Prospects

Liquid membrane separation has been widely used in petroleum, chemical, metallurgy, atomic energy, wastewater treatment, medicine, and biological fields because of its high efficiency, fast speed, simplicity, and energy efficiency. However, there are still many challenges to be addressed, such as the lack of continuous, efficient, and energy-saving equipment for the preparation and extraction of liquid membrane, swelling, and demulsification. IS-RPB performs well in emulsion preparation and extraction due to its advantages of short operation time, high efficiency, easy scaling up, and continuous operation; moreover, it is applicable to any (including both W/O and O/W) emulsion and has promising applications in the industry. Despite its numerous advantages and potential applications, more efforts are needed to:

1. Develop high-performance surfactant that plays a critical role in the removal of solutes because of its potential impact on the stability and swelling of the liquid membrane.
2. Develop novel packing and pre-mixing units. The packing has important effects on the mixing and mass transfer process of the emulsion. Also, the role of pre-mixing units in emulsification should not be ignored. It is of great significance to develop lightweight packing and high-performance pre-mixing units.
3. Extend the application of ELM. In the context of high global greenhouse gas emissions, the combination of ELM separation and high gravity technology will provide a novel means for CO_2 removal [53].

References

[1] Li NN, Somernet NJ. Separation hydrocarbons with liquid membranes. US 3410794, 1968.
[2] Bloch R, Finkelstein A, Kedem O. Metal ion separation by dialysis through solvent membranes. Industrial & Engineering Chemistry Process Design and Development, 1967,6:231-237.
[3] Ward WJ, Robb WL. Carbon dioxide-oxygen separation: facilitated transport of carbon dioxide across a liquid film. Science, 1967,156:1481-1486.
[4] Rappert M, Draxler J, Marr R. Liquid-membrane-permeation and its experiences in pilot plant and industrial scale. Separation Science and Technology, 1988,23:1659-1666.
[5] Zhang MD, Zhang LJ, Liu G. Advances on technique and application of liquid membrane separation progress. Chemical World, 2015,8:506-511.
[6] Kulkarni PS, Bellary MP, Ghosh SK. Study on membrane stability and recovery of uranium (VI) from aqueous solutions using a liquid emulsion membrane process. Hydrometallurgy, 2002,64:49-58.
[7] Du SW, Liu WF. Research and application of emulsion liquid membrane. Contemporary Chemical Industry, 2015,44:101-104.
[8] Boyadzhier L, Bezenshen E. Carrier mediated extraction, application of double emulsion technique for mercury removal from wastewater. Journal of Membrane Science, 1983, 14:13-18.
[9] Wang XS. Advances in liquid membrane separation. Chemical Industry and Engineering Progress, 1990,23:1-6.
[10] Zhang RH. Liquid Membrane Separation Technology. Nanchang: Jiangxi People's Publishing House, 1984.
[11] Davoodi-Nasab P, Rahbar-Kelishami A, Safdari J, et al. Evaluation of the emulsion liquid membrane performance on the removal of gadolinium from acidic solutions. Journal of Molecular Liquids, 2018,262:97-103.
[12] Krull FF, Fritzmann C, Melin T. Liquid membranes for gas/vapor separations. Journal of Membrane Science, 2008,325:509-519.
[13] Li GJ, Pan JZ. Method of combining high pressure with impinging stream applied in emulsion and homogenization. Machinery Design & Manufacture, 2010,2:72-74.
[14] Zhang J, Guo K. Experimental study about flow of liquid in rotating packed bed. Journal of Chemical Engineering of Chinese Universities, 2000,14:378-381.
[15] Ouyang CB, Liu YZ, Qi GS. A new type reaction facility: rotating packed bed technology and its application. Science & Technology in Chemical Industry, 2002,10:50-53.
[16] Liu YZ, Jiao WZ, Qi GS. Preparation and properties of methanol-diesel oil emulsified fuel under high-gravity environment. Renew Energy, 2011,36:1463-1468.
[17] Jiao WZ, Luo S, He Z, et al. Emulsified behaviors for the formation of methanol-diesel oil under high-gravity environment. Energy, 2017,141:2387-2396.
[18] Shi J, Yuan Q, Gao CJ. Handbook of Membrane Technology. Beijing: Chemical Industry Press, 2001.
[19] Dai YY. New Extraction Technology and Application. Beijing: Chemical Industry Press, 2007.
[20] Li KB, Jin SD. The kinetics and conditions for removal of ammonia nitrogen from waste water by emulsion liquid membrane. Acta Scientiae Circumstantiae, 1996,16:412-417.
[21] Li JP. Preparation of Nano Barium Sulfate Using Impinging Steam-rotating Packed Bed. Taiyuan: North University of China, 2002.
[22] Li C, Li ZP, Gao ZM. Micromixing characteristics of an opposed-jet reactor. Journal of Beijing University of Chemical Technology (Natural Science Edition), 2009,36(6):2-4.
[23] Chen JF, Zou HK. Synthesis of nanomaterials by high-gravity reactive precipitation method. Modern Chemical Industry, 2001,9-12.
[24] Zhou LH. Measurement of Liquid Holdup and Residence Time in a Rotating Packed Bed. Taiyuan: North University of China, 2010.
[25] Zhang J. Experimental and Simulation Studies of the Liquid Flow and Mass Transfer in a Rotating Packed Bed. Beijing: Beijing University of Chemical Technology, 1996.

[26] Guo K. Observation and Measurement of Liquid Flow in the Packing of High-gravity Rotator. Beijing: Beijing University of Chemical Technology, 1996.
[27] Gou F, Zheng C, Guo K, et al. Hydrodynamics and mass-transfer in cross-flow rotating bed. Chemical Engineering Science, 1997,52:3853-3859.
[28] Patnaik PR. Liquid emulsion membranes: principles, problems and applications in fermentation processes. Biotechnology Advances, 1995,13:175-208.
[29] Wan YH, Wang XD, Zhang XJ. Progress in the study of surfactant used for emulsion liquid membrane. Chemical Industry and Engineering Progress, 1998,17:5-12.
[30] Chiha M, Samar MH, Hamdaoui O. Extraction of chromium (VI) from sulphuric acid aqueous solutions by a liquid surfactant membrane (LSM). Desalination, 2006,194:69-80.
[31] Ahmada AL, Kusumastuti A, Dereka CJC, et al. Emulsion liquid membrane for heavy metal removal: an overview on emulsion stabilization and destabilization. Chemical Engineering Journal, 2011,171:870-882.
[32] Liu GG, Lv WY, Xue XL. The extraction of acetic acid with emulsion liquid membrane. Environmental Chemistry, 2002,21:385-388.
[33] Qiao YJ, Jin YZ, Zhao QN. Investigation on the stability of emulsion liquid membrane with petroleum ether as membrane phase. Environmental Pollution & Control, 2008,30(8):19-23.
[34] Manea M, Chemtob A, Paulis M, et al. Miniemulsification in high-pressure homogenizers. AIChE Journal, 2008,54:289-297.
[35] Shen JQ, Yin WP, Zhao YX. A study on separation of alanine using emulsion liquid membranes. Chinese Journal of Biotechnology, 1996,12:410-415.
[36] Wan YH, Wang XD, Zhang XJ. Study of the treatment of wastewater containing high concentration of phenol by liquid membrane. South China University of Technology, 1998,26:29-32.
[37] Wang ZH, Fu JF. Viscosity effects on the swelling of liquid surfactant membrane. CIESC Journal 1992,43:148-152.
[38] Clinart PD, Elepine ST, Rouve G. Water treatment in emulsion liquid membrane processes. Journal of Membrane Science, 1984,20:167-169.
[39] Nakashio F. Recent advances in separation of metals by liquid surfactant membranes. Journal of Chemical Engineering of Japan, 1993,26:123-133.
[40] Chen HP. The latest progress in the research and application of the demulsification methods. Speciality Petrochemicals, 2012,29:71-76.
[41] Zolfaghari R, Fakhru'L-Razi A, Abdullah LC, et al. Demulsification techniques of water-in-oil and oil-in-water emulsions in petroleum industry. Separation and Purification Technology, 2016, 170:377-407.
[42] Lin S, Wang CS. Treatment of high strength phenolic wastewater by a new two-step method. Journal of Hazardous Materials, 2002,90:205-216.
[43] Yang RL. IS-RPB Emulsion Liquid Membrane for Treatment of Phenol Waste Water. Taiyuan: North University of China, 2004.
[44] Yang RL, Liu YZ, Qi GS. Study about treating phenolic effluent through emulsion liquid membrane by impinging stream-rotating packed bed. Applied Chemical Industry, 2004,33(3):31-33.
[45] Liu YZ, Zhang DY, Jiao WZ. Further development of phenol removal by emulsion liquid membrane with new pattern technique. Chemical Engineer, 2005,3:1-5.
[46] Wang G, Lu Y, Wang L. Research on treatment of aniline contained dyeing wastewater. Environmental Protection Science, 2007,33:8-61.
[47] GB/T 8978—1996. Comprehensive wastewater discharge standard. Beijing: China Standards Press, 1996.
[48] Liu BT, Yang Y, Zhang L. Experimental studies on the treatment of aniline wastewater by activated carbon fiber. Advanced Materials, 2011,282:64-67.
[49] Zhang GM. Advanced Oxidation Technology for Water Treatment. Harbin: Harbin Institute of Technology Press, 2007.

[50] Emtiazi G, Satarii M, Mazaherion F. The utilization of aniline, chlorinated aniline and aniline blue as the only source of nitrogen by fungi in water. Water Research, 2001,35:1219-1224.
[51] Li QT. IS-RPB Study on the Treatment of Aniline Wastewater by Emulsion Liquid Membrane in IS-RPB. Taiyuan: North University of China, 2014.
[52] Li QT, Liu YZ, Qi GS. Treatment of aniline wastewater by emulsion liquid membranes. Modern Chemical Industry, 2013,33:76-79.
[53] Jindaratsamee P, Ito A. Separation of CO_2 from the CO_2/N_2 mixed gas through ionic liquid membranes at the high feed concentration. Journal of Membrane Science, 2012,423-424:27-32.

CHAPTER 7

Adsorption

Contents

7.1 Overview	243
7.2 Adsorption and separation technologies	245
7.2.1 Theoretical basis of adsorption	245
7.2.2 Adsorption process	253
7.3 High-gravity adsorption	259
7.3.1 High-gravity adsorption device	260
7.3.2 Mechanism of intensification of high-gravity adsorption process	261
7.4 Application examples	264
7.4.1 Treatment of resorcinol wastewater	264
7.4.2 Treatment of toluene-containing exhaust gas	270
References	280

7.1 Overview

Adsorption is the adhesion of atoms, ions, or molecules of a gas, liquid, or dissolved solid to the surface of a solid adsorbent, whereby one or more components of the fluid (adsorbate) are attracted to the surface of the porous solid (adsorbent) in contact with the fluid. Adsorption can be broadly classified into physical adsorption and chemical adsorption according to the intermolecular interactions between adsorbate and adsorbent. Physical adsorption, such as adsorption of gas on activated carbon, occurs through the weak van der Waals interaction between adsorbate and adsorbent with low adsorption heat, and in this case, desorption is more likely to occur, while chemical adsorption occurs through the strong chemical bonds between adsorbate and adsorbent and it is an irreversible process with high adsorption heat.

The basic principle of the adsorption process is to use a porous solid to selectively adsorb one or more components from a fluid mixture. Desorption is the reverse process of adsorption in which adsorbed components are removed from the adsorbent in order to regenerate the adsorbent and recover the adsorbate for reuse. In the chemical industry, the adsorption process consisting of several successive adsorption-desorption cycles is an important unit operation for separating gas or liquid mixtures. Adsorption can also be classified into gas/solid adsorption and liquid/solid adsorption according to the adsorbate to be adsorbed. The gas/solid adsorption process is intended to absorb the organic vapor in industrial exhaust gas using two adsorption columns, one for adsorption and the other one for desorption and regeneration. Regeneration is often achieved by heating, and the desorbed adsorbate (gas) is condensed for recovery and the regenerated adsorbent is cooled at room temperature for reuse in subsequent processes, forming an adsorption-desorption cycle. The liquid/solid adsorption process is often used for removing organic pollutants in wastewater and decolorizing cooking oil and petroleum products. After oil separation, flotation, and sand filtration, the wastewater flows sequentially through the adsorbent layer for purification. According to the contact between liquid and adsorbent, the adsorption equipment can be divided into an adsorption tank, fixed bed, fluidized bed, and moving bed.

The parameters for evaluating the adsorption process mainly include the removal rate (purification rate) of the adsorbate, processing capacity (the flow rate of the gas or liquid mixture per unit volume of the equipment), and energy consumption, which are affected by adsorbent properties (e.g., specific surface area, adsorption capacity, selectivity, and mechanical strength), equipment, operating temperature, and pressure. The adsorbent with larger specific surface area and porosity has a higher equilibrium adsorption capacity and in this case, less adsorbent is required. Typical adsorbents include natural adsorbent, activated carbon, zeolite molecular sieves, and activated alumina. Traditional industrial adsorbers include fixed bed, moving bed, and fluidized bed. The fixed bed is featured by simple structure, low cost, and low adsorbent wear, but the adsorption process is a batch process requiring alternating adsorption and regeneration steps and local overheating is likely to occur because of the poor heat conductivity of the bed. The moving bed has high processing capacity and the adsorbent is recyclable, and it is applicable to the case where there is a high adsorbent/gas ratio but less applicable to pollution control. The adsorbent for the moving bed should have high wear resistance. In the fluidized bed, the adsorbent particles on the sieve plate are strongly agitated by the high-velocity airflow, leading to rapid mass and heat transfer, uniform temperature distribution, stable operation, but high wear loss. However, it is rarely used for purification of exhaust gases because of the low saturation of the absorbent as a result of backmixing. Lower temperatures and higher pressures are favorable for adsorption, while the opposite conditions are favorable for desorption. Therefore, temperature and pressure play critical roles in separation processes.

In comparison to other separation processes, the adsorption process has the unique advantages of low energy consumption, high product purity, the ability to remove trace substances, and low operating temperature. Now, it is widely used in the chemical industry, pharmaceutical industry, food industry, and light industry. It is also used in environmental protection for dehydration, deodorization and decolorization of gas or liquid, recovery of volatile organic compounds (VOCs), adsorption of trace components in the gas, separation of some substances that are difficult to separate by distillation, and treatment of exhaust gas and wastewater.

To sum up, each adsorber has its advantages and disadvantages that make it more or less suited to a particular application. The fixed bed requires alternating adsorption and desorption processes and local overheating is likely to occur in the desorption process. The moving bed requires a high adsorbent/gas ratio and is less applicable to pollution control. The fluidized bed is susceptible to the backmixing of the adsorbent and is rarely used for exhaust gas purification. Thus, there is a practical need to develop more efficient adsorption technologies for pollution control and exhaust gas purification. The high-gravity adsorption technology has emerged as a highly promising adsorption technology, whereby the adsorption process takes place in a high-gravity field created by the high-speed rotation of the packing in the rotating packed bed (RPB). The adsorbent is loaded as the packing in the rotor and the gas is brought into intimate contact with the rotating packing, which can not only effectively improve the adsorption rate and the gas-solid heat transfer rate but can also avoid local overheating in the regeneration process and increase the desorption rate. The time needed for adsorption and regeneration is also greatly reduced. This novel separation technology is expected to have more potential applications in pollution control.

This chapter discusses the mechanisms and applications of high-gravity adsorption processes and the devices used for adsorption and desorption. Both gas/solid adsorption and liquid/solid adsorption are covered, and some application examples are presented.

7.2 Adsorption and separation technologies

In this section, we first discuss the adsorption theory, thermodynamic equilibrium, and dynamics involved in high-gravity adsorption processes, and then discuss the mass transfer characteristics and adsorbent and adsorption equipment.

7.2.1 Theoretical basis of adsorption

7.2.1.1 Adsorption mechanism

Adsorption refers to the selective attachment of one or more components of a fluid (liquid or gas) phase onto the surface of a porous solid adsorbent in contact with the fluid phase. The adsorption process can be classified into physical adsorption and chemical adsorption, and their differences are shown in Table 7.1.

Physical adsorption is widely used in the industry. It is a reversible process and the adsorbent can be readily regenerated for reuse. It relies on the van der Waals force between adsorbate and adsorbent, including induction force, electrostatic force, and dispersion force between atoms and molecules. The van der Waals force plays a significant role in the adsorption and diffusion rates, adsorption heat, and adsorption activation energy. Table 7.2 summarizes the van der Waals forces of several common organic compounds.

Table 7.1 Differences between physical absorption and chemical absorption [1].

Properties	Physical absorption	Chemical absorption
Adsorption force	van der Waals force, hydrogen bond	Chemical bond
Adsorption heat /(kJ/mol)	42-62	80-400
Adsorption rate	No need for activation, diffusion controlled, and high adsorption rate	Need of activation, overcoming energy barrier, and low adsorption rate
Selectivity	No or very poor	Very high
Reversibility	Reversible	Irreversible
Adsorption spectra	Changes in the intensity of adsorption peaks or beam displacement	Appearance of new characteristic absorption peaks
Number of adsorption layers	Single or multiple	Single
Stability	Unstable and prone to desorption	Relatively stable
Adsorption equilibrium	Easy	Difficult

Table 7.2 van der Waals forces of some organic compounds.

Compounds	Dipole moment/ (10^{-3} C·m)	Polarizability/ (10^{-24} C/m^2)	Electrostatic force/(kJ/mol)	Induction force/(kJ/mol)	Dispersion force/(kJ/mol)
Benzene	0.00	10.50	0.00	0.00	100.00
Toluene	1.43	11.80	0.10	0.90	99.00
Paraxylene	0.00	14.23	0.00	0.00	100.00
Phenol	1.60	11.15	—	—	—

7.2.1.2 Adsorption equilibrium

Desorption occurs simultaneously with adsorption. The apparent adsorption rate will be zero when the adsorption rate is equal to the desorption rate, and adsorption equilibrium will be reached when the concentrations of adsorbate in the solution and on the adsorbent surface remain unchanged. Gas adsorption equilibrium and liquid adsorption equilibrium are discussed in the following sections.

(1) Gas adsorption equilibrium

Gas adsorption equilibrium is reached when the adsorption rate of gas adsorbate molecules on the adsorbent surface is equal to the desorption rate under specific temperature and pressure conditions. According to the number of components of the adsorbate, single- and multi-component adsorption equilibrium are discussed. The single-component adsorption equilibrium aims to determine the relationship between the adsorption capacity per unit mass of adsorbent and the partial pressure of the single component in the gas phase (or the molar concentration of each solute per unit volume of the fluid phase) at a given temperature, while the multi-component adsorption equilibrium aims to determine the relationship between the adsorption capacity per unit mass of adsorbent for each component and the partial pressure of each component in the gas phase (or the molar concentration of each solute per unit volume of the fluid phase). Both single- and multi-component adsorption equilibrium can be described by adsorption isotherm equations. The multi-component adsorption equilibrium is an appropriate extension of the single-component adsorption equilibrium.

① Single-component adsorption equilibrium isotherm

Brunauer has proposed five types of single-component gas adsorption isotherms, as shown in Fig. 7.1. These curves represent different relationships between the adsorption capacity of the adsorbent and the partial pressure of the adsorbate in the gas phase or the intermolecular forces between adsorbate and adsorbent surface molecules. Type I isotherm indicates the monolayer adsorption of adsorbate molecules on the adsorbent surface, Type II isotherm indicates the multilayer adsorption of adsorbate molecules on the adsorbent surface, Type III isotherm is characterized by a convexity toward the p/p_0 axis over the entire range, Type IV and Type V isotherms correspond to Type II and Type III isotherms with capillary condensation.

A number of single-component adsorption isotherm equations have been proposed, including the Langmuir equation based on dynamic equilibrium, Gibbs equation based on classical thermodynamics, and the Dubinin-Polanyi equation based on the adsorption potential theory [2]. The last two equations that are seldom used in practice are not discussed in detail here. The Langmuir equation is routinely used to describe the monolayer adsorption of adsorbate molecules on a uniform adsorbent surface at a constant temperature. It is the simplest isotherm model for describing both physical and chemical adsorption. The Langmuir equation can be expressed as follows:

$$\frac{q}{q_m} = \frac{k_L p}{1 + k_L p} \tag{7.1}$$

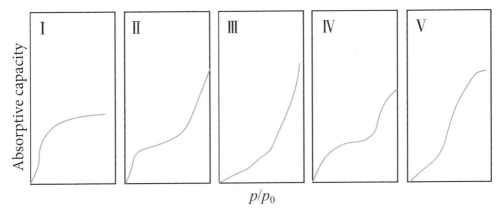

Figure 7.1 Typical single-component gas adsorption isotherms.

where q—the adsorption capacity of the adsorbent, kg/kg;
q_m—the maximum adsorption capacity of the adsorbent, kg/kg;
p—the partial pressure of the gas-phase adsorbate, atm(1atm = 101325Pa);
k_L—Langmuir constant, m³/kg.

Different versions of the Langmuir equation are used depending on adsorbate properties and adsorption conditions.

Henry equation is applicable to the case where the coverage ratio of the adsorbate on the adsorbent surface ($\theta = q/q_m$) is below 10%:

$$q = k_H p \tag{7.2}$$

where k_H—Henry coefficient.

Freundlich equation is applicable to the case where the adsorption heat decreases logarithmically with increasing coverage ratio of the adsorbate at a constant temperature on a heterogeneous surface:

$$q = k_F p^{1/n} \tag{7.3}$$

where k_F—Freundlich constant, m³/kg;
$1/n$—adsorption capacity, which is determined experimentally.

Langmuir-Freundlich equation is used in the dissociation of the adsorbate, each molecule has two active sites in the adsorption and desorption processes, and the adsorption and desorption rate is proportional to $(1-\theta)^2$ and θ^2, respectively, where θ is the coverage ratio of the adsorbate on the adsorbent surface. The equation is expressed as:

$$\frac{q}{q_m} = \frac{(k_L p)^{1/2}}{1 + (k_L p)^{1/2}} \tag{7.4}$$

BET equation is applicable to multilayer adsorption. The adsorption in the first layer is governed by van der Waals interactions at the active sites, and dynamic equilibrium is reached in other layers with the same adsorption heat. It is applicable to the case where $p/p_0 = 0.05\text{-}0.30$. The equation is expressed as:

$$\frac{p}{q(p_0 - p)} = \frac{1}{k_b q_m} + \frac{k_b - 1}{k_b q_m} \frac{p}{p_0} \tag{7.5}$$

where p_0—the saturated vapor pressure of the gas-phase adsorbate at the adsorption temperature, atm;
p—the partial pressure of the gas-phase adsorbate, atm;
q—the adsorption capacity of the adsorbent, kg/kg;
q_m—the saturated adsorption capacity of the first monolayer, kg/kg;
k_b—BET constant, m³/kg.

② Multicomponent adsorption equilibrium isotherm

The gas or liquid to be treated is a multi-component mixture and these components compete with each other for adsorption. The multi-component adsorption equilibrium isotherm can be obtained by thermodynamic methods and extended from the single-component adsorption isotherm, such as the Langmuir equation, BET, potential theory, and Grant-Manes equation.

The extended Langmuir equation is widely used for describing multi-component gas adsorption, which is established under ideal adsorption conditions, and the coverage area of one component is not affected by other components:

$$\frac{q_i}{q_{mi}} = \frac{k_i p_i}{1 + \sum_{j=1}^{n} k_j p_j} \tag{7.6}$$

where q_i—the adsorption capacity of the adsorbent for component i, kg/kg;
q_{mi}—the maximum single-layer adsorption capacity for component i, kg/kg;
p_i—the partial pressure of component i of the gas-phase adsorbate, atm;
k_i—Langmuir constant of component i, m³/kg.

The BET equation assumes that the ratio of the Langmuir constant between $(n + 1)$th and nth layers is a constant for all components of the mixture. The extended potential theory assumes single-layer molecular adsorption without interference in the adsorption of different components. The Grant-Manes equation is based on the potential theory. These equations are only applicable to specific situations and rarely used in practice [2].

(2) Liquid adsorption equilibrium

The liquid adsorption isotherms can be divided into four categories according to the change in the slope of the curve closest to the origin, as shown in Fig. 7.2.

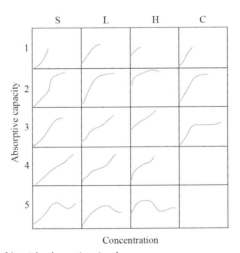

Figure 7.2 Classification of liquid adsorption isotherms.

S-shaped isotherm: The adsorbed molecules are perpendicular to the adsorption surface, and the isotherm is concave toward the concentration axis.

L-shaped isotherm: This isotherm is a typical Langmuir adsorption isotherm characterized by a decreasing slope as concentration increases. The longitudinal axes of adsorbed molecules are parallel to the adsorbent surface, and sometimes the forces between adsorbed ions are very strong.

H-shaped isotherm: This isotherm is characterized by the high affinity of the solute for an adsorbent. It is concave toward the adsorption capacity axis, and low-affinity ions are substituted with high-affinity ions.

C-shaped isotherm: This isotherm is characterized by an initial linear part indicative of a constant partition of the solute between solution and adsorbent. The solute is more easily adsorbed into the solid adsorbent than the solvent, and the initial slope remains independent of adsorbate concentration until a maximum adsorption is achieved.

Compared to the gas adsorption process, the liquid adsorption process is much more complex as it involves not only temperature, concentration, and adsorbent properties, but also solvent and solute properties. Typical adsorption equilibrium isotherm equations include the Langmuir equation, Freundlich equation, Temkin equation, and Redlich-Peterson equation. Langmuir equation and Freundlich equation are suited to the adsorption of low-concentration solutions, and the characteristic curve or equation based on the adsorption potential theory is suited for the adsorption of mixed solutions.

Langmuir equation assumes that only monolayer adsorption occurs on the adsorbent surface, which is expressed as follows [3]:

$$q = q_m \frac{k_L c_e}{1 + k_L c_e} \tag{7.7}$$

$$\frac{c_e}{q} = \frac{1}{q_m} c_e + \frac{1}{k_L q_m} \tag{7.8}$$

where c_e—equilibrium concentration, mmol/L;

q—adsorption capacity, mg/g;

k_L—adsorption equilibrium constant, and the larger the k value is, the higher the adsorption capacity will be;

q_m—the maximum adsorption capacity of the adsorbent, kg/kg, which is related to the adsorption potential but not to the temperature in theory.

Freundlich equation is an empirical equation based on adsorption on a heterogeneous surface, which is expressed as follows:

$$q = k_F c_e^{1/n} \tag{7.9}$$

where c_e—equilibrium concentration, mmol/L;
$\quad q$—adsorption capacity, mg/g;
$\quad k_F$—adsorption equilibrium constant.

The $1/n$ value indicates the effect of concentration on the adsorption capacity. The smaller the $1/n$ value is, the higher the adsorption capacity will be. A $1/n$ value in the range of 0.1-0.5 indicates easy adsorption, and a $1/n$ value higher than 2 indicates difficult adsorption.

The Temkin equation indicates that the adsorption heat decreases linearly with increasing adsorption capacity and the simplified expression is as follows:

$$q = A + B \lg c_e \tag{7.10}$$

where c_e—equilibrium concentration, mmol/L;
$\quad q$—adsorption capacity, mg/g;
$\quad A, B$—constants.

If the plot of q against $\lg c_e$ is a straight line, there would be a good agreement between predicted and experimental results.

Redlich-Peterson equation:

$$q = k_R c_e / \left(1 + a_R c_e^a\right) \tag{7.11}$$

where c_e—equilibrium concentration, mmol/L;
$\quad q$—adsorption capacity, mg/g;
k_R, a_R, a—empirical constants.

7.2.1.3 Adsorption kinetics

Previous studies on adsorption kinetics have suggested that the adsorption and desorption rates are influenced by a variety of factors. Similar to adsorption equilibrium, the adsorption and desorption rates also depend on the interactions between adsorbent and adsorbate, as well as temperature and pressure. The adsorption kinetic model is a mathematical model relating the adsorption of adsorbate onto the adsorbent surface to the main influencing factors under specific assumptions. The most widely accepted adsorption models include pore diffusion model, equilibrium model, dusty gas model, and linear driving force model [4]. The pore diffusion model assumes adsorbent particles as consisting of a solid phase interspersed with very small pores. The adsorbate diffuses into the pores and adsorption occurs at the internal surfaces. This process can be described by the partial differential equation. The equilibrium model assumes that there is no mass transfer resistance and that the adsorption process can reach equilibrium instantaneously. The dusty gas model assumes adsorbent particles as a pseudo-component of the mixture and it is suited for describing multi-component mass transfer in an external force field. The linear driving force model assumes that the adsorption rate of the solid adsorbent is proportional to the adsorbate concentration gradient between external and internal adsorbent, which is simple and easy to analyze. As the first three models are difficult to solve with strict conditions, they are rarely used in engineering applications. We will focus on the linear driving force model that is more widely used.

(1) Pseudo-first-order kinetic model [5]

The pseudo-first-order kinetic model proposed by Lagergren is a simple model describing the initial surface adsorption stage:

$$\lg(q_e - q_t) = \lg q_e - \frac{k_1}{2.303}t \tag{7.12}$$

where q_t—adsorption capacity at time t, mg/g;
q_e—adsorption capacity at equilibrium, mg/g;
k_1—first-order adsorption rate constant, min^{-1}.

The $\lg(q_e - q_t)$ is plotted against t based on the q_e value and the time-varying q_t values obtained under optimal conditions, and the fitting degree is determined by the determination coefficient R^2. The k_1 value can be obtained from the intercept and slope of the fitted straight line. This model is applicable to the case where surface adsorption dominates rather than the adsorption of porous adsorbents.

(2) Pseudo-second-order kinetic model [6]

The pseudo-second-order kinetic model assumes that the rate-limiting step is the chemical adsorption involving valency forces through sharing or exchange of electrons between adsorbent and adsorbate, and it is expressed as follows:

$$\frac{t}{q_t} = \frac{1}{k_2 q_e^2} + \frac{1}{q_e}t \tag{7.13}$$

where q_t—adsorption capacity at time t, mg/g;
q_e—adsorption capacity at equilibrium, mg/g;
k_2—second-order adsorption rate constant, g/(mg · min).

The t/q_t is plotted against t based on the q_e value and the time-varying q_t values obtained under optimal conditions. The fitting degree is determined by the determination coefficient R^2. The k_2 value can be obtained from the intercept and slope of the fitted straight line. This model can be used to describe the whole adsorption process and the adsorption rate, and it is applicable to the adsorption of most porous adsorbents.

(3) Weber-Morris model (intraparticle diffusion model)

This model can be used to determine the rate-limiting steps in different stages of the adsorption process [7], which is expressed as follows:

$$q_t = k_{WM} t^{1/2} + C \tag{7.14}$$

where q_t—adsorption capacity at time t mg/g;
   k_{WM}—intraparticle diffusion rate constant;
   C—a constant related to the thickness and boundary layer.

If the plot of q_t against $t^{1/2}$ passes through the origin, the intraparticle diffusion is the only rate-limiting step of the adsorption process; otherwise, the adsorption process consists of three stages: external diffusion, intraparticle diffusion, and adsorption equilibrium; external and intraparticle diffusion of particles play dominant roles.

7.2.2 Adsorption process

7.2.2.1 Adsorption and mass transfer process

Adsorption is the selective attachment of one or more components of a fluid (liquid or gas) phase onto the surface of a porous solid adsorbent in contact with the fluid phase, which is widely used in deep drying, decolorization of food and drugs, air separation, and exhaust gas treatment. Various devices and adsorbents can be utilized for different adsorption processes.

(1) Adsorption and separation process

The selection of a proper adsorbent is based on the properties of the components to be separated from the mixture and the separation requirements. Based on the operation mode, adsorption can be classified into batch and continuous adsorption. The desorption and separation processes are schematically shown in Fig. 7.3, which mainly consists of the feed system, adsorption-desorption system, and separation and recovery system. The feed system is responsible for pre-treatment, measurement, and delivery of the feed. The adsorption-desorption system is used for adsorption and desorption/regeneration. For continuous adsorbers, adsorption and desorption take place continuously in an adsorber. For batch adsorbers, desorption is conducted after the completion of adsorption in an adsorber, or adsorption and desorption take place sequentially in two adsorbers. The adsorbent is regenerated for subsequent reuse and the desorbed adsorbate is transferred to the separation and recovery system, where the adsorbate is recovered by distillation or other methods according to its properties and operating conditions.

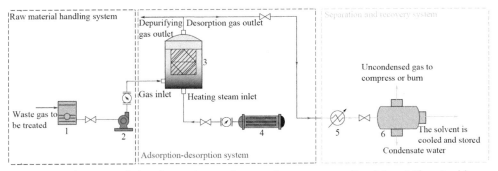

Figure 7.3 Schematic of the desorption and separation processes (1—dehumidifier, 2—blower, 3—adsorption and desorption column, 4—steam heater, 5—condenser, 6—gas-liquid separator).

(2) Mass transfer

As far as liquid/solid and gas/solid adsorption processes are concerned, the overall mass transfer process consists of three consecutive steps: external diffusion, intraparticle diffusion, and surface adsorption.

In external diffusion, the adsorbate diffuses from the bulk solution to the external surface of the adsorbent through molecular diffusion and convective diffusion.

In intraparticle diffusion, the adsorbate diffuses from the external surface into the pores of the adsorbent.

In surface adsorption, the adsorbate binds on the active sites of the internal pores of the adsorbent until equilibrium is established between adsorption and desorption.

Any of the three steps has an impact on the adsorption process and the slowest step will be the rate-limiting step. For chemical adsorption, surface adsorption that occurs slowly is the rate-limiting step, while for physical adsorption, surface adsorption occurs almost instantaneously and external diffusion and intraparticle diffusion are supposed to be the rate-limiting steps. The diffusion of the adsorbate from the bulk solution to the external surface of the adsorbent is a typical mass transfer process between fluid and solid surface, which is related to fluid properties, geometric properties of particles, temperature, pressure, and various other operating conditions. The diffusion of the adsorbate from the external surface into the pores of the adsorbent is closely related to the micropore structure of the adsorbent. The adsorbate diffuses either along the cross-section of the pore or on the surface of the pore. Pore size plays a critical role in the former case. Molecular diffusion dominates at pore sizes much larger than the mean free path of adsorbate molecules, Knudsen diffusion dominates at pore sizes much smaller than the mean free path of adsorbate molecules, and both of them occur at non-uniform pore sizes, which is referred to as transition diffusion. The intraparticle diffusion is also related to adsorbate properties and operating conditions [8].

Several techniques have been proposed to improve the mass transfer. For chemical adsorption, modification of the adsorbent with affinity groups can significantly increase surface adsorption, for physical adsorption, almost no adsorption occurs as adsorbate molecules diffuse from the bulk solution to the external surface of the adsorbent, which is regarded as ineffective adsorption. The shear force caused by the external force is expected to reduce the time of ineffective adsorption and increase the diffusion rate.

7.2.2.2 Types and applications of adsorbents

Carbon- [9], oxygen- [10] and polymer-based adsorbents [11] are among the most widely used adsorbents. Carbon-based adsorbents (e.g., activated carbon and graphene) have low polarity and hydrophilicity, oxygen-containing adsorbents (e.g., silica gel, zeolite molecular sieves, and metal oxides) have high polarity and certain hydrophilicity, and polymer-based adsorbents mainly include polymer resins. Activated carbon is the most widely used industrial adsorbent, which is prepared by carbonization and activation using natural materials (e.g., coconut shell, fruit shell, and wood), asphalt, and pulverized coal as raw materials. The adsorbents for use in industrial applications should have (1) high adsorption capacity (large specific surface area and pore structure—in general, the specific surface area is about 300-1200 m^2/g), (2) high mechanical strength and wear resistance, (3) high selectivity, (4) long service life, and (5) simple preparation and low cost.

A large number of studies have been conducted on the development and use of adsorbents. An important research trend is the development of high-performance adsorbents for industrial use, such as natural organic or inorganic adsorbents, synthetic adsorbents, and biological adsorbents. Much research effort is also being made to develop environmentally friendly adsorbents with high adsorption capacity and reusability using waste materials and to prepare composite adsorbents in order to fully exploit the potential of different adsorbents. Surface modification of adsorbents is necessary to improve the adsorption efficiency and selectivity.

7.2.2.3 Types and applications of adsorbers

The most commonly used adsorbers include fixed bed, moving bed, fluidized bed, and rotating bed, and their differences are visible in terms of adsorbent arrangement and operating conditions. These adsorbers can be operated in a batch or a continuous mode.

(1) Fixed bed [12,13]

In a fixed bed adsorber, adsorbent particles are uniformly distributed on a porous support and remain stationary throughout the adsorption process. The fluid flows through the adsorbent bed from top to bottom (or from bottom to top) in the axial direction. As the adsorption process comes to an end, the fluid should not be fed anymore and the adsorbent is regenerated for reuse. Take the use of activated carbon for adsorption of organic solvent in the air as an example. Water vapor is introduced to remove organic solvent (adsorbate) adsorbed on the activated carbon (adsorbent). The adsorbent is regenerated and the resulting gas mixture (water vapor and organic vapor) is condensed for separation and recovery of the adsorbate. The fixed bed has the advantages of simple structure and low adsorbent wear, but the adsorption process is a batch process requiring periodic adsorption and desorption, which is very complicated and labor intensive. The fixed bed is cumbersome with low adsorption capacity. The low thermal conductivity of the adsorbent makes it difficult for heating or variable temperature regeneration, and local overheating is also likely to occur in case of large thermal effect. In order to achieve continuous adsorption and high adsorption capacity, three or more fixed beds can be connected in series in practical applications, as shown in Fig. 7.4.

(2) Moving bed [14,15]

In a moving bed adsorber, the adsorbent moves from top to bottom and the gas flows from bottom to top, leading to intensive countercurrent contact between the gas and the adsorbent. The gas mixture is introduced through the gas inlet at the middle of the

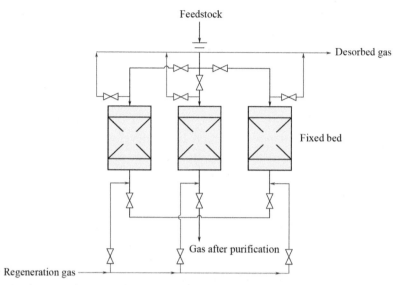

Figure 7.4 A schematic diagram of the structure of the fixed bed.

moving bed and contacts countercurrently with the moving adsorbent. The adsorbent goes through cooling, adsorption, concentration, steam stripping, and regeneration processes. The adsorption and desorption processes take place sequentially in the same adsorber, which makes it particularly appealing for stable and continuous purification of exhaust gas in large quantities. For example, an activated carbon moving bed can be used for the treatment of flue gas containing sulfur dioxide, and adsorption of sulfur dioxide and catalytic oxidation are performed in the same adsorber (Reinluft process).

As shown in Fig. 7.5, the moving bed consists of:

Cooler: the cooler comprises of a bundle of pipes which are used to cool the adsorbent in order to improve its adsorption capacity.

Adsorption sector (I): The adsorbent is brought into countercurrent contact with the gas flowing upward. One or more components of the gas mixture are adsorbed on the adsorbent and the purified gas is discharge from the top of the bed.

Concentration sector (II): The adsorbent contacts countercurrently with ascending gas. The desorbed components are replaced, resulting in the concentration effect.

Stripping sector (III): The adsorbent is heated and stripped off by stripping steam. A portion of the stripped gas is discharged as product or concentrated harmful gas, and the rest is refluxed to the concentration sector.

Desorption sector: The saturated adsorbent is desorbed for regeneration.

Regenerator: The temperature is raised for further desorption and regeneration of the adsorbent. The resulting adsorbent is transferred to the top of the adsorber for the next adsorption process.

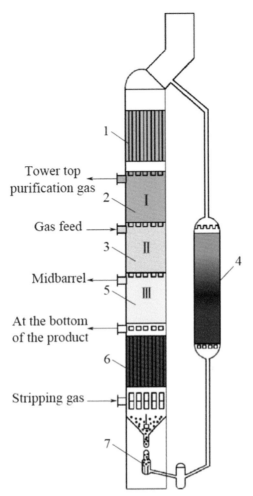

Figure 7.5 The structure of a moving bed adsorber (1—cooler, 2—adsorption sector, 3—concentration sector, 4—regenerator, 5—stripping sector, 6—desorption sector, 7—outlet valve).

The moving bed permits continuous adsorption and effectively improves the utilization rate of the adsorbent and the adsorption capacity of the adsorber. It can be applied to the case where there is a high adsorbent/gas ratio, but it is seldom used for pollution control. The adsorbent has high wear resistance, but high energy consumption is required due to additional power consumption.

(3) Fluidized bed [16,17]

In a fluidized bed adsorber, the gas flow rate is high, leading to intimate contact between two phases and consequently high heat and mass transfer. The fluidized bed is applicable to the treatment of exhaust gas continuously released in large quantities from stable sources. As shown in Fig. 7.6, the gas mixture to be separated flows

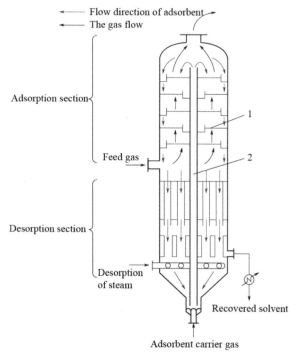

Figure 7.6 The structure of the fluidized bed adsorber (1—plate, 2—gas lifting pipe).

through the bed at a high velocity to suspend adsorbent particles. The adsorber can be divided into adsorption and desorption sectors. The exhaust gas is introduced from the lower end of the adsorption sector and then flows through the adsorption layers sequentially, and it is discharged from the upper end of the adsorption sector. The adsorbent is carried to the top of the adsorption sector through the center pipe by the carrier gas and then moves downward along the plates. On each plate, the solid adsorber is suspended in a fluidized state by the upward-flowing gas. The exhaust gas is purified by the adsorbent. After adsorption, the saturated adsorbent is heated and regenerated in the desorption sector, and the desorbed solvent is condensed and recovered for reuse.

The fluidized bed is characterized by a high mass and heat transfer rate, uniform bed temperature, and high stability. However, the adsorbent is in a fluidized form, and the transport of the adsorbent to the top of the adsorber causes severe wear of the adsorbent and high power and heat consumption. Therefore, the adsorbent should have high wear resistance and mechanical strength. The low saturation of adsorbent particles as a result of high backmixing with the gas makes the bed less suitable for exhaust gas purification.

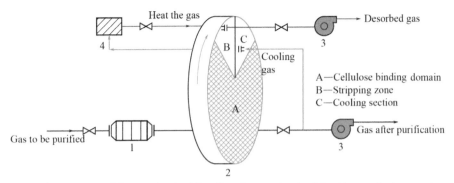

Figure 7.7 The structure of the rotating bed (1—filter, 2—rotating bed, 3—blower, 4—heat exchanger).

(4) Rotating bed [18]

The rotating bed has emerged as a novel and efficient adsorber for the treatment of large quantities of low-concentration exhaust gas. As shown in Fig. 7.7, the rotating bed consists of three parts: adsorption sector, cooling sector, and desorption sector.

Adsorption sector: As the dry dedusted gas flows through the adsorption layer, it comes into contact with the adsorbent that rotates at a speed of 1-6 r/h.

Desorption sector: The gas in the rotor is concentrated and the adsorbent is desorbed for regeneration by the heat flow provided by the heat exchanger. The purified gas is transferred to the next sector.

Cooling sector: The desorbed adsorbent is cooled to room temperature and then transferred to the adsorption sector for recycling.

The rotating bed is characterized by continuous operation, large adsorption capacity, automatic control, low gas pressure drop, and compacted structure. Its potential disadvantages include large equipment size and high power loss. Deceleration gearing is needed and the seal between the rotor and the connection pipe is very complicated.

7.3 High-gravity adsorption

The high-gravity adsorption process takes place in an RPB with adsorbent as the packing. In physical adsorption, the high-gravity field can effectively increase the diffusion of phases and reduce the ineffective adsorption layer, which in turn can enhance the van der Waals interaction and hydrogen bonding force. In chemical adsorption, it can effectively increase the reaction rate between adsorbate and adsorbent. The porous solid adsorbents used in high-gravity adsorption processes, such as activated carbon, silica gel, and molecular sieves, should have high mechanical strength and wear resistance because they have to rotate at a high speed throughout the adsorption process.

For both high-gravity gas/solid and liquid/solid adsorption processes, adsorption and desorption can be carried out alternatively in the RPB, but it is also possible to establish a continuous adsorption and desorption process by connecting two or more RPBs in series. As internal diffusion is the controlling step of the initial stage of adsorption, the overall adsorption efficiency is significantly improved by increasing the internal diffusion rate [19]. The high centrifugal force increases the wetted area of the packing and the liquid-solid contact area. The mass transfer coefficient of RPB is 3-6 times higher than that of the fixed bed [20]. The RPB can also provide more active adsorption sites to facilitate internal diffusion and mass transfer [21].

The high-gravity adsorption process is advantageous over traditional adsorption processes in the following aspects. First, more adsorption sites are available and the high driving force resulting from the concentration gradient between the surface and pores increases the adsorption capacity of activated carbon and reduces the time needed to reach adsorption equilibrium. Second, the rotation of the packing can prevent the short circuit of the fluid in the adsorbent layer and the "wind dead zone" to increase the adsorption capacity of the bed. Third, the rotation of the packing increases the desorption rate and reduces the desorption time. In general, the higher the high-gravity factor is, the better the desorption effect will be.

7.3.1 High-gravity adsorption device
7.3.1.1 Rotating packed bed for liquid/solid adsorption

The RPB used for liquid/solid adsorption consists of motor, rotor, liquid distributor, liquid inlet and outlet, rotor shaft, shaft seal, and shell, as shown in Fig. 7.8 [22]. The rotor is a hollow cylinder with a central cavity where the liquid distributor is installed. The liquid distributor is positioned concentrically with the rotor with a gap between them, and the rotor is loaded with adsorbent particles supported on wire mesh. The dynamic balance of the rotor is tested and adjusted if necessary to ensure high stability

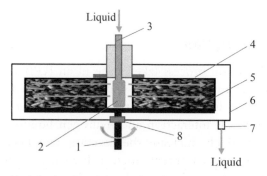

Figure 7.8 RPB for liquid/solid adsorption (1—rotor shaft, 2—liquid distributor, 3—liquid inlet, 4—rotor, 5—activated carbon, 6—interior wall of the casing, 7—liquid outlet, 8—shaft seal). *RPB*, rotating packed bed.

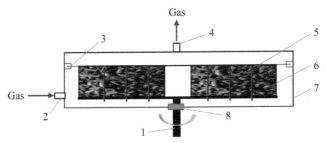

Figure 7.9 RPB for gas/solid adsorption (1—rotor shaft, 2—gas inlet, 3—airtight seal, 4—gas outlet, 5—rotor, 6—adsorbent, 7—interior wall of the casing, 8—shaft seal). *RPB*, Rotating packed bed.

during rotation. The motor speed can be adjusted via a frequency converter. The liquid flow rate is measured by a flowmeter. The liquid is uniformly sprayed onto the inner edge of the rotor using the liquid distributor, and then it flows through the packing from the inner edge to the outer edge of the rotor under the high centrifugal force generated by the high-speed rotating packing. The liquid is collected at the shell and exits the bed through the liquid outlet.

7.3.1.2 Rotating packed bed for gas/solid adsorption

The RPB used for gas/solid adsorption also consists of motor, rotor, liquid distributor, gas inlet and outlet, rotating shaft, shaft seal, gas seal, and shell, as shown in Fig. 7.9 [23,24]. The rotor is a hollow cylinder loaded with adsorbent particles supported on wire mesh, and the dynamic balance is tested and adjusted if necessary to ensure high stability during rotation. The seal is mounted between rotor and casing in order to separate inlet and outlet gas streams. The motor speed can be adjusted via a frequency converter. The gas flow rate is measured by a flowmeter. Gas is introduced into the RPB through the gas inlet at the bottom of the RPB, and then it flows upward from bottom to top for mass transfer with the adsorbent. Finally, it exits the bed through the gas outlet.

7.3.2 Mechanism of intensification of high-gravity adsorption process

The most important feature of the high-gravity adsorption process is to use the centrifugal force generated by high-speed rotation to change the micromorphology, characteristic scale, and flow mode of the fluid. The adsorbent rotates synchronously with the rotor, leading to fundamental changes in the contact and mass transfer between fluid and adsorbent. The renewal and diffusion rates of the adsorbate on the adsorbent surface are accelerated. The high-gravity field plays a role in intensifying the external diffusion, pore diffusion, and adsorption equilibrium. Adsorbate molecules can easily diffuse into the pores of the adsorbent for adsorption and mass transfer.

7.3.2.1 Intensification of liquid/solid adsorption processes

In RPB, the adsorbent will rotate synchronously with the rotor at a high speed and the liquid is subjected to high shear stress, leading to a substantial increase in mass transfer in liquid/solid adsorption processes [25,26].

Changes in liquid: Liquid is subjected to high shear stress in a high-gravity field, forming micro/nano-scale liquid elements (e.g., droplets, filaments, and films) with a small volume but a large specific surface area. Consequently, the liquid is well dispersed on the adsorbent surface and in the voids between adsorbent particles. The liquid condenses immediately as it contacts with the adsorbent, and it is adsorbed on the adsorbent surface. Because of the high centrifugal force, micro/nano-scale liquid elements are formed once again. The liquid is highly dispersed and exists as a dispersed phase rather than as a continuous phase, which is completely different from immersion or submerged adsorption.

Changes in adsorbent: Adsorbent is loaded into the rotor as the packing and rotates at a high speed during the adsorption process. The adsorbent is immobile in a fixed bed but rotates continuously in an RPB. The liquid appears on the adsorbent surface and then disappears instantaneously due to the high centrifugal force, leading to rapid surface renewal and elimination of the external diffusion resistance of the adsorbate from the bulk fluid to the adsorbent surface.

Changes in the contact between liquid and adsorbent and the environment: first, the adsorbent is not immersed in liquid, but instead the liquid is dispersed on the adsorbent surface as micro/nano-scale elements and soon separated from the adsorbent under the high centrifugal force; second, the high-speed rotation leads to more uniform distribution and rapid renewal of the liquid on the adsorbent surface, as well as a higher frequency of contact between adsorbent and adsorbate and a higher concentration of adsorbate on the adsorbent surface. There is a high driving force for pore diffusion and the adsorption rate is increased; third, more effective adsorption sites are available and the liquid coverage area is increased on the adsorbent surface; finally, the liquid is repeatedly dispersed and then condensed in the rotating packing, which contributes greatly to adsorption.

To sum up, the essence of high-gravity liquid/solid adsorption processes is the rapid renewal of the liquid on the adsorbent surface and the elimination of the external diffusion resistance of the adsorbate from the bulk fluid to the adsorbent surface. A high concentration of adsorbate is maintained on the adsorbent surface. As a result, there is a high concentration gradient between external and internal surfaces of adsorbent particles, which increases the driving force for mass transfer and consequently the adsorption rate at equilibrium. The high-gravity plays a role in all three stages of the adsorption process: external diffusion, pore diffusion and adsorption equilibrium.

7.3.2.2 Intensification of gas/solid adsorption processes

The high-gravity gas/solid adsorption is conducted in an RPB using solid adsorbent particles as the packing. These adsorbent particles rotate at a high speed throughout the adsorption process and the gas flows on the adsorbent surface or in the voids between adsorbent particles. Similarly, substantial changes are expected in the gas, adsorbent, and mass transfer between gas and adsorbent [27,28].

Changes in gas: Gas is introduced into the RPB by pressure and flows through the packing in the axial direction. As the gas is also subjected to the drag force resulting from the adsorbent particles and rotor and the centrifugal force resulting from the rotation of the packing, the gas flow velocity is the sum of the axial, circumferential, and radial components. The gas flows upward through the packing in a helical, inverted conical path, and the resulting flow pattern is much more complex than that in a fixed bed. Boundary-layer separation occurs as the gas flows over the surface of rotating adsorbent particles, leading to the formation of vortex that can facilitate the mass transfer between adsorbent particles and the adsorbate in the gas.

Changes in adsorbent: Adsorbent is loaded into the rotor as the packing and rotates synchronously with the rotor at a high speed during the adsorption process, and the gas flows over the adsorbent surface. There is a high relative velocity between adsorbent and gas. Under the influence of gas flow and centrifugal force, adsorbent particles are squeezed against each other, leading to slight rotation, displacement, or rolling over.

Changes in the contact between gas and adsorbent and the environment: The high-speed rotation of adsorbent particles leads to a large relative velocity with respect to the gas, which can increase the frequency of contact between gas and solid phases and the external diffusion of the gas on the adsorbent surface. The gas on the adsorbent surface is rapidly renewed, leading to a high adsorbate concentration on the adsorbent surface and the diffusion of more adsorbate into the pores of the adsorbent. The particle surface where the angle between the normal direction of the surface and the gas flow direction is 180-360 degrees is the windward side, and the particle surface where the angle is 0-180 degrees is the leeward side. There is a difference in the gas/adsorbent contact between the windward and leeward sides. A stationary point is formed as the gas flows on the windward side and the kinetic energy is converted into static pressure energy. The gas pressure is high and the adsorption rate in the pores is increased. On the leeward side, vortices are formed due to boundary-layer separation. In this case, the gas pressure is low, which is favorable for the removal of gas in the pores and subsequently the diffusion of new adsorbate into the pores for adsorption.

To sum up, the essence of high-gravity gas/solid adsorption processes is the rapid renewal of the gas on the surface of the adsorbent that rotates at a high speed and the elimination of the external diffusion resistance of the adsorbate from the bulk gas to the adsorbent surface. A high concentration of adsorbate is maintained on the adsorbent surface, which can greatly increase the adsorption rate and reduce the adsorption time.

Table 7.3 Effective coverage area and coverage ratio of different adsorbers.

Adsorbers	Effective coverage area/(m²/g)	Coverage ratio/%
Fixed bed for gas/solid adsorption	218.36	48.42
RPB for gas/solid adsorption	233.74	54.09
Fixed bed for liquid/solid adsorption	249.12	41.78
RPB for liquid/solid adsorption	446.04	51.67

During the desorption process, the contact between heated stream and adsorbent particles in a high-gravity field leads to rapid heat transfer and renewal of the gas on the adsorbent surface, so that the adsorbate adsorbed on the adsorbent surface can be rapidly desorbed and then carried out of the bed with the gas stream. The high adsorbate concentration difference between the surfaces and pores of the adsorbent can increase the driving force for the desorption of the adsorbate adsorbed in the pores. In a high-gravity field, the gas and adsorbent are rotating at high speed, leading to an increase in the driving force for both external and intraparticle diffusion and the contact time and area between the two phases, but a reduction in the desorption time.

The coverage ratio and the effective coverage area of the adsorbate on the adsorbent surface indicate the number of effective adsorption sites and the adsorption capacity of the adsorbent. Taking the adsorption of phenol in wastewater and that of toluene in gas using activated carbon as an example. Under the same operating conditions, the effective coverage area and the coverage ratio of the adsorbate on the activated carbon surface are shown in Table 7.3. The effective coverage area and the coverage ratio of the RPB are higher than those of the fixed bed. More specifically, the effective coverage area and the coverage ratio are increased by 7% and 11.7% for gas/solid adsorption and by 79% and 23.7% for liquid/solid adsorption, respectively. These results demonstrate that the high-gravity technology can significantly increase the effective adsorption area and eliminate the short-circuit phenomenon of the fluid flowing in the adsorbent layer and the "wind dead zone" area, which can provide more adsorption sites for adsorption.

7.4 Application examples
7.4.1 Treatment of resorcinol wastewater

This section describes the application of high-gravity liquid/solid adsorption for treatment of resorcinol wastewater. Resorcinol is an endocrine-disrupting chemical that may cause reproductive and developmental abnormalities. Routine treatment methods include ultrasonic degradation, photocatalysis, biological oxidation, and adsorption. We have investigated the thermodynamics and kinetics in high-gravity adsorption processes for treatment of resorcinol wastewater using activated carbon as the adsorbent [29,30], and the effects of high-gravity factor, liquid flow rate and pH on the removal rate.

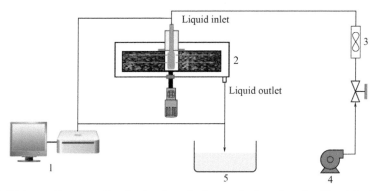

Figure 7.10 High-gravity adsorption for treatment of resorcinol wastewater (1—detector, 2—rotating bed, 3—flow meter, 4—centrifugal pump, 5—liquid storage tank).

The wastewater containing 1000 mg/L resorcinol was treated using activated carbon (particle size: 2 mm, specific surface area: 670 m²/g) as the adsorbent in RPB (packing height: 40 mm, outer diameter: 65 mm, inner diameter: 30 mm), as shown in Fig. 7.10. The resorcinol solution stored in a liquid storage tank is pumped into the bed by a centrifugal pump and the flow rate is measured using a flow meter. Then, the liquid is uniformly sprayed on the inner edge of the packing through a liquid distributor, and as it flows through the packing in the radial direction, it is sheared into micro-sized droplets and brought into intimate contact with activated carbon for mass transfer. Finally, the resorcinol is stored in the liquid storage tank and the wastewater is recycled in the RPB.

7.4.1.1 Adsorption thermodynamics

Under conditions of liquid flow rate = 50 L/h, high-gravity factor = 41.3, initial resorcinol concentration in wastewater = 900 mg/L, and pH = 5, the thermodynamics for the high-gravity adsorption process of resorcinol is described by van't Hoff equation [31]:

$$\ln \frac{1}{c_e} = \ln k_0 + \left(-\frac{\Delta H}{RT}\right) \qquad (7.15)$$

where c_e—equilibrium concentration, mol/L;
  ΔH—isosteric adsorption enthalpy, kJ/mol;
    T—temperature, K;
    R—gas constant, 8.314 J/(mol · K);
   k_0—van't Hoff equation constant.

Fig. 7.11 shows the effect of temperature on the removal rate of resorcinol. It is seen that the removal rate of resorcinol decreases from 96.86% to 94.44% with the increase of temperature from 20°C to 50°C. At adsorption equilibrium, there is only a

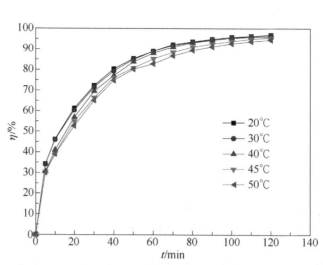

Figure 7.11 Effect of temperature on the removal rate of resorcinol.

marginal difference in the removal rate (<2.42%), indicating that temperature has no significant effect on the adsorption rate of resorcinol. This is characteristic of physical adsorption. As the adsorption process is exothermic, a lower temperature is expected to be more favorable. The plot of $\ln(1/c_e)$ against $1/T$ shows that $\Delta H = 14.65$ kJ/mol, indicating that the adsorption heat is 14.65 kJ/mol. This is within the range of the adsorption heat for physical adsorption (2.1-20.9 kJ/mol) [32] and thus indicates that the adsorption of resorcinol by activated carbon in a high-gravity field is dominated by physical adsorption or surface adsorption (van der Waals' force).

7.4.1.2 Adsorption kinetics

The removal rates of resorcinol obtained in the fixed bed, stirred tank, and RPB are shown in Fig. 7.12. Under the same operating conditions, the removal rate of the RPB (96.86%) is 16.3% and 23.7% higher than that of the fixed bed (80.56%) and stirred tank (73.16%), respectively, and batch adsorption can be carried out in a single adsorber.

Table 7.4 shows that the correlation coefficients R^2 are high for the pseudo-second-order models of RPB, stirred tank, and fixed bed, and the corresponding adsorption rate constants k_2 are 2.576×10^{-3}, 1.823×10^{-3} and 1.584×10^{-3} g/(mg · min), respectively. The adsorption rate constant of the RPB is 1.413 and 1.626 times that of the stirred tank and fixed bed, respectively, implying that the high-gravity field can significantly enhance the mass transfer between liquid and solid phases.

The adsorption kinetics of resorcinol is fitted by the Weber-Morris (intraparticle diffusion) model, and the plots of q_t against $t^{1/2}$ are shown in Fig. 7.13. The initial stage of the curve represents the boundary diffusion (film diffusion), and the subsequent stage of the curve represents the intraparticle diffusion. It is found that q_t is not

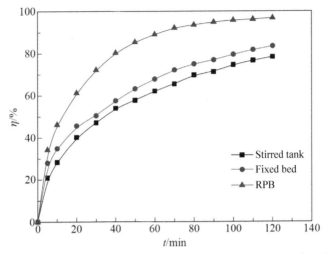

Figure 7.12 The removal rates of resorcinol obtained by fixed bed, stirred tank, and RPB. *RPB*, rotating packed bed.

Table 7.4 The pseudo-second-order models for the fixed bed, stirred tank and RPB (rotating packed bed).

Model	Adsorber	Linear equation	R^2	Adsorption rate constant/[g/(mg · min)]
Pseudo-second-order model	Fixed bed	$y = 0.04016x + 1.018$	0.9894	1.584×10^{-3}
	Stirred tank	$y = 0.03883x + 0.8272$	0.9871	1.823×10^{-3}
	RPB	$y = 0.03423x + 0.4549$	0.999	2.576×10^{-3}

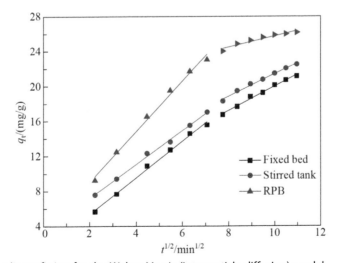

Figure 7.13 The linear fitting for the Weber-Morris (intraparticle diffusion) model.

linearly related to $t^{1/2}$, indicating that intraparticle diffusion is not the only rate-limiting step of the adsorption process. Both boundary diffusion and intraparticle diffusion dominate. The initial stage is the film diffusion process in which intraparticle diffusion is the rate-limiting step. The driving force is large at the very beginning of the adsorption process because of the large concentration difference between adsorbate and adsorbent, and resorcinol is easily adsorbed on the surface of activated carbon. Subsequently, film diffusion becomes the rate-limiting step. As saturated adsorption is reached on the surface of activated carbon, resorcinol diffuses into the pores of activated carbon. The increase in film diffusion resistance leads to a decrease in the film diffusion rate. As the resorcinol concentration decreases in the solution, the driving force decreases and the film diffusion rate becomes lower until an equilibrium is reached. Therefore, at the later stage, film diffusion is the rate-controlling step, which is consistent with the adsorption mechanism of resorcinol by activated carbon [33] and the pseudo-second-order model.

7.4.1.3 Effects of operating parameters
(1) Effect of liquid flow rate on the removal rate of resorcinol

Increasing the liquid flow rate can increase not only the coverage ratio of micro/nano liquid elements (droplets, filaments, and films) on the surface of activated carbon and the contact area between liquid and solid phases but also the renewal rate of wastewater on the surface of activated carbon. As a result, the surface diffusion resistance is reduced and the mass transfer rate is increased. As shown in Fig. 7.14, the higher the liquid flow rate is, the higher the removal rate of resorcinol will be. However, the variation of the total adsorption rate with time is approximately the

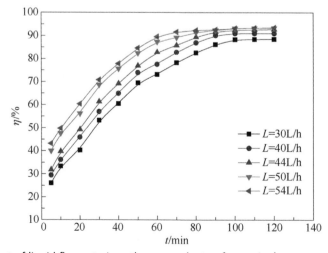

Figure 7.14 Effect of liquid flow rate L on the removal rate of resorcinol.

same and the slope decreases over time irrespective of the liquid flow rate. It is concluded that increasing the liquid flow rate at any period contributes to improving the total adsorption rate. In this regard, the high-gravity adsorption process has high operation flexibility.

It should be noted that when the wastewater flows through the activated carbon packing in a highly dispersed state, increasing the liquid flow rate will increase the adsorption rate. However, the wastewater will flow in a continuous or submerged state at extremely high liquid flow rates. In this case, the mass transfer mechanism changes fundamentally, and the adsorption rate is decreased to a level close to that of fixed bed or stirred tank.

(2) Effect of high-gravity factor on the removal rate of resorcinol

The effect of high-gravity factor β on the removal rate of resorcinol is shown in Fig. 7.15. Increasing the high-gravity factor induces a higher rotation speed and thus a higher centrifugal force acting on the liquid. The liquid is dispersed into finer droplets, filaments, and films and distributed more uniformly. The resorcinol in wastewater can easily reach the surface of activated carbon and is quickly renewed, and the mass transfer resistance is significantly reduced, which can effectively enhance the adsorption of resorcinol. The variation of the removal rate of resorcinol with time is similar and the slope of the curve decreases gradually irrespective of the high-gravity factor. However, it should also be noted that as the high-gravity factor increases, the residence time of the wastewater on the surface of activated carbon is significantly reduced, potentially leading to lower mass transfer with activated carbon and consequently lower adsorption efficiency. These conflicting effects should be taken into account in the selection of an appropriate high-gravity factor.

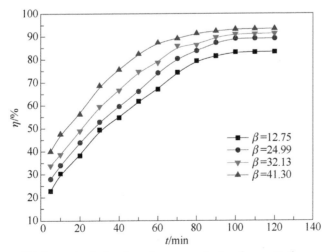

Figure 7.15 Effect of high-gravity factor β on the removal rate of resorcinol.

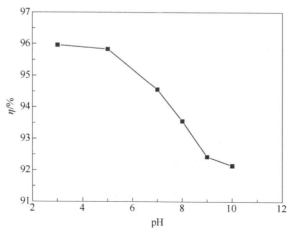

Figure 7.16 Effect of pH on the removal rate of resorcinol.

(3) Effect of pH on the removal rate of resorcinol

Fig. 7.16 shows that pH has a significant effect on the removal rate of resorcinol, and it changes little in acidic and neutral solutions but decreases significantly with increasing pH in alkaline solutions. This is because resorcinol is a weak acid and its solubility increases with the decrease in pH [34]. At pH <7, resorcinol exists in the form of molecules in the solution and it is easily adsorbed by activated carbon, while at pH >7, resorcinol is converted into resorcinol salt with a higher solubility, and thus it is desorbed from the activated carbon and the removal rate of resorcinol is reduced significantly. This is also the theoretical basis for the generation of activated carbon with strong alkali.

7.4.2 Treatment of toluene-containing exhaust gas

Toluene can not only cause air and environmental pollution but can also harm human health because of its toxic and carcinogenic potential. Very often, high-concentration toluene gas is treated by combustion, while medium- and low-concentration toluene gas is treated by adsorption. However, traditional adsorption methods are less effective and the utilization rate of adsorbent is low. We have demonstrated that the high-gravity adsorption process is effective for the treatment of toluene [23]. The adsorption and desorption of toluene-containing gases were investigated in the RPB and fixed bed using activated carbon as the adsorbent. The equipment setup and operating parameters are shown in Table 7.5, and the process is shown in Fig. 7.17. The toluene gas generated by a gas generator is measured using a flowmeter and then introduced into the RPB using a blower. The gas flows through activated carbon from bottom to top, and finally, it leaves the RPB from the top outlet. Portable Heated Flame Ionization Detector (HFID) VOCs online detectors are installed at both inlet and outlet. At the

Table 7.5 The structure and operating parameters of the RPB (rotating packed bed).

Parameters	Value	Parameters	Value
Adsorption column outer diameter/mm	300	Initial concentration/(mg/m^3)	0-3000
Adsorption column height/mm	800	Rotation speed/(r/min)	0-900
Operating pressure/kPa	103	Gas flow rate/(m^3/h)	0-50
Operating temperature/K	298-490		

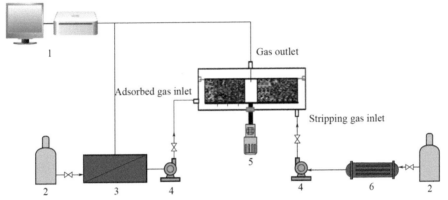

Figure 7.17 The adsorption and desorption processes in an RPB (1—HFID detector, 2—nitrogen cylinder, 3—VOCs gas generator, 4—blower, 5—RPB, 6—gas heater). *RPB*, rotating packed bed.

completion of adsorption, the inlet valve is closed and the nitrogen valve is opened. Nitrogen is heated by an air heater and then introduced into the saturated adsorption bed using a blower for desorption. Adsorption and desorption are carried out alternately.

7.4.2.1 Comparison of adsorption efficiency of rotating packed bed and fixed bed

To compare the adsorption efficiency of RPB and fixed bed, an integrated adsorption bed operating either in a fixed bed or an RPB mode is used in this study. Specifically, the adsorption column will not rotate in the fixed bed mode but rotate in the RPB mode.

(1) Breakthrough curve

The breakthrough curve is used to characterize the adsorption of toluene by activated carbon in the RPB and fixed bed, based on which relevant parameters, including saturated adsorption capacity, bed utilization rate, and adsorption rate constant, can be obtained. The initial stage of the adsorption process is characterized by the diffusion of adsorbate molecules to the surface of the adsorbent, which is called the ineffective adsorption layer. The length of the ineffective adsorption layer can be determined from the breakthrough curve.

The saturated adsorption capacity is an indicator of the adsorption capacity of an adsorbent, which is expressed as follows:

$$q_e = \frac{Qc_0}{m}\left(t_e - \int_0^{t_e} \frac{c_e}{c_0}dt\right) \tag{7.16}$$

where Q—gas flow rate, m³/h;
q_e—equilibrium adsorption capacity, mg/g;
c_0—initial gas concentration, mg/m³;
t_e—adsorption equilibrium time, h;
c_e—outlet concentration at equilibrium, mg/m³;
m—adsorbent loading, g.

The height of the mass transfer zone is an important physical parameter for evaluating the effective utilization and adsorption performance of the adsorption bed:

$$z_a = z\left[\frac{t_e - t_b}{t_e - (1-E)(t_e - t_b)}\right] \tag{7.17}$$

where z_a—the height of the mass transfer zone, cm;
z—the height of the adsorption column bed, cm;
t_e—adsorption equilibrium time, min;
t_b—breakthrough time, min;
E—the ratio of the adsorbent with adsorption capacity in the mass transfer zone to all adsorbents with adsorption capacity, which is generally in the range of 0.4-0.5. It is taken to be 0.5 in this book.

The bed utilization rate γ [35] is an important parameter for evaluating adsorption performance and selecting adsorption conditions:

$$\gamma = \frac{t_b}{t_0} \times 100\% \tag{7.18}$$

where t_b—breakthrough time, min;
t_0—time needed to reach 5% breakthrough concentration, min.

The breakthrough curves and saturated adsorption capacity of the fixed bed and RPB are shown in Fig. 7.18, where the flow rate of toluene is controlled at 40 m³/h, the initial concentration of toluene is 3500 mg/m³, the activated carbon loading is 300 g, and the high-gravity factor is 60. The breakthrough time of the RPB (60 min) is 42.9% longer than that of the fixed bed (42 min). The adsorption equilibrium time is 135 min in the fixed bed and 175 min in the RPB, and the saturated adsorption capacity is increased by 2.2 times. The breakthrough curve of the RPB shifts to the right and is much steeper. This is mainly because the gas flows upward in a helical, inverted conical path in the adsorbent packing due to the drag force from the

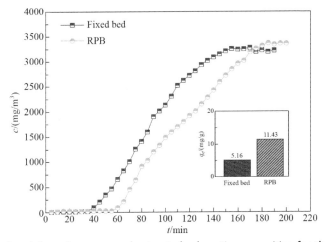

Figure 7.18 The breakthrough curves and saturated adsorption capacities for the fixed bed and RPB. *RPB*, rotating packed bed.

Table 7.6 Comparison of adsorption of toluene using activated carbon in the fixed bed and RPB (rotating packed bed).

Adsorber	Breakthrough time/min	Adsorption equilibrium time/min	Saturated adsorption amount/(mg/g)	Height of mass transfer zone/cm	Bed layer utilization rate/%
Fixed bed	42	135	5.16	5.20	58.53
RPB	60	175	11.43	4.97	61.85
Increase rate/%	+42.9	+29.6	+121.5	−4.4	+5.7

adsorbent particles and rotor and the centrifugal force from the rotation of the packing. The breakthrough time is prolonged and the contact between activated carbon and toluene is enhanced. There is a higher driving force for mass transfer and diffusion, which can increase the adsorption of toluene molecules on the adsorption sites of activated carbon and consequently the saturated adsorption capacity.

The breakthrough time and bed utilization rate for the adsorption of toluene using activated carbon in the fixed bed and RPB are shown in Table 7.6.

Under the same operating conditions, RPB has a higher saturated adsorption capacity and bed utilization rate but a lower height of the mass transfer zone. This is because the high-speed rotation of activated carbon in the RPB can enhance mass transfer between toluene and the adsorption layer on the surface of activated carbon, and reduce the diffusion time of the adsorbate in the ineffective adsorption zone. The high mass transfer rate in turn leads to the adsorption of more toluene on the surface of activated carbon. Thus, less activated carbon is used for the adsorption of the same amount of toluene and a lower height of the mass transfer zone is required.

More effective adsorption sites are available to improve the adsorption rate and the bed utilization rate.

(2) Adsorption rate constant

The adsorption rate constant is calculated from the semi-empirical gas adsorption model proposed by Chen [36] based on adsorption probability. The outlet concentration is obtained as follows:

$$c_b = \frac{c_0}{1 + \exp[k'(t_0 - t)]} \tag{7.19}$$

where c_0—initial gas concentration, mg/m³;
c_b—concentration at the breakthrough point, mg/m³;
k'—adsorption rate constant, min⁻¹;
t_0—time needed to reach 5% breakthrough concentration, min.

It can be converted into the following straight-line formula:

$$t = t_0 + \frac{1}{k'} \ln \frac{P}{1-P} \tag{7.20}$$

where P is the penetration fraction, which is defined as c_b/c_0 and is the measure of the velocity of the adsorbate passing through the adsorption layer.

Fig. 7.19 shows the plots of $\ln[P/(1-P)]$ against t fitted by the Yoon-Nelson model for the fixed bed and RPB, respectively, and Table 7.7 gives the corresponding fitting parameters. The results show that the adsorption rate constant of the RPB is 7% higher than that of the fixed bed, indicating that RPB can more effectively improve the adsorption of toluene by activated carbon.

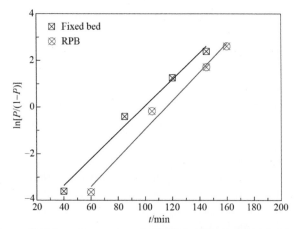

Figure 7.19 The plots of $\ln[P/(1-P)]$ against t for the adsorption of toluene by activated carbon in the fixed bed and RPB.

Table 7.7 Related parameters for the Yoon-Nelson model.

Adsorber	Fitted equation	R^2	Rate constant/min^{-1}
Fixed bed	$\ln[P/(1-P)] = 0.05683t - 5.6282$	0.98	0.057
RPB	$\ln[P/(1-P)] = 0.06134t - 7.08332$	0.98	0.061

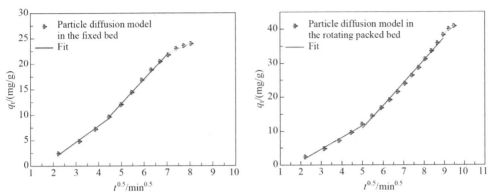

Figure 7.20 The Weber-Morris diffusion model for the adsorption of toluene by activated carbon in the fixed bed and RPB. *RPB*, rotating packed bed.

(3) Weber-Morris diffusion model

The Weber-Morris diffusion model is used to describe the adsorption process of toluene by activated carbon in the RPB. As shown in Fig. 7.20, the multi-linear plot (broken line) indicates that the adsorption process is controlled by both external and intraparticle diffusion. The adsorption process is divided into three stages. The first stage is characterized by surface adsorption of activated carbon, in which toluene is diffused from the liquid phase to the surface of activated carbon. This process is determined by the relative velocity between gas and solid phases. The second stage is characterized by intraparticle diffusion of toluene, in which toluene is diffused into the pores of activated carbon. This process is determined by the number of pores of activated carbon and the driving force resulting from the concentration gradient of toluene between the surface and pores. The third stage is characterized by slow adsorption until equilibrium, which is determined by the renewal rate of toluene in the pores. The fitted parameters and the contribution rates of each stage to the total adsorption capacity of the fixed bed and RPB are shown in Tables 7.8 and 7.9, respectively.

According to the diffusion model, the adsorption of toluene by activated carbon in the fixed bed and RPB is the combination of both surface diffusion and intraparticle diffusion, but the latter plays a dominant role. The external diffusion rate of the RPB is slightly higher than that of the fixed bed, but its intraparticle diffusion rate is 1.4 times that of the fixed bed. In the RPB, intraparticle diffusion contributes more to the total adsorption capacity than surface diffusion. Thus, RPB can effectively improve

Table 7.8 Fitted parameters for the Weber-Morris diffusion model.

Adsorber	Stage 1		Stage 2		Stage 3	
	k_{WM1}	R_1^2	k_{WM1}	R_1^2	k_{WM1}	R_1^2
Fixed bed	3.21	0.99	4.79	0.99	2.16	0.92
RPB	3.46	0.98	6.69	0.99	4.69	0.92

Table 7.9 Contribution rate of each stage to the total adsorption capacity of the fixed bed and RPB (rotating packed bed).

Adsorber	Stage 1 Surface diffusion/%	Stage 2 Intraparticle diffusion/%	Stage 3 Adsorption until equilibrium/%
Fixed bed	40.15	50.70	9.15
RPB	29.43	58.39	12.18

the pore adsorption of activated carbon, which is attributed to the particular contact mode between activated carbon and toluene in the RPB. The external diffusion rate is increased because the high-speed rotation of the packing makes it easier for toluene to be adsorbed onto the surface of activated carbon. The high external diffusion rate leads to a large difference in toluene concentration between surfaces and pores of activated carbon, which increases the driving force for the diffusion of toluene to the pores of activated carbon and consequently the intraparticle diffusion rate. Because of the high-efficient mass transfer in the RPB, the equilibrium adsorption rate is about two times that of the fixed bed. According to the Weber-Morris diffusion model, RPB can improve the mass transfer in the adsorption process as it can increase the external diffusion, as well as the intraparticle diffusion rate by increasing the driving force for the diffusion of toluene into the pores of activated carbon. More adsorption sites are available in the pores of activated carbon to increase the adsorption performance of activated carbon.

(4) Adsorption sites of toluene

In order to further investigate the high adsorption capacity of activated carbon in the RPB, we have explored the effective coverage area and the coverage ratio of toluene on the surface of activated carbon in the RPB and fixed bed under different conditions of high-gravity factor. The coverage ratio is defined as the ratio of the surface coverage area of toluene to the specific surface area of activated carbon, which is indicative of the contact probability and adsorption performance. The coverage ratio of toluene on the surface of activated carbon is calculated by the following formula:

$$x = \frac{S_{cx}}{S_{BET}} = \frac{6.023 \times 10^{23} A_m q_x}{1000 M_w S_{BET}} \tag{7.21}$$

Table 7.10 The coverage area and the coverage ratio of toluene on the surface of activated carbon.

Parameters	Fixed bed	RPB (rotating packed bed)		
		$\beta = 5$	$\beta = 20$	$\beta = 50$
Surface coverage area/(m²/g)	349.24	359.57	388.57	428.05
Surface coverage ratio/%	52.01	53.55	57.78	63.77

where x—surface coverage ratio, %;
$\quad S_{cx}$—surface coverage area, m²/g;
$\quad S_{BET}$—specific surface area, m²/g;
$\quad A_m$—projected area of a molecule (calculated by molecular diameter), m²;
$\quad q_x$—saturated adsorption capacity, mg/g;
$\quad M_w$—molar mass of the adsorbate, g/mol.

Table 7.10 summarizes the surface coverage area and the coverage ratio of toluene on the surface of activated carbon. Both the surface coverage area and the coverage ratio of the RPB are higher than that of the fixed bed, and they increase with an increasing high-gravity factor. At $\beta = 50$, the surface coverage ratio of toluene in the RPB is 22.6% higher than that in the fixed bed, leading to a higher contact area between activated carbon and toluene in the RPB and more adsorption sites.

7.4.2.2 Comparison of desorption efficiency of rotating packed bed and fixed bed

Desorption is carried out when the allowable adsorption amount or the saturated adsorption capacity is reached. In the desorption step, adsorbent is regenerated for reuse in order to avoid the waste of the adsorbent and reduce secondary pollution and operating cost. The heated medium is introduced into the bed to facilitate the detachment of adsorbate from the surfaces and pores of activated carbon. The desorbed adsorbate in the medium is separated and recovered.

(1) Effect of desorption temperature

Fig. 7.21 shows the variation of the desorption rate of activated carbon in the fixed bed with temperature using nitrogen as the desorption gas. The desorption process can be divided into three stages. The desorption rate increases rapidly in the first stage, and it decreases significantly in the second stage but slightly in the last stage. A higher desorption temperature leads to a higher desorption rate. Increasing the nitrogen temperature can accelerate the desorption rate of toluene on the saturated activated carbon due to the low specific heat capacity of nitrogen. As the temperature of nitrogen is increased from 120°C to 180°C, the maximum toluene concentration desorbed from the saturated activated carbon is increased from 1800 to 2300 mg/m³, implying that a higher nitrogen temperature favors the desorption of activated carbon. Less toluene is retained in the saturated activated carbon and the regeneration rate is high. The desorption time is reduced from 265 to 220 min. Temperature plays an important role in desorption. Although a higher temperature leads to a higher desorption rate of toluene and a shorter adsorption time, it causes higher energy consumption and damage to the pore structure and surface properties of activated carbon. The adsorption performance and service life of regenerated activated carbon may be reduced. Therefore, the desorption temperature of 180°C is selected for the fixed bed and RPB.

(2) Comparison of desorption performance of fixed bed and RPB

The desorption performance of fixed bed and RPB is compared at a desorption temperature of 180°C. Fig. 7.22 reveals that the two desorption curves are approximately the

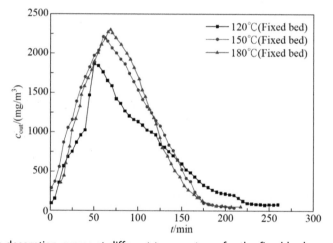

Figure 7.21 The desorption curves at different temperatures for the fixed bed.

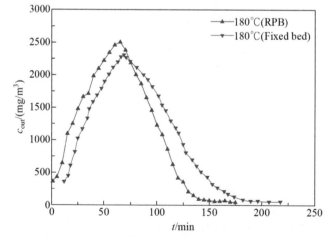

Figure 7.22 Desorption curves of the fixed bed and RPB at 180°C. *RPB*, rotating packed bed.

same in shape, but the desorption curve of the RPB shifts to the left, thus indicating a shorter desorption time and a higher desorption rate. The slopes are similar in the first stage, but the slope of the RPB in the second is steeper with a shorter tail, indicating that the desorption rate of the RPB is larger and there is less residual adsorbate.

(3) Comparison of desorption rate of fixed bed and RPB

The desorption rate is described as follows:

$$R = R_0 + Ae^{-kt} \qquad (7.22)$$

where R—desorption rate, %;

R_0—desorption rate at equilibrium, %;

k—desorption speed, $\min^{-1}$;

A—constant, %;

t—desorption time, min.

Fig. 7.23 shows the variation of desorption rate with time under different conditions of high-gravity factor. All correlation coefficients are higher than 0.99. Table 7.11 shows the parameters obtained under different conditions of high-gravity factor.

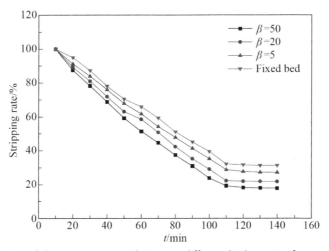

Figure 7.23 Variation of desorption rate with time at different high-gravity factors.

Table 7.11 Parameters for the fixed bed and RPB (rotating packed bed) at different high-gravity factors.

Parameters	Fixed bed	RPB		
		$\beta = 5$	$\beta = 20$	$\beta = 50$
R_0/%	31.58	27.69	21.96	18.25
$k/\min^{-1}$	0.00757	0.00843	0.00928	0.01023

It is seen from Table 7.11 that the desorption rate of the RPB is higher than that of the fixed bed. Also, the desorption rate increases with increasing high-gravity factor. At a high-gravity factor of 50, the desorption rate of the RPB is 1.35 times that of the fixed bed. The high-gravity field leads to an increase in the desorption rate and a decrease in the regeneration time of the adsorbent, which is favorable for the adsorption-desorption process and the service life of activated carbon. The high-gravity adsorption technology as a new adsorption technology is capable of adsorbing pollutants in wastewater and exhaust gas. It can not only improve the utilization rate of the adsorbent but can also reduce the adsorption time and improve the adsorption and desorption rates.

Compared with traditional adsorbers, RPB has the advantages of flexible operation, low labor intensity, small equipment volume, and convenient start-up and shutdown. The high-gravity adsorption process is characterized by high energy efficiency, high product purity, high capacity to remove trace substances, and low operating temperature. This technology would have wider applications in chemical, pharmaceutical, and food industries for the treatment of wastewater and exhaust gas.

References

[1] Zhao ZG. Application Principle of Adsorption. Beijing: Chemical Industry Press, 2005.
[2] Shi J, Wang JD, Yu GC. Handbook of Chemical Engineering. Beijing: Chemical industry press, 1996.
[3] Wang XF, Zhang HP, Xiao XY. Adsorption equilibrium and dynamics of phenol on bamboo charcoals. Journal of Functional Materials, 2005,36(5):746-749.
[4] Glueckauf E, Coates JI. Theory of chromatography. Part IV. The influence of incomplete equilibrium on the front boundary of chromatograms and on the effectiveness of separation. Journal of the Chemical Society, 1947,149(10):1315-1321.
[5] Kong LM, Zhang T, Wang PD. Equilibrium and kinetics of phenol adsorption from petrochemical wastewater with activated carbon fiber. CIESC Journal, 2015,66(12):4874-4882.
[6] Ho YS, Mckay G. Pseudo-second order model for sorption processes. Process Biochemistry 1999,34(5):451-465.
[7] Weber WJ, Morris JC. Proceedings of International Conference Water Pollution Symposium Pergamon: Oxford Pergamon Press, 1962.
[8] Zhu JW, Wu YY. Separate Process. Beijing: Chemical Industry Press, 2019.
[9] Liu ZJ, Huang YF, Liu JH. Research progresses in removal of volatile organic compounds by adsorption on activated carbon. Natural Gas Chemical Industry, 2014,39(2):75-79.
[10] Wang RS, Li HM, Feng W. Defluorination performance of activated alumina/silica gel and its application. Acta Scientiae Circumstantiae, 1992,12(3):333-340.
[11] Chen XK, Ning PS, Ding ZM. Research progress on organic wastewater treatment using adsorption resin. Thermosetting Resin, 2015,6:55-64.
[12] Zhao WR. Fixed bed honeycomb VOCs adsorption and desorption device and method. CN 105944500B, 2018.
[13] Li ML, Chen L, Yan ZC. Study on modifying of cassava adsorbent and fixed bed adsorption. Science and Technology of Food Industry, 2010,31(4):195-197.
[14] Li LQ. Moving bed adsorption device. CN 201016969Y, 2008.
[15] Bai JL, He YH, Wang JW. Experimental study on removal of mercury emission from the flue gases by using moved bed adsorption apparatus. Clean Coal Technology, 2018,24(4):114-119.

[16] Zhang LQ, Jiang HT, Li B. Kinetics of SO_2 adsorption by powder activated carbon using fluidized bed reactor. Journal of China Coal Society, 2012,37(6):1046-1050.
[17] Duan WL, Song WL, Luo LA. Experimental study on VOCs adsorption in a two-stage circulating fluidized bed. The Chinese Journal of Process Engineering, 2004,4(3):210-214.
[18] Dong N, Lin YJ, Cui YD. A new process for controlling VOCs-zeolite runner adsorption concentration + catalytic combustion. Environment and Development, 2017,29(7):118-119.
[19] Lin CC, Liu HS. Adsorption in a centrifugal field: basic dye adsorption by activated carbon. Industrial & Engineering Chemistry Research, 2011,16(1):161-167.
[20] Lin CC, Chen YR, Liu HS. Adsorption of dodecane from water in a rotating packed bed. Journal of the Chinese Institute of Chemical Engineers, 2004,35(5):531-538.
[21] Chang CF, Lee SC. Adsorption behavior of pesticide methomyl on activated carbon in a high-gravity rotating packed bed reactor. Water Research, 2012,46(9):2869-2880.
[22] Wu XN, Liu YZ, Jiao WZ. Adsorption of phenol on activated carbon in rotating packed bed. Chinese Journal of Energetic Materials, 2016,24(5):509-514.
[23] Guo Q, Liu YZ, Qi GS. Adsorption and desorption behaviour of toluene on activated carbon in a high-gravity rotating bed. Chemical Engineering Research and Design, 2019,143(3):47-55.
[24] Liu Y Z, Guo Q, Qi G S. High-gravity moving bed for gas adsorption. CN 108404601A, 2018.
[25] Guo F, Liu YZ, Guo Q. Adsorption of resorcinol on activated carbon under supergravity field. Chinese Journal of Energetic Materials, 2018,26(4):339-345.
[26] Guo Q, Liu YZ, Qi GS. Study of low temperature combustion performance for composite metal catalysts prepared via rotating packed bed. Energy, 2019,179(7):431-441.
[27] Guo Q, Liu YZ, Qi GS. Application of high-gravity technology NaOH-modified activated carbon in rotating packed bed (RPB) to adsorb toluene. Journal of Nanoparticle Research, 2019,21(8):175.
[28] Guo Q, Liu YZ, Qi GS. Behavior of activated carbons by compound modification in high-gravity for toluene adsorption. Adsorption Science & Technology, 2018,36(3-4):1018-1030.
[29] Guo F. Study on Adsorption Process of Resorcinol on Activated Carbon Under High-gravity. Taiyuan: North University of China, 2018.
[30] Guo F, Liu YZ, Guo Q. Desorption behavior and kinetics of resorcinol from activated carbon in rotating packed bed. The Chinese Journal of Process Engineering, 2018,18(3):503-508.
[31] Garcia-Delgado RA, Cotoruelo-Minguez LM, Rodriguez JJ. Equilibrium study of single solute adsorption of anionic surfactants with polymeric XAD resins. Separation Science, 1992,27(7):975-987.
[32] Suresh S, Srivastava VC, Mishra IM. Isotherm, thermodynamics, desorption, and disposal study for the adsorption of catechol and resorcinol onto granular activated carbon. Journal of Chemical & Engineering Data, 2011,56(4):811-818.
[33] Suresh S, Srivastava VC, Mishra IM. Study of catechol and resorcinol adsorption mechanism through granular activated carbon characterization, pH and kinetic study. Separation Science & Technology, 2011,46(11):1750-1766.
[34] Liao Q, Sun J, Gao L. The adsorption of resorcinol from water using multi-walled carbon nanotubes. Colloids & Surfaces A: Physicochemical & Engineering Aspects, 2008,312(2):160-165.
[35] Lee SW, Park HJ, Lee SH. Comparison of adsorption characteristics according to polarity difference of acetone vapor and toluene vapor on silica alumina fixed-bed reactor. Journal of Industrial & Engineering Chemistry, 2008,14(1):10-17.
[36] Chen L. Research of Monolithic Activated Carbon Foam and Its Application. Beijing: Beijing University of Chemical Technology, 2010.

CHAPTER 8

Gas-solid separation

Contents

8.1 Overview   283
    8.1.1 Sources of PM2.5   284
    8.1.2 Emission standard of PM2.5   286
    8.1.3 Separation technologies of PM2.5   286
8.2 Key technologies and principles of high-gravity gas-solid separation   289
    8.2.1 Key problems and theoretical analysis   289
    8.2.2 Principles of high-gravity intensification of gas-solid separation   289
8.3 Performance of high-gravity gas-solid separation   291
    8.3.1 Research advances   291
    8.3.2 Separation performance of cross-flow and countercurrent rotating packed bed   295
    8.3.3 Separation performance of low-concentration PM2.5 in the gas   299
    8.3.4 Separation of different components   307
8.4 Application examples   308
    8.4.1 Removal of dust particles and tar in semi-water gas   308
    8.4.2 Removal of dust particles and tar in coal-lock gas   309
    8.4.3 Removal of dust particles in boiler flue gas   310
    8.4.4 Removal and recovery of fine particles in chemical tail gas   310
8.5 Prospects   311
References   312

8.1 Overview

A gas containing small solid particles is essentially a heterogeneous mixture in which solid particles are the dispersed phase and gas is the continuous phase. Gas-solid separation is an important unit operation for the separation of solid particles from a gas mixture based on one or more physical processes. Particles larger than 100 μm will quickly settle out of the gas. Ultrafine particles smaller than 0.1 μm are almost undetectable with current methods and are beyond the capability of current gas-solid separation technologies that generally deal with solid particles of 0.1-100 μm [1,2]. It is much easier to separate particles larger than 10 μm than particles of 0.1-10 μm. Particles with an aerodynamic diameter equal to or less than 10 μm are known as PM10 or "respirable" particles, and particles with an aerodynamic diameter equal to or less than 2.5 μm are known as PM2.5 or fine particulate matter [3,4].

PM2.5 is characterized by small volume, large specific surface area, high activity, and lightweight. It can be suspended in the air for a long period of time and travel along with the air. It may also contain some toxic substances, such as polycyclic aromatic hydrocarbons, viruses, and bacteria, that may adversely impact human health, atmospheric environment, and industrial production. Previous experimental studies have demonstrated that particles with aerodynamic diameters of 2.5-10 μm can be deposited in the upper respiratory tract, such as nose, pharynx, larynx, trachea, and bronchi, and particles with aerodynamic diameters less than 2.5 μm can settle in the alveoli and even circulate via the blood system, resulting in obstructive pulmonary disease and cardiovascular disease [5]. Therefore, PM2.5 is considered one of the most harmful air pollutants. Long-term suspension of PM2.5 in the air can reduce atmospheric visibility by absorbing or scattering the light, leading to significant environmental impacts, such as acid rain, haze, and global climate change [6]. A substantial amount of PM2.5 with complex components, such as tar, naphthalene, and benzene, may be generated in chemical production processes, such as coal gasification and coking, which may have an impact on the continuous and stable operation and sometimes suspend the production.

Various effective technologies are readily available to collect total particulate matter, but significant challenges remain for the separation of PM2.5. In recent years, air pollution caused by total suspended particles (TSP) and inhalable particulate matter (PM10) has been declining in China, but the emission of PM2.5 has been increasing. Therefore, how to reduce PM2.5 has become the focus of recent research of gas–solid separation in China.

8.1.1 Sources of PM2.5

The PM2.5 in the atmosphere is derived from either natural sources (e.g., volcanic eruption, forest fire, and sandstorm) [7] or anthropogenic activities (e.g., combustion and industrial process).

8.1.1.1 Combustion

Flue gas from coal-fired boilers. Two types of fine particles can be generated in the combustion of pulverized coal, namely submicron particulate matter (PM1.0) and residual ash particles. The former is much greater in number but lighter in weight than the latter. A recent report shows that industrial boilers and coal-fired power plants are the dominant anthropogenic sources of atmospheric particulate matter in nine metropolises in China [8]. China now has approximately 47×10^4 coal-fired boilers with an annual production of 410×10^6 tons of dust, which accounts for 32% of the total emissions of particulate matter of the country [9].

Flue gas from biomass boilers. The fine particles generated in the combustion of straw consist mainly of carbonaceous compounds from incomplete combustion and

Table 8.1 Main sources and compositions of ship particulate matter.

No.	Sources	Compositions
1	Incomplete combustion of fuel oil and lubricating oil and agglomeration of small particles produced	Submicron
2	Ash residues originated from the ash initially present in the fuel oil after combustion	0.2–10 μm
3	Secondary synthesis of gas pollutants produced by combustion, such as sulfuric acid/sulfate and nitric acid/nitric acid salt aerosols resulting from the oxidation of SO_2 and NO_x produced by combustion	Micron

inorganic compounds from gasification, nucleation, condensation and flocculation of volatile elements, such as K, Cl, and N. Note that their particle sizes are generally less than 1 μm. The particles in the flue gas from biomass boilers are primarily floating dust particles with small particle size, lightweight, high diffusivity, and a large amount of tar and water vapor.

Exhaust gas from ship diesel engines. China has become a global maritime powerhouse, and shipping emissions are now one of the main sources of air pollution in many coastal and riverside areas. The particles in the exhaust gas from ship diesel engines are derived from diverse sources and have complex compositions (Table 8.1) [10]. It is found that 98%, 94%, and 92% of particles have a diameter equal to or less than 10 μm (PM10), 2.5 μm (PM2.5), and 1 μm (PM1), respectively, and PM1 accounts for approximately 78%–80% of the total particulate matter. There is often a high content of sulfur in fuel oil for use in most ships, resulting in a higher level of PM2.5 emissions compared to vehicles. For a medium- or large-sized container ship that uses fuel oil with 3.5% of sulfur and the output power is 70% of the maximum power, the total amount of PM2.5 emitted each day is equal to that of 500,000 trucks with National IV emission standard [11,12].

8.1.1.2 Emissions from industrial processes

A substantial amount of particle-containing gas (e.g., feed gas and exhaust gas) is generated in chemical, metallurgy, and construction industries. For example, PM2.5 (e.g., fine dust particles and tar) is generated in large quantities in coal gasification, coking, and pyrolysis processes, accounting for over 30% of the total inhalable particulate matter [13]. The coke oven gas used for the production of methanol contains PM2.5, tar, naphthalene, etc. The semi-water gas generated in the production of synthetic ammonia contains many impurities such as PM2.5 and tar. The feed gas from the calcination of pyrite for the production of sulfuric acid contains dust, arsenic, selenium, etc. The presence of impurities in feed gas can significantly affect continuous and stable production and cause an increase in system resistance and energy consumption [14]. The presence of PM2.5 in fuel gas, such as the fuel gas generated in the combustion of purge waste gas in the urea industry and the dry exhaust gas containing the product particulate matter in the drying process of the product, can cause not only air pollution but also a significant waste of resources of product particles in the gas.

8.1.2 Emission standard of PM2.5

In China, there has been a growing emphasis on the control of PM2.5 emissions. The *National Ambient Air Quality Standards* issued in 1996 specifies that the emission limit of PM10 is 40 μg/m^3 (annual average) or 50 μg/m^3 (daily average), but no emission limit of PM2.5 is specified. In 2012, the revised *National Ambient Air Quality Standards* (GB 3095—2012) specifies that the level I (AQI = 0-50) limit of PM2.5 is 15 μg/m^3 (annual mean) or 35 μg/m^3 (daily mean). In the United States, the emission limit of PM2.5 is set to 15 μg/m^3 (annual mean) or 65 μg/m^3 (daily mean) in 1997, and a strict emission limit of PM2.5 is also set in Australia and Europe. However, it is noted that current ambient air quality standards of PM2.5 have focused on the emission limit of the total smoke and no standards are established for the emission limit, chemical composition, and toxicity of PM10 and PM2.5 emitted from the combustion [15].

China has set rigorous emission standards for coal-fired power plants to control PM2.5 emissions at the source. In 2011, the Ministry of Environmental Protection and the General Administration of Quality Supervision, Inspection and Quarantine jointly issued the revised *Emission Standard of Air Pollutants for Thermal Power Plants* (GB 13223−2011) [16], in which the emission limit of smoke (dust) from coal-fired boilers is set to 30 mg/m^3. In 2014, the National Development and Reform Commission, the Ministry of Environmental Protection, and the National Energy Administration jointly issued the *Action Plan for the Upgrade and Reconstruction of Energy Saving and Emission Reduction of Coal-fired Power Plants* (2014-2020) [17], in which the emission limit of smoke (dust) from newly built coal-fired power plants is set to below 10 mg/m^3. In 2015, they again issued the *Work Plan on Full Implementation of Ultra-low Emission and Energy-conservation Reconstruction of Coal-fired Power Plants* [18], in which the emission limit of smoke (dust) is set to 5 mg/m^3. The standard for the emission of particulate matter from industrial boilers and coal-fired power plants becomes increasingly stringent.

8.1.3 Separation technologies of PM2.5

Current separation technologies of PM2.5 are broadly divided into dry and wet separation technologies [15]. Dry separation processes are used to separate dust particles from gas based on their differences in physical properties, such as density, particle size, and conductivity, and commonly used equipment includes gravity dust collector, inertial dust collector, cyclone, filter, and electrostatic precipitator, each of which has its own advantages and disadvantages. The removal rate is generally high for coarse particles larger than 10 μm but exceedingly low for PM2.5. For example, the removal rate of particles less than 1 μm is only 27% for high-performance cyclones. At present, bag filters, electrostatic precipitators, and electrostatic bag filters that combine the advantages of both bag filters and electrostatic precipitators are widely used. Bag filters use fabric filtration to separate dust particulates from dust-laden gases, and the removal rate of particles of 1-5 μm is greater than 99%. Bag filters are even capable of removing particles of 0.1-1 μm with a cut size of 0.04 μm. Despite their high removal efficiency of PM2.5, bag filters have the disadvantages of low filtration speed, large pressure drop, large volume, periodic cleaning and replacement, and especially bag blockage in the removal of particles from wet or oil gas. The working principle of electrostatic precipitators is moderately simple. The gas-borne particles are negatively charged by the high-voltage discharge electrodes and then attracted to positively charged collector plates. The removal efficiency is high (>99%) for coarse particles larger than 10 μm but low for submicron particles which could not be effectively charged. The removal rate of particles with a specific resistance lower than $10^4 \Omega \cdot cm$ or larger than $10^{11} \Omega \cdot cm$ is low. Electrostatic precipitators normally require high investment and could not be used for explosive gases or under high humidity conditions. Electrostatic bag filters are capable of treating ultra-low emissions of particles in flue gas from coal-fired boilers, but are less applicable to gases with complex compositions such as fly ash, mist, tar, and other viscous substances, or conditions of high temperature or pressure.

Wet separation processes are used to separate dust particles from gas based on the inertial impaction between the dust-laden gas and the washing liquid (usually water) and diffusion of dust particles. Some gaseous pollutants could also be removed. Compared to dry separation processes, wet separation processes have a wider range of applications, simple structure, and high stability, and allow simultaneous cooling and absorption of harmful gases. Common wet separation equipment includes spray tower, packed bed scrubber, spray scrubber, rotating-stream tray scrubber, and Venturi scrubber [19,20].

The spray tower uses nozzles to break water into small droplets, which in turn impact with dust particles and then condense and settle. The dust particles in the dust-laden gas are captured by the water. The spray tower has a simple structure, low resistance, and convenient operation, which can be used for gases with a high concentration of dust particles. However, it is bulky, and the removal rate of PM2.5 is only 50% with a cut size of 2-3 μm. Therefore, it is mainly used to treat gases with a high concentration of coarse particles.

In a packed bed scrubber, the liquid is sprayed onto the packing via the liquid distributor mounted at the top of the scrubber, forming a liquid film on the packing surface. As the liquid flows downward through the packing by gravity, it comes into countercurrent contact with the gas that flows upward through the packing. Dust particles are captured into the liquid. The scrubber has the advantages of high processing capacity, low-pressure drop, and high-operating flexibility. However, the removal rate of PM2.5 is low because of the small gas-liquid contact area as a result of the low dispersion of the liquid. The cut size is 1 µm and the removal rate of PM2.5 is only 30%. Blockage occurs for gases containing suspended particles.

The working principle of spray scrubbers is to break liquid into small droplets to increase the capture probability of dust particles and the agglomeration of small particles into large ones. Traditional spray scrubbers have only one layer with a dust removal rate of 60%-80%, and the use of multi-layer spray scrubbers can significantly improve the removal rate. For a three-layer spray scrubber, when the spraying pressure is 0.6 MPa and the gas flow rate is 0.6 m/s, the removal rate of PM15 reaches 98.3% with a cut size of 3 µm, and the pressure drop is close to that for empty spraying.

The rotating-stream tray scrubber consists of a scrubber housing and a rotating-stream tray with paddle-wheel type fan. The liquid is uniformly distributed onto each blade via the blind plate in the middle to form a thin liquid film. The gas flows upward from the lower part to the upper part of the scrubber in a spiral path, which breaks the liquid on the blade into small droplets. Dust particles are trapped in the liquid and thrown onto the wall. Then, the liquid droplets with dust particles are collected into the liquid collection tank by gravity. The scrubber has many advantages, such as simple structure, high efficiency, low-pressure drop, high-operating flexibility, and low probability of blockage. It has a higher gas flow rate and better liquid dispersion than the packed bed scrubber, but the removal rate is low for particles smaller than 1 µm. The cut size is 1-2 µm, and the removal rate of PM2.5 is only 40%.

Venturi scrubber performs well in removing PM2.5 with a cut size of 0.1-0.3 µm and a removal rate higher than 99%. However, a high removal rate would be achieved at the expense of high-pressure loss as the particle size decreases, and the gas phase resistance is up to 3000-20,000 Pa, which requires more blowers and results in higher power consumption and additional investment and operating cost.

Current wet technologies are not as economical and effective as expected in the removal of PM2.5. In recent years, high-gravity technology as an emerging process intensification technology has shown great promise in the removal of PM2.5. It uses high-speed rotating packing to increase the inertia force of small particles in the gas and to produce a large shear force to break the liquid into micro/nanodroplets, filaments, and films. Compared to traditional columns, a larger gas/liquid relative velocity is obtained in the high centrifugal field created by the high-speed rotation of the packing. The inertial impaction between micro/nano-liquid droplets and small particles is coupled with the forced impaction for efficient capture of PM2.5. The gas pressure drop is only 350-800 Pa, which is similar to that in a spray tower but much smaller than that in a Venturi scrubber. The cut size is 0.01-0.08 µm, and the capture rate is 100% for PM10 and 99.7% for PM2.5, respectively. The liquid flows through the packing at an acceleration that is dozens and even hundreds of times higher than the gravitational acceleration, which can strongly scour the packing and thus prevent the blockage of the packing. Therefore, the packing has a high self-cleaning capacity. This makes the high gravity wet scrubbing economically competitive. In particular, it is also advantageous in the purification of gas and difficult working conditions for the dry methods.

8.2 Key technologies and principles of high-gravity gas-solid separation

8.2.1 Key problems and theoretical analysis

In the wet separation process of particulates from the gas, the particulates to be removed are likely to creep or flow around large liquid droplets at low liquid flow rates because of their small size, lightweight, and low inertia, which make them difficult to be captured by liquid droplets. According to the theories of fluid mechanics and particle sedimentation, the Stokes number (St) can be used to indicate the flow of PM2.5 in a gas with the fluid, and its physical meaning is the ratio of particle inertia to diffusion:

$$St = \frac{C_c \rho_p d_p^2 u_0}{18 \mu \, d_c} \tag{8.1}$$

where C_c—Cunningham correction coefficient;
ρ_p—particle density, kg/m³;
d_p—particle size, m;
u_0—gas/liquid relative velocity, m/s;
μ—gas viscosity, Pa · s;
d_c—droplet size, m.

Smaller St values indicate that particulates can completely migrate along the streamline of the gas flow field, and the impaction probability with liquid droplets decreases with increasing gas flow rate. Increasing the St value can increase the inertia force acting on particulates; thus, they are more likely to deviate from the streamline and collide with liquid droplets. For example, increasing the St value from 0.29 to 1.58 leads to an increase in the removal rate from 11% to 59%. Formula (8.1) shows that the capture rate of PM2.5 can be increased by increasing the gas/liquid relative velocity u_0 and/or decreasing the droplet size d_c.

8.2.2 Principles of high-gravity intensification of gas-solid separation

The high-gravity technology can increase the relative velocity between gas and liquid phases and the impact force between liquid droplets and particles. In a high-gravity rotating packed bed (RPB), the high-speed rotation of the packing allows the gas to flow at a velocity greater than the tangential velocity of the packing and the liquid to flow at a velocity less than the tangential velocity of the packing, resulting in a large gas/liquid relative velocity. The u_0 value is an order of magnitude larger than that in a column, and accordingly, the St value is increased by an order of magnitude. As a result, the collision between liquid droplets and particles is increased, which can increase the removal rate of particles in the gas.

The high-gravity technology can also reduce droplet size and increase droplet number. The sizes of liquid droplets in RPB are measured by particle image velocimetry (PIV), and the results reveal that the mean diameter is 0.2–2 mm for wire mesh packing and most liquid droplets are 0.4–0.5 mm in diameter, as shown in Fig. 8.1 [21]. However, the diameters of liquid droplets in a spray tower range from 0.25 to 4.0 mm, most of which are distributed in the range of 1.7–3.0 mm (Fig. 8.2, where m is the uniformity index) [22,23]. Thus, the particle size in the spray tower is one order of magnitude larger than that in RPB. As a result, the St value of the RPB is

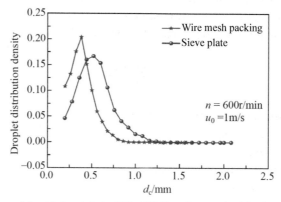

Figure 8.1 Distribution of liquid droplets in RPB. *RPB*, Rotating packed bed.

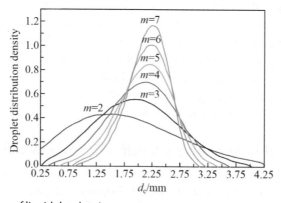

Figure 8.2 Distribution of liquid droplets in spray tower.

increased by an order of magnitude, indicating that more droplets are formed in RPB, and they are smaller in size and more likely to be captured.

The high-gravity technology allows for multi-time and multi-stage capture of particles. In RPB, the liquid is sprayed on the first-stage packing from the liquid distributor and then well dispersed in the radial direction by the strong centrifugal force. The liquid droplets collide with each other, forming larger droplets that will be dispersed again in the second-stage packing. The repeated dispersion-coagulation-dispersion processes of liquid droplets in the packing lead to the capture of particles. The detailed capture process can be described as follows. In the "formation stage," the high shear force generated from the high-speed rotating packing can reduce the surface tension of the liquid and break the liquid into micro/nano-droplets, filaments, and films. In the "development stage," the liquid is dispersed in the gas in packing voids, which leads to intimate contact with particulates and subsequently the wetting of particulates. In the "disappearance stage," the rotation of the packing leads to the collision of micro/nano-liquid droplets with particulates and other liquid droplets, and therefore rapid aggregation, coagulation, and disappearance of the liquid and the capture of particulates in the liquid. In the "multi-stage capture" stage, the aggregation, coagulation, and disappearance of the liquid and the capture of particulates in the liquid occur instantaneously in packing voids. As the liquid and particles pass through each packing layer, they undergo the same formation, development, and disappearance stages until they leave the last layer. These four stages occur in spatial and temporal sequences, which can greatly enhance the capture of particulates by liquid droplets.

It is concluded that the formation, development, and disappearance of the micro/nano-liquid are the mechanisms of how high gravity intensifies the wet separation of PM2.5. Previous studies on the mass transfer mechanism of RPB have demonstrated that the height of the mass transfer unit is 20-40 mm and the number of the packing layer can reach 7 and more [24]. The total removal rate for a seven-stage RPB can reach 91.8% and 99.2% if the removal rate at each stage is 30% and 50%, respectively.

Traditional wet processes that utilize a large amount of water to "wash" the gas are not effective enough in removing PM2.5. The main technical challenges are: (1) the liquid is thousands of times larger than PM2.5, making it less likely to contact and collide with each other, and (2) the liquid is dispersed into droplets only once, and as liquid droplets settle by gravity, they will condense and coagulate into larger droplets, resulting in a decrease in the contact and collision with PM2.5.

8.3 Performance of high-gravity gas-solid separation

8.3.1 Research advances

In China, the researchers in the field of high-gravity wet gas-solid separation are mainly from the North University of China, Beijing University of Chemical Technology, and

South China University of Technology, and their research topics include gas–liquid contact mode, packing characteristics, equipment design and modification, liquid surface properties, and solid particle properties.

The gas–liquid contact mode in RPB is divided into countercurrent: cross-flow and cocurrent contact. In their investigation on the high-gravity wet removal of fly ash from coal-fired power plants in a countercurrent RPB [25], Zhang et al. reported that the dust removal rate reached over 99.9%, the outlet dust concentrations were generally less than 50 mg/m^3, and the pressure drops were 600–1250 Pa. The dust removal rate increased with increasing rotation speed and liquid–gas ratio, and the liquid–gas ratio was only 0.21 L/m^3. Liu [26] investigated the effects of gas–liquid contact mode (cocurrent and countercurrent) on the removal of fly ash from coal-fired power plants. The results demonstrate that the pressure drop of cocurrent RPB is 1130 Pa lower than that of countercurrent RPB under the same operating conditions. When the gas flow rate is 100 m^3/h, the liquid–gas ratio is 4 L/m^3, and the rotation speed is 1050 r/min, the outlet dust concentration is lower than 150 mg/m^3 for both beds. Therefore, a countercurrent RPB is preferred for low liquid rates. Under appropriate operating conditions, the fractional efficiency is 99.98% for particles with an average particle size of 0.3 μm, and all particles larger than 2 μm are removed in the outlet gas. The inlet dust concentration has little effect on the total removal rate, but the particle size distribution has an effect on the fractional efficiency. Fu [27] studied the removal of PM2.5 (fly ash) in countercurrent and cross-flow RPBs, where the inlet dust concentration is 16 g/m^3 and the average particle size is 2.25 μm. The results demonstrate that the fractional efficiency of cross-flow and countercurrent RPB is 94.3% and 96.1% for particles of 2.5 μm, and 79.4% and 85.3% for particles of 1.0 μm, respectively. Under the same operating conditions, their dust removal rates are approximately the same, but the pressure drop of the cross-flow RPB is 51% of that of the countercurrent RPB. Therefore, the cross-flow RPB has the advantages of high processing capacity, low pressure drop, and high dust removal rate.

The packing characteristics have an important influence on dust removal. Li et al. [28] investigated the purification of oil smoke in the catering industry using steel corrugated wire mesh packing with a porosity of 0.97 and nylon wire mesh packing with a porosity of 0.88. The results reveal that when the gas flow rate is 20 m^3/h, the liquid flow rate is 0.5 m^3/h, and the inlet oil smoke concentration is 18 mg/m^3, the fractional efficiency of steel corrugated mesh packing is much higher than that of nylon mesh packing and the dust removal rate is 95%. Song et al. [29] investigated the removal of dust in a countercurrent RPB with RS corrugated wire mesh packing with a porosity of 0.97 and nylon wire mesh packing with a porosity of 0.88, respectively. The results demonstrate that the fractional efficiency of RS corrugated wire mesh packing is much higher than that of nylon wire mesh packing. Because of its higher interfacial tension with water, the RS corrugated wire mesh packing is more easily wetted by water. Its higher porosity also allows the formation of a larger number of smaller droplets, which is favorable for the capture of particles. When the gas flow rate is 200 m^3/h, the liquid flow rate is 0.5-2.0 m^3/h, and the rotation speed is 900-1500 r/min, the capture rate is over 99% and the cut size is 0.02-0.3 μm. Wang [30] investigated the removal of PM2.5 in a cross-flow RPB with structured or random packing. When the inlet concentration is 200 mg/m^3, the average particle size is 46.95 μm, and the liquid-gas ratio is 1.25 L/m^3, a higher removal rate is obtained with the use of the structured packing, and at the same removal rate the loading of structured packing is 16.67% lower than that of random packing. Huang et al. [31] investigated the effects of rotation speed, liquid flow rate, and the number of axial plane mesh packing layers on the dust removal rate in a multi-stage high-gravity RPB. The results demonstrate that the removal efficiency of the three-layer packing is much higher than that of the one-layer packing. These studies have suggested that structured packing and steel corrugated wire mesh packing are better than random packing and nylon wire mesh packing in terms of removal efficiency, respectively. However, the mechanism of how packing properties (e.g., specific surface area, porosity, and surface tension) influence dust removal remains to be elucidated.

The configuration of RPB plays a critical role in increasing the interphase mass transfer in the gas-solid separation process. Wei [32] used a novel counter airflow shear-rotating packed bed (CAS-RPB) consisting of two (upper disk and lower) concentric disks that can rotate independently of each other for removal of small particles in the gas. There are two groups of concentric circular packing rings of increasing radius that are attached alternately to the upper and lower disk with a gap between adjacent rings and between the free end of each ring and the disk. The packing is made from multi-layer wire mesh or random fine porous packing. The high-speed rotation of the packing produces high shear force and disturbance to gas and liquid phases, which is favorable for the capture of PM2.5. When the inlet dust concentration is 2000 mg/m^3, the liquid-gas ratio is 0.8 L/m^3, and the high-gravity factor is 101, the outlet dust concentration is only 3.06 mg/m^3, the pressure drop is 460-500 Pa, the cut size is 0.01 μm, and the removal rate of PM1.0 is 92%, which are better than that in a cross-flow RPB. Liu et al. [33] invented a multi-stage cross-flow RPB for integrated desulfurization and dedusting in ships. Two or more rotors are used with one stator sandwiched between two adjacent rotors. As the gas flows from the rotor to the stator, the flow direction is changed and the gas is subjected to a large shear force, leading to strong turbulence and redistribution of the gas in the stator. The direct consequence is that the relative velocity and collision frequency between gas and liquid phases are increased, which in turn enhance the coagulation and agglomeration of fine particles in the gas. The multi-stage cross-flow RPB has many advantages, such as high dust removal and desulfurization rate, low energy consumption, free from bumping and tilting of the ship, and convenient start up and shut down.

In wet gas–solid separation processes, water is often used as the liquid. Because of the differences in the hydrophilic and lipophilic properties of solid particles and packing, the surface properties of the liquid have some effects on the separation of solid particles. In their investigation on the effects of four wetting agents on dust removal in an RPB, Fu [27] reported that sodium dodecylbenzene sulfonate has the highest wetting ability, and other conditions being equal, the use of 0.03% wetting agent as the liquid leads to a 1.0% increase in the total removal rate, and a 3.43% and 8.44% increase in the fractional efficiency of PM2.5 and PM1.0, respectively. Wei [32] investigated the effect of liquid viscosity on dust removal in a CAS-RPB using glycerin as additive. It is found that increasing the liquid viscosity leads to a decrease in the total dust removal rate irrespective of the rotation direction of the two rotors. However, the total dust removal rate is decreased more significantly from 95.63% to 78.87% when the two rotors rotate in the same direction, but it is decreased by only 5.17% when the two rotors rotate in the opposite directions.

In wet gas–solid separation processes, the properties of PM2.5 in the gas have an important influence on the separation of PM2.5 and recycling of cleaned gas. Li [34] investigated the removal of ultrafine dust particles in RPB, where the fly ash from power plants was used to simulate the carbonaceous components of ultrafine dust particles, and aluminum powder, oxalic acid, and ammonium sulfate were used to simulate the metal components, soluble organic and inorganic components, respectively. When the inlet dust concentration is 6 g/m^3, the average particle size is $\leqslant 10$ μm, the liquid-gas ratio is 1.67 L/m^3, and the rotation speed is 1200 r/min, the removal rates of all components are greater than 95%, and the outlet concentrations are all lower than 0.25 mg/m^3. Liu et al. [35–37] proposed high-gravity equipment and technologies for the removal of fine particles with complex components, such as PM2.5, tar, toluene, and naphthalene, in chemical feed gas such as coal-lock gas, semi-water gas, and coke oven gas. By optimizing the structure and operating parameters of the high-gravity equipment, PM2.5, tar and naphthalene in the feed gas can be removed in an integrated manner. For example, the concentrations of tar and fine dust particles in coal lock gas can be reduced from 300-600 mg/m^3 to below 300 mg/m^3, which solves the problem of unstable operation of the compressor in the presence of dust particles and tar. In the past, the coal-lock gas discharged under normal pressure could be transferred to the pressurization system and has to be combusted directly. The high-gravity wet scrubbing technology has a wider range of applications for the separation and recovery of complex components, especially toxic and harmful components.

The high-gravity dust removal technology is based on the separation of the gas-liquid-solid mixed fluid in RPB. The liquid is subjected to repeated dispersion-coagulation-dispersion processes in the packing for capture of dust particles, and the internal flow field is very complex. Mathematical models are used to describe the removal process of dust particles, based on which RPB structure can be optimized in a labor- and cost-efficient manner. However, most previous models have focused on the relatively simple flow field. The mathematical model proposed by Claudia et al. [38] for wet electrostatic scrubbers takes into account inertial impaction, directional interception, Brownian diffusion, electrostatic interactions, and thermophoresis, and the effects of concentration, liquid/gas contact time, droplet size, gas/water relative velocity, and droplet charge on dust removal rate are investigated. The model predicts that a collection efficiency as high as 99.5% can be reliably obtained in few seconds under optimal conditions. Mohan et al. [39,40] established a correlation formula for dust removal in a spray scrubber. Kim et al. [41] considered diffusion, interception, and impaction in the dust removal model in a gravitational wet scrubber. Lim et al. [42] reported that impaction is the main mechanism for the capture of dust particles in reverse jet scrubbers and the removal rate increases with increasing particle size. Pulley [43] proposed a mathematical model considering the inertial impaction in Venturi scrubbers. However, there are few reports on the mathematical models for high-gravity wet scrubbing. Li [34] established an empirical model for high-gravity wet scrubbing that relate particle motion, interception efficiency, and Brownian effect to the capture, fractional efficiency, and total removal efficiency of particles, which could well predict the removal rate of PM2.5 with an error of $\pm 2\%$ between simulated and experimental results. Pan et al. [44] proposed a droplet size-location jointing distribution probability model to describe the three-dimensional motion of droplet population in a cross-flow RPB considering the gas flow field and droplet collision. The effects of gas and liquid flow rates on droplet motion were analyzed. It is found that increasing the gas flow rate can break the liquid into smaller droplets and consequently increase the interfacial area, while increasing the liquid flow rate can increase the specific surface area of the liquid phase. This model provides some important insights into the mechanism of how high-gravity intensifies wet scrubbing.

8.3.2 Separation performance of cross-flow and countercurrent rotating packed bed

High-gravity wet scrubbing as an emerging dust removal technology has attracted much research and public interest due to its unique advantages. Cross-flow and countercurrent RPBs are often used for this purpose. Here, we describe their dust removal efficiency, fractional efficiency, and gas pressure drop for the capture of PM2.5. Fu [27] compared the total dust removal efficiency and fractional efficiency between cross-flow and countercurrent RPBs with similar rotor in the removal of dust simulated using fly ash (the particle sizes measured using a laser particle size analyzer were in the range of 1-3 μm with an average particle size of 2.25 μm). The particle size distribution is given in Table 8.2.

8.3.2.1 Gas pressure drop and residence time

The liquid flow path is the same for the two RPBs. Specifically, the liquid is sprayed from the liquid distributor and then flows radially from the inner edge of the packing to the outer edge of the packing. However, it is worth noting that the gas flow path differs between the two RPBs. In a cross-flow RPB [Fig. 8.3(A)] [45], gas is introduced from the bottom and then flows through the packing in the axial direction, forming a cross-flow contact with the liquid. The flow cross-sectional area remains constant $\left(\lceil\pi(r_2^2 - r_1^2)\rceil\right)$. In a countercurrent RPB [Fig. 8.3(B)] [45], the gas flows radially from the outer edge to the inner edge of the packing, forming a countercurrent contact with the liquid. The flow cross-sectional area is gradually decreased from $2\pi H r_2$ to $2\pi H r_1$.

The residence time of the gas in a cross-flow RPB is determined mainly by rotation speed and packing thickness, and the pressure drop of the gas flowing through the packing is lower than that in a countercurrent RPB. In the countercurrent RPB, the residence time of the gas is determined mainly by rotation speed and packing size in the diameter direction $(r_2 - r_1)$. The gas flows through the rotating packing in a direction opposite to the centrifugal force, and because of the centrifugal pressure drop and the gradual decrease in the cross-sectional area, the total pressure is greater than the gas pressure drop in the cross-flow bed.

Table 8.2 Particle size distribution of simulated dust particles.

Particle size range/μm	Content/ %	Accumulative/ %	Particle size range/μm	Content/ %	Accumulative/ %
0–0.5	1.32	1.32	1.7–2.0	13.54	73.59
0.5–0.7	4.63	5.95	2.0–2.3	8.63	82.22
0.7–1.0	8.45	14.40	2.3–2.5	6.42	88.64
1.0–1.3	10.82	25.22	2.5–3.0	5.92	94.56
1.3–1.5	16.36	41.58	3.0–5.0	3.56	98.12
1.5–1.7	18.47	60.05	5.0–6.0	1.88	100

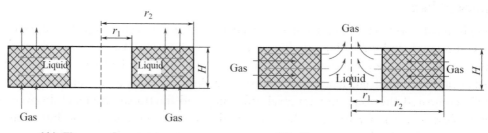

(A) The cross-flow structure (B) The countercurrent structure

Figure 8.3 Gas and liquid flow paths in cross-flow and countercurrent RPBs. *RPB*, Rotating packed bed.

Table 8.3 Comparison of gas pressure drop between cross-flow and countercurrent RPBs (rotating packed bed) under optimal operating conditions.

RPB	High-gravity factor	Liquid spraying density/ [m³/(m² · h)]	Gas velocity/ (m/s)	Gas pressure drop/Pa
Cross-flow RPB	163	6.3	2.6	140
Countercurrent RPB	80	4.8	1.6	250

The difference in gas flow path between countercurrent and cross-flow RPBs leads to different gas/liquid contact modes and consequently different gas/liquid relative velocity and residence time of the gas. The gas/liquid relative velocity has an effect on the Stokes number and the inertial impaction of dust particles, and the residence time of the gas determines the gas/liquid contact time and a longer gas/liquid contact time favors the capture of dust particles.

Resistance is an important indicator for evaluating the performance of the dust collector, which is indicative of the pressure loss of the gas flowing through the dust collector. As greater resistance implies higher power consumption for the blower, resistance is also an important indicator for evaluating the power consumption and operating cost. Generally, the dust collector with a system pressure drop greater than 2000 Pa is a high-resistance dust collector, such as high-energy Venturi scrubber. Despite its high removal efficiency for PM2.5, the energy consumption is extremely high because the pressure drop is greater than 3000 Pa (8000 Pa most of the time). It is seen in Table 8.3 that the gas pressure drop is lower than 300 Pa for both cross-flow and countercurrent RPBs under optimum operating conditions, indicating that RPB is a low-energy dust collector. Also, the pressure drop of the cross-flow RPB is 78% lower than that of the countercurrent RPB.

The operating parameter optimization results reveal that RPB is a low-energy dust collector. Under the same operating conditions, countercurrent RPB is slightly more efficient than cross-flow RPB and the removal rate of PM2.5 μm is 99.5%, implying that RPB is capable of effectively removing PM2.5 from the gas.

8.3.2.2 Separation performance

The effects of high-gravity factor β, liquid spraying density q, and superficial gas velocity u on the total dust removal efficiency and fractional efficiency in cross-flow and countercurrent RPBs are compared under similar operating conditions.

8.3.2.2.1 Total dust removal efficiency

Table 8.4 compares the total dust removal efficiencies between cross-flow and countercurrent RPBs under their respective optimal operating conditions. A higher removal rate can be obtained in the countercurrent RPB at lower levels of high-gravity factor and liquid spraying density.

Table 8.4 Comparison of total dust removal efficiencies between cross-flow and countercurrent RPB (rotating packed bed).

RPB	High-gravity factor	Liquid spraying density/[m³/(m² · h)]	Superficial gas velocity/(m/s)	Dust removal efficiency/%
Cross-flow RPB	163	6.3	2.6	99.6
Countercurrent RPB	80	4.8	1.6	99.8

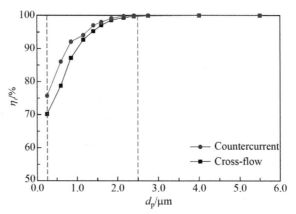

Figure 8.4 Comparison of fractional efficiency between cross-flow and countercurrent RPBs. *RPB*, rotating packed bed.

8.3.2.2.2 Fractional efficiency

Fig. 8.4 compares the fractional efficiency between cross-flow and countercurrent RPBs under their respective optimal operating conditions. The fractional efficiency of the countercurrent RPB is slightly higher than that of the cross-flow RPB. For particles larger than 2.5 μm, the removal rate is 99.87% and 99.90%, and the fractional efficiency $\eta_{0.25}$ is 70.1% and 75.7%, respectively, which are higher than that of current dust collectors. Note that the fractional efficiency of the countercurrent RPB is higher than that of the cross-flow RPB.

8.3.2.2.3 Fitting of fractional efficiency

The relationship between fractional efficiency and dust particle size for most dust collectors can be expressed as follows [46]:

$$\eta_i = 1.0 - \exp(-md_p^n) \qquad (8.2)$$

where m, n—characteristic coefficients.

The characteristic coefficient m indicates the extent of fractional efficiency. The larger the m value is, the higher the fractional efficiency will be. And the characteristic

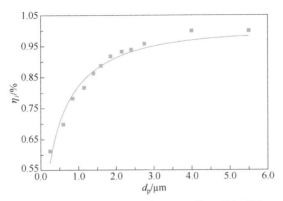

Figure 8.5 The fitting curve of fractional efficiency for cross-flow RPB. RPB, rotating packed bed.

Table 8.5 The characteristic coefficients of fractional efficiency.

Dust collector	m	n
Cyclone	0.002–0.10	0.80–2.00
Wet scrubber	0.09–0.60	0.50–0.80
Electrostatic precipitator	0.80–2.50	0.20–0.50
Cross-flow RPB (rotating packed bed)	1.72	0.51
Countercurrent RPB	2.02	0.50

coefficient n indicates the effect of dust particle size on the fractional efficiency. The smaller the n value is, the flatter the fractional efficiency curve and the higher the fractional efficiency will be.

For cross-flow RPB, the regression analysis reveals that $m = 1.72$ and $n = 0.51$ ($R^2 = 0.96$) when the high-gravity factor is 163, the spraying density is 6.3 m^3/(m$^2 \cdot$ h), and the gas flow velocity is 2.6 m/s, as shown in Fig. 8.5.

Table 8.5 summarizes the characteristic coefficients of the fractional efficiency for commonly used dust collectors. The fractional efficiency of RPB is higher than that of wet scrubbers but similar to that of electrostatic precipitators, indicating that RPB can efficiently remove PM2.5 in the gas. The m value of the countercurrent RPB is higher than that of the cross-flow RPB, indicating that countercurrent RPB has higher fractional efficiency. Their n values are similar, indicating that the two RPBs have comparable removal efficiency.

8.3.3 Separation performance of low-concentration PM2.5 in the gas

The coal gas contains a wide variety of impurities such as dust particles and sulfides that could not be completely removed by wet dedusting and desulfurization processes. High concentrations (several tens to several hundreds of milligrams per cubic meter) of small dust particles and tar are still present in the purified gas. After the coke oven gas is cleaned by wet scrubbing and sulfur, ammonia, benzene, and naphthalene are removed, the concentration of dust particles and tar is 30-100 mg/m^3. The presence of PM2.5 can block the compressor and sometimes it has to be shut down for maintenance. The dust concentration is 150-300 mg/m^3 in the boiler flue gas after conventional dust removal processes, which does not meet the emission limit. A trace amount of dust particles is still present in purified feed gas and flue gas that could not be removed using conventional technologies. Therefore, it is of practical significance to develop technically and economically viable technologies for the purification of low-concentration dust particles. Wang [30] used high-gravity wet scrubbing to remove low-concentration (200 mg/m^3) dust particles from the gas and investigated the effects of high-gravity factor, gas volume, liquid volume, packing type, and inlet dust concentration on dust removal efficiency.

8.3.3.1 Separation process

The fly ash generated in a power plant is used to simulate dust-laden gas. Prior to the experiment, large particles are filtered out and ground into smaller ones with a particle size range of 0.01-5 μm and an average particle size of 2.45 μm. The inlet concentration is about 200 mg/m^3. The separation process is shown in Fig. 8.6 [30,47]. The dust-laden gas is introduced into RPB and then flows through the rotating packing from bottom to top. The liquid is pumped into RPB and dispersed on the inner edge of the packing via a liquid distributor. As it flows from the inner edge to the outer edge of the packing, the liquid is broken into small droplets, filaments, and films by

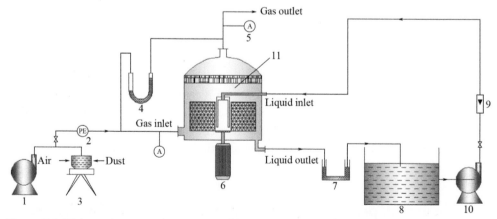

Figure 8.6 High-gravity separation process of low-concentration dust particles (1—blower, 2—gas flowmeter, 3—screw feeder, 4—U-shaped tube differential pressure gauge, 5—sampling port, 6—motor, 7—liquid sealer, 8—liquid storage tank, 9—liquid flowmeter, 10—liquid pump, 11—high-gravity RPB). *RPB*, rotating packed bed.

the high shear force as a result of the high-speed rotation of the packing, and comes into contact with dust particles repeatedly on the surface and in the voids of the packing. Section 8.2 describes the whole process (formation, development and disappearance of liquid droplets, filaments, and films). The liquid entrained in the purified gas is removed using a demister and then discharged, and the liquid is discharged from the outlet into the water tank for subsequent reuse.

8.3.3.2 Effects of various factors on dust removal efficiency
8.3.3.2.1 Effect of high-gravity factor
Under the same conditions (inlet dust concentration = 200 mg/m^3, gas flow rate = 300 m^3/h, and liquid flow rate = 0.3 m^3/h), the high-gravity factor β is varied by adjusting the rotation speed. Fig. 8.7 shows that the dust removal rate first increases and then decreases slightly with increasing high-gravity factor, and it reaches a maximum of 92.13% at $\beta = 163$. Increasing the high-gravity factor can increase the shear force of the rotating packing on the liquid, so that the liquid is broken into smaller droplets, filaments, and films, making it easier for the capture of dust particles. Also, the higher the high-gravity factor is, the higher the centrifugal force acting on dust particles will be. This makes it easier for dust particles to deviate from the gas streamline and increase their collision frequency with liquid droplets and consequently the removal rate of dust particles.

8.3.3.2.2 Effect of gas flow rate
The effect of gas flow rate on dust removal efficiency is shown in Fig. 8.8. The dust removal rate first increases and then decreases with increasing gas flow rate. Increasing the gas flow rate would increase the gas/liquid relative velocity and consequently the impaction frequency and intensity between liquid droplets and dust particles. Increasing

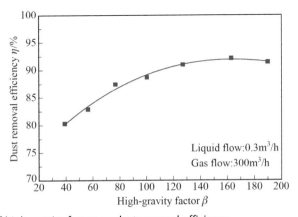

Figure 8.7 Effect of high-gravity factor on dust removal efficiency.

the gas flow rate can also increase gas/liquid turbulence so that dust particles can be better wetted for coagulation and capture [48]. However, the gas/liquid contact time is extremely short at gas flow rates higher than 400 m³/h, leading to a decrease in the dust removal efficiency.

8.3.3.2.3 Effect of liquid flow rate

The effect of liquid flow rate on dust removal efficiency is shown in Fig. 8.9. The removal rate first increases and then decreases with increasing liquid flow rate, and it reaches a maximum of 93.18% at a liquid flow rate of 0.5 m³/h under conditions of $\beta = 163$ and gas flow rate = 400 m³/h. Increasing the liquid flow rate would increase the liquid holding capacity of the packing [49]. As a result, more liquid will be broken into small droplets and thus more dust particles will be captured. Increasing the liquid flow rate also increases the wetting extent of the packing and the gas/liquid contact

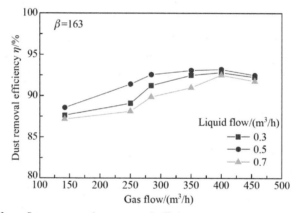

Figure 8.8 Effect of gas flow rate on dust removal efficiency.

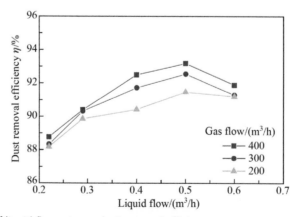

Figure 8.9 Effect of liquid flow rate on dust removal efficiency.

area, which contribute to wetting and trapping dust particles. However, the liquid holding capacity of the packing will be very high at liquid flow rates higher than 0.5 m³/h, and the kinetic energy of the gas and the shear force from the high-speed rotation of the packing are not sufficiently high to break the liquid into smaller droplets, leading to a reduction in the removal rate.

8.3.3.2.4 Effect of packing characteristics

Packing is the core component of the RPB, and its structure and volume have important impacts on the flow pattern. The high-speed rotation of the packing can break the liquid into micro droplets. Thus, a good match of the sizes of liquid droplets and dust particles is the key for the high-efficient removal of dust particles. Here, the effects of structured wire mesh packing and Pall ring random packing are compared. The dust removal rate as a function of the high-gravity factor for structured wire mesh packing and Pall ring random packing is shown in Fig. 8.10, where the inlet dust concentration is 200 mg/m³, the gas flow rate is 300 m³/h, and the liquid flow rate is 0.3 m³/h. The removal rate increases with increasing high-gravity factors for both types of packing. At the same level of high-gravity factor, a higher removal rate can be obtained using the structured wire mesh packing due to its higher porosity and specific surface area [50]. At low levels of high-gravity factor, the shear force from the high-speed rotation of the packing is not sufficiently high to break the liquid, and in this case, the advantages of the structured packing are particularly evident. At high levels of high-gravity factor, it is easy to break the liquid and the gas/liquid contact is maximized, and the difference between the two packing becomes less pronounced.

The dust removal rate as a function of the gas flow rate is shown in Fig. 8.11. The dust removal rate increases with increasing gas flow rate for both structured wire mesh

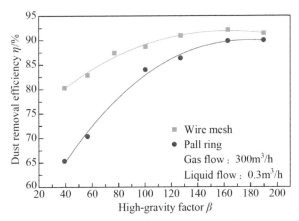

Figure 8.10 Effect of high-gravity factor on the dust removal rate for structured wire mesh packing and Pall ring random packing.

packing and Pall ring random packing, and again higher removal rates are obtained using the structured packing. The voids of random packing are not uniform in size, and gas short circuit is more likely to take place in large voids [51], resulting in a reduction in gas/liquid contact time and making it more difficult for the capture of dust particles. In contrast, the voids of structured packing are more uniform in size, and intricate channels are more likely to form as the packing rotates at a high speed, which can increase gas/liquid turbulence and contact and thus lead to more efficient removal of dust particles.

The dust removal rate as a function of the liquid flow rate for structured wire mesh packing and Pall ring random packing is shown in Fig. 8.12. The dust removal rate increases with increasing liquid flow rate for both types of packing, but higher

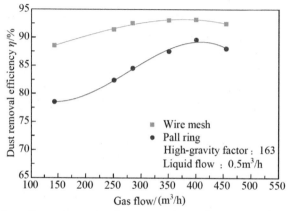

Figure 8.11 Effect of gas flow rate on the dust removal rate for structured wire mesh packing and Pall ring random packing.

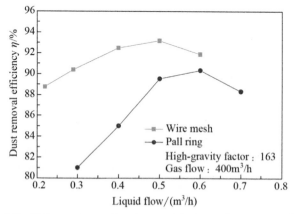

Figure 8.12 Effect of liquid flow rate on the dust removal rate for structured wire mesh packing and Pall ring random packing.

removal rates are obtained using structured packing. The structured packing has a larger specific surface area for the contact between dust particles and the liquid, which is favorable for the capture of dust particles. The larger porosity also allows more liquid to be attached to the internal surface of the packing, which increases the effective gas/liquid contact area. The structured packing made of metal has better wettability than random packing made of plastic material, which allows the capture of more dust particles within a short period of time. When the high-gravity factor is 163, the gas flow rate is 400 m³/h, the liquid flow rate is 0.6 m³/h, and the dust removal rate of the random packing is 88.5%. The liquid-gas ratio of the structured packing at the maximum dust removal rate is 1.25 L/m³, which is 16.67% lower than that of the random packing. Therefore, less water and energy would be consumed, which can reduce not only energy consumption and operating cost but also equipment size of the whole dust removal system.

8.3.3.2.5 Effect of inlet dust concentration

The effect of inlet dust concentration on the dust removal rate is shown in Fig. 8.13. The dust removal rate decreases more slowly with decreasing inlet dust concentration, and the decreasing rate is reduced rapidly at inlet dust concentrations lower than 30 mg/m³. As the inlet dust concentration decreases from 200 to 30 mg/m³, the removal rate decreases from 92.05% to 81.25% and the outlet dust concentration is 5.60–15.9 mg/m³. As the inlet dust concentration decreases to 12 mg/m³, the removal rate is 58.33% and the outlet dust concentration reaches the ultra-low emission standard of 5 mg/m³, which indicates that the high-gravity wet scrubbing technology is capable of removing low-concentration particles.

8.3.3.2.6 Determination of fractional efficiency

Under conditions of inlet dust concentration = 200 mg/m³, high-gravity factor = 163, gas flow rate = 400 m³/h, and liquid flow rate = 0.3, 0.5, or 0.6 m³/h, dust samples

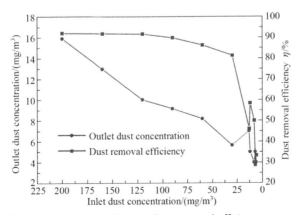

Figure 8.13 Effect of inlet dust concentration on dust removal efficiency.

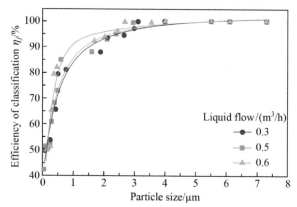

Figure 8.14 The fractional efficiency curve as a function of particle size.

were collected at the inlet and outlet of the RPB to determine the fractional efficiency. Fig. 8.14 shows that the high-gravity technology has higher removal efficiency for larger dust particles. The removal rate is 97.11%-99.84% for particles of about 3 μm, and particles larger than 4 μm can be completely removed because they will experience higher gravity and centrifugal force because of the higher mass inertia force, and thus they are more likely to deviate from the gas streamline.

Fig. 8.14 shows that the cut size (particle size corresponding to the fractional efficiency of 50%) is about 0.08 μm for RPB, which is significantly lower than that for spray tower and high-efficiency Venturi scrubber, indicating that RPB is highly effective in removing fine particles. Also, the removal rate of PM2.5 is as high as about 95%. In traditional wet removal processes, there is a mismatch in size between liquid and particles, resulting in a lower impaction probability between them and consequently a lower removal rate of particles. In the high-gravity field created by the high-speed rotation of the packing, the repeated impaction and condensation of the liquid lead to continuous and stable removal of particles.

8.3.3.2.7 Scanning electron microscopy of inlet and outlet dust

In order to demonstrate the ability of the high-gravity technology to remove low-concentration small particles, scanning electron microscopy (SEM) was performed on dust samples collected at the inlet and outlet of the RPB. Fig. 8.15 shows that most particles are spherical in shape and some small particles are attached to larger ones. Some flake- and rod-shaped particles are also found, which are membrane fibers. The particle size is 1-5 μm for inlet particles and about 1 μm for outlet particles, indicating that particles larger than 3 μm are almost completely removed. This is in good agreement with the fractional efficiency results.

The removal rate of PM2.5 reaches up to 95% at inlet dust concentrations of about 200 mg/m^3, and the cut size is significantly smaller than that of the spray tower and

(A) SEM image of inlet dust   (B) SEM image of outlet dust

Figure 8.15 SEM images of inlet (A) and outlet (B) dust particles. *SEM*, Scanning electron microscopy.

high-efficiency Venturi scrubber, which meets the deep purification requirement for low-concentration PM2.5. The pressure drop is lower than 490 Pa. Therefore, high-gravity technology is a high-efficiency and low-energy dust removal technology. RPB is characterized by small floor space, rapid and convenient start-up and shut-down, simple operation and maintenance. It can be easily coupled with existing dust removal technologies for small and medium industrial boilers. For instance, RPB can be placed behind the water dust scrubber to achieve the ultra-low emission of fine particles in the flue gas from industrial boilers.

8.3.4 Separation of different components

Li [34] adopted the high-gravity technology to remove four key components of PM2.5 in the simulated gas using water as the absorbing liquid. The main conclusions are as follows: (1) The effects of rotation speed, gas–liquid ratio, inlet dust concentration, inlet dust temperature, and packing type on the removal rate of carbonaceous components were investigated. Under optimal conditions, the removal rate reaches over 98% and the outlet concentration is reduced to below 0.18 mg/m^3. (2) The effects of rotation speed, gas–liquid ratio, inlet dust concentration, and inlet dust temperature on the removal rate of metal components were investigated. Under optimal conditions, the removal rate reaches over 97% and the outlet concentration is reduced to below 0.2 mg/m^3. (3) The effects of rotation speed, gas–liquid ratio, inlet dust concentration, and inlet dust temperature on the removal rate of soluble organic components were investigated. Under optimal conditions, the removal rate reaches over 96% and the outlet concentration is reduced to below 0.25 mg/m^3. (4) The effects of rotation speed, gas–liquid ratio, inlet dust concentration, and inlet dust temperature on the removal rate of soluble inorganic components were investigated. Under optimal conditions, the removal rate reaches over 98% and the outlet concentration is reduced to below 0.15 mg/m^3. The removal rate of mixed components can reach

96% and higher and the outlet dust concentration can be reduced to below 0.24 mg/m³. (5) An empirical correlation is proposed by fitting the experimental data, which can predict the removal efficiency of the high-gravity wet method with temperature, gas-liquid ratio, and rotation speed. There is a good agreement between simulated and experimental results with an error of ± 2%, indicating that the proposed empirical correlation can well predict the removal rate of PM2.5.

8.4 Application examples

The high-gravity wet dust removal technology is advantageous over many conventional technologies in terms of operating cost and removal efficiency. In this section, we present several application cases of high-gravity wet dust removal.

8.4.1 Removal of dust particles and tar in semi-water gas

The production process of synthetic ammonia mainly includes gasification, purification, compression, and synthesis. The semi-water gas produced in the gasification process is desulfurized, converted, purified, and compressed, but it still contains dust particles and tar that can block the compressor valve and thus affect the continuous and stable operation of the compressor. The high-gravity technology was applied to remove dust particles and tar in the semi-water gas generated in a compressor (Fig. 8.16), where the synthetic ammonia production is 200,000 ton/year, the gas flow rate is 20,700 m³/h, the water circulation volume is 10 m³/h, and the gas pressure drop is 800 Pa.

The removal rate of dust particles and tar reaches over 85%, and that of hydrogen sulfide reaches over 80%. The successful implementation of this project contributes greatly to the purification of semi-water gas and ensures continuous stable operation and low corrosion. The concentration of dust particles and tar in the purified gas is reduced below 5 mg/m³. No accumulation of dust particles and tar occurs at the valve

Figure 8.16 The high-gravity dust removal equipment.

of the compressor 30 days after implementation of the project, and the compressor operates continuously and stably for half a year, indicating that the high-gravity gas-solid separation technology is an economical and effective method for purification of semi-water gas, and it contributes greatly to the stable production of synthetic ammonia in an energy-efficient manner.

8.4.2 Removal of dust particles and tar in coal-lock gas

The concentrations of dust particles and tar are 100-1000 mg/m^3 in the coal-lockgas, which may block subsequent sections and affect the normal operation of the system. The coal-lock gas discharged under normal pressure could be transferred to the pressurization system and has to be combusted directly, which causes not only environmental pollution but also waste of gas resources. In the West-to-East Natural Gas Transmission Project of China, Lurgi pressurized gasification has been used in a company in Xinjiang Uygur Autonomous Region with an annual production of 1.2 million tons of methanol and 0.8 million tons of dimethyl ether. As the concentrations of dust particles and tar are 500-900 mg/m^3 in the purified coal-lock gas, the compressor has to stop for maintenance only after one week of operation due to the blockage of the value. In order to completely solve this problem, a high-gravity gas-solid separator (Fig. 8.17) is used for removal of dust particles, tar and other impurities in the coal-lock gas before it is introduced into the compressor. The water used for purification is the water used in the gasification process. The removal rate of oil and dust is 90% and 95% at liquid-gas ratios of 0.4-0.5 L/m^3, respectively, and the gas phase resistance is <1000 Pa, which ensures long-term stable operation of the compressor and reduces environmental pollution caused by combustion of coal-lock gas.

Figure 8.17 The high-gravity gas-solid separator for purification of coal-lock gas.

Table 8.6 Particle size distribution of dust particles in the flue gas at the outlet.

Particle size/μm	⩽30	⩽11.5	⩽4.5	⩽1.3
Content/%	100	90	50	10

Figure 8.18 The high-gravity separation device.

8.4.3 Removal of dust particles in boiler flue gas

The flue gas discharged from a boiler in a urea production plant is purified. The concentration of fine dust particles at the outlet is about 150-200 mg/m^3, and the particle size distribution is shown in Table 8.6.

The flue gas contains only dust particles, 50% and 10% of which are smaller than 4.5 and 1.3 μm, respectively. Assuming that all particles larger than 1.3 μm can be removed, the concentration of residual dust particles is still higher than 15 mg/m^3 in theory. Thus, there is a practical need to capture dust particles equal to or smaller than 1 μm. A high-gravity RPB with multi-layer packing and form-drag baffles (Fig. 8.18) is adopted, which can enhance the tangential slip velocity of the gas in the radial direction and break liquid into smaller droplets to match with dust particles (⩽1 μm). The liquid-gas ratio is 0.3-0.5 L/m^3 and the pressure drop is 800 Pa. The final outlet dust concentration is lower than 10 mg/m^3, which meets the local ultra-low emission standard of flue gas.

8.4.4 Removal and recovery of fine particles in chemical tail gas

The tail gas generated in the production process of fertilizers often contains a large number of fine particles, which can cause not only environmental pollution but also significant resource waste because these particles may be valuable chemical products. It is necessary to recover these particles and ensure ultra-low emission of tail gas. In a fertilizer plant with an annual production of 900,000 tons of nitrophosphate fertilizer, which accounts for 90% of the total nitrophosphate fertilizer production, a substantial amount of tail gas is generated in the production of calcium ammonium nitrate because of the unstable operating temperature and pressure, and the dust particles contain 2500 mg/m^3 calcium ammonium nitrate. Because of the small size and high concentration of particles in the tail gas, conventional methods are less effective in removing these particles and the equipment is easily blocked and could not operate stably over a long period of time. The production equipment is mounted on a 22-m platform and the space is limited in the workshop (the clear height is only 5 m). It is inappropriate to use conventional dust collectors which occupy a large space. The high-gravity RPB that can be installed in a narrow and small space is a more suitable choice. Over the past years, it has demonstrated high dust removal efficiency. The outlet dust concentration is only 5 mg/m^3 and the dust removal efficiency reaches 99%. The water consumption is 20%-40% of that for conventional dust collectors, and the used liquid can be transferred back to the production process for reuse.

8.5 Prospects

The high-gravity gas–solid separation technology makes substantial theoretical and technical contributions by facilitating the conversion from traditional "gas scrubbing" to "micro-liquid capture." The disturbance of the gas and the dispersion and condensation of the liquid are enhanced, resulting in higher gas/liquid relative velocity and making liquid droplets smaller. Thus, there is a good match between liquid droplets and particles, which increases their impaction frequency and the removal rate of particles. This technology provides a new approach to the control and management of fine particles in the gas. The cut size is 0.01-0.08 μm, the liquid-gas ratio is 0.2-0.5 L/m^3, the water consumption is only 10%-20% of that of traditional wet scrubbing methods. The gas pressure drop is 350-800 Pa, which implies lower power consumption. The equipment volume is about 10% of the traditional tower. This technology has low investment and operating cost and high-operating flexibility. The gas load is 30%-120% of the normal load. The rotating packing has a self-cleaning capacity, which can prevent scaling and blockage and thus ensure long-term stable operation. It can be used under complex operating conditions and in the presence of oil, dust, moisture, and other particles.

In China, the high-gravity gas–solid separation technology has been widely used to remove particles in industrial gases (e.g., coal gas, flue gas, and dry tail gas) in various national projects, such as West-to-East Natural Gas Transmission Project and National Coal Chemical Industry Demonstration Project. Given the diversity of particles and the complex operating conditions, it is necessary to improve the technology for complex operating conditions and pollutant components. The operating conditions and unit structural parameters should be further optimized. We expect that this technology will provide an economically and technically viable method for the control of the emission of fine particles into atmosphere.

References

[1] Liu XJ. Study on hydrocyclone radial jet structure of parameter optimization. Wuhan: Huazhong University of Science and Technology, 2012.
[2] Li MJ. Theoretical analysis and experimental study of new-style dust precipitator. Beijing: China Ship Research Institute, 2004.
[3] David YHP, Chen SC, Zuo ZL. PM2.5 in China: measurements, sources, visibility and health effects, and mitigation. Particuology, 2014,13(2):1-26.
[4] He KB, Yang FM, Duan FK. Atmospheric particulates and regional combined pollution. Beijing: Science Press, 2011.
[5] Gao ZY. Study on impacts in human health exposure to fine particulate matter and genetic susceptibility. Shanghai: Fudan University, 2010.
[6] Feng SL, Gao D, Liao F, et al. The health effects of ambient PM2.5 and potential mechanisms. Ecotoxicology and Environmental Safety, 2016,128(6):67-74.
[7] Ministry of Ecology and Environment of the People's Republic of China. 2016 China Environmental Status Bulletin.
[8] Xu X. Research on electrostatic precipitation of fine particle in high-temperature gas. Hangzhou: Zhejiang University, 2016.
[9] Wang YP. Initial analysis of ultra low emission technology of industrial pulverized coal boiler. Energy and Energy Conservation, 2017,1:77-79.
[10] Natale FD, Carotenuto C. Particulate matter in marine diesel engines exhausts: emissions and control strategies. Transportation Research Part D: Transport and Environment, 2015,40(10):166-191.
[11] Huo H, Zhang Q, He KB, et al. Vehicle-use intensity in China: current status and future trend. Energy Policy, 2012,43(4):6-16.
[12] Natural Resources Defense Council. White Paper of Shipping and Port Emission Control in China 2014.
[13] Research Group of China's Total Coal Consumption Control Program and Policy Research Project. Contribution of Coal Use to Emission Control in China. 2014.
[14] Niu YX. Study on coke-oven gas purification process. Journal of Library and Information Science, 2010,20(32):191-193.
[15] Yang LJ. Control technology of fine particulate pollution from combustion source. Beijing: Chemical Industrial Press, 2011.
[16] GB 13271—2014. Emission Standard of Air Pollutants for Boiler.
[17] GB 13223—2011. Emission Standard of Air Pollutants for Thermal Power Plants.
[18] National Development and Reform Commission, Ministry of Environmental Protection, and National Energy Administration. Action Plan for the Upgrade and Reconstruction of Energy Saving and Emission Reduction of Coal-fired Power Plants (2014—2020).
[19] Guo CR. Selection of wet scrubber. Knowledge Economy, 2013,14:08.
[20] Zhang LD, Li XB, Wang Q. Performance of dust removal in a multi-layer spray column scrubber. Chemical Industry and Engineering Progress, 2017,3(7):2375-2380.
[21] Li HT. Study on flow characteristics of liquid in countercurrent rotating packed bed. Taiyuan: North University of China, 2019.
[22] Zhu J, Wu ZY, Ye SC. Drop size distribution and specific surface area in spray tower. CIESC Journal, 2014,65(12):4709-4715.
[23] Zhu J, Liu ZH, Yang YF. Applying the Rosin-Rammler function to study on drop size distribution in spray tower. Journal of Chemical Engineering of Chinese Universities, 2015,29(5):1059-1064.
[24] Su XP, Liu YZ, Zhang ZC. Mass transfer property of high effective rotating distillation bed. Modern Chemical Industry, 2011,31(2):77-80.
[25] Zhang YH, Liu LS, Liu YZ. Experimental study on flue gas dedusting by hyper gravity rotary bed. Environmental Engineering, 2003,21(6):42-43.
[26] Liu W. Research on co-current flow dust removal in rotating packed bed. Beijing: Beijing University of Chemical Technology, 2004.
[27] Fu J. Study on technology of wet dust collection under high-gravity. Taiyuan: North University of China, 2015.

[28] Li JH, Liu YZ. Experimental study of removal dust from flue gas by high-gravidity technology and its mechanism. Chemical Production and Technology, 2007,14(2):35-37,67.
[29] Song YH, Chen JF, Fu JW. Research on particle removal efficiency of the rotating packed bed. Chemical Industry and Engineering Progress, 2003,22(5):499-502.
[30] Wang T. Application of high-gravity wet dust removal technology in the removal of dust particles from industrial boiler. Taiyuan: North University of China, 2017.
[31] Huang DB, Deng XH, Tian DL. Experimental research on removal of micron level dust by high-gravity rotating packed bed. Chemical Engineering (China), 2011,39(3):42-45.
[32] Wei SK. Study on dust removal performance of counter airflow shear rotating packed bed. Taiyuan: North University of China, 2018.
[33] Liu Y Z, Yuan Z G, Qi G S. An integrated device for desulphurization and dust removal for shipping flue gas. CN 104492210B, 2016.
[34] Li ZY. Study on the novel technology of ultrafine dust removal in a rotating packed bed. Beijing: Beijing University of Chemical Technology, 2015.
[35] Liu Y Z, Qi G S, Jiao W Z. A high-gravity device and method for removal of fine particles in the gas. CN 105642062B, 2018.
[36] Liu Y Z, Qi G S, Jiao W Z. A high-gravity device and method for treatment of tar coal lock gas. CN105561713B, 2018.
[37] Liu YZ, Jiao WZ, Yuan ZG. A novel counter airflow shear-rotating packed bed. CN 103463827B, 2015.
[38] Claudia C, Francesco DN, Amedeo L. Wet electrostatic scrubbers for the abatement of submicronic particulate. Chemical Engineering Journal, 2010,165(1):35-45.
[39] Mohan BR, Biswas S, Meikap BC. Performance characteristics of the particulates scrubbing in a counter-current spray-column. Separation and Purification Technology, 2008,61(1):96-102.
[40] Mohan BR, Jain RK, Meikap BC. Comprehensive analysis for prediction of dust removal efficiency using twin-fluid atomization in a spray scrubber. Separation and Purification Technology, 2008,63(2):269-277.
[41] Kim HT, Jung CH, Oh SN, et al. Particle removal efficiency of gravitational wet scrubber considering diffusion, interception, and impaction. Environmental Engineering Science, 2001,18(2):125-136.
[42] Lim KS, Lee SH, Park HS. Prediction for particle removal efficiency of a reverse jet scrubber. Aerosol Science, 2006,37(12):1826-1839.
[43] Pulley RA. Modelling the performance of Venturi scrubbers. Chemical Engineering Journal, 1997,67 (1):9-18.
[44] Pan CQ, Deng XH. Motion model of droplet in cross-rotating packed bed. CIESC Journal, 2003,54(7):918-922.
[45] Qi GS. Removal of hydrogen sulfide in the gas using high-gravity wet oxidation method. Taiyuan: North University of China, 2012.
[46] Licht W. Air pollution control engineering: basic calculations for particulate collection. Florida: CRC Press, 1988.
[47] Wang T, Qi GS, Liu YZ. Removal of low concentration dust from gas by wet dust collection technology under high-gravity. The Chinese Journal of Process Engineering, 2017,17(1):721-729.
[48] Byeon SH, Lee BK, Mohan BR. Removal of ammonia and particulate matter using a modified turbulent wet scrubbing system. Separation and Purification Technology, 2012,98(98):221-229.
[49] Meikap BC, Biswas MN. Fly-ash removal efficiency in a modified multi-stage bubble column scrubber. Separation and Purification Technology, 2004,36(3):177-190.
[50] Guo Q, Qi GS, Liu YZ. Comparative study of mass transfer performance of different structured packings in rotating packed bed. Chemical Industry and Engineering Progress, 2016,35(3):741-747.
[51] Yuan ZG, Song W, Jin GL. Comparison of pressure-drop and mass-transfer characteristics of two kinds of packing in countercurrent rotating packed bed. Chemical Engineering (China), 2015,43(7):7-11.

Index

Note: Page numbers followed by "*f*" and "*t*" refer to figures and tables, respectively.

A

ABS. *See* Acrylonitrile-butadiene-styrene (ABS)
Absorption, 23
 high-gravity absorption, 24–30
  gas-liquid mass transfer theory, 24–25
  high-gravity technology for intensification of mass transfer, 25–30
 of other volatile gases, 70–71
Absorption-desorption process, 75–76
Acrylon Chemical Plant of Petro China Fushun Petrochemical Company, 95–96
Acrylonitrile (AN), 95
Acrylonitrile-butadiene-styrene (ABS), 95
Activated alumina, 244
Activated carbon, 244, 255, 264, 275
Activity coefficient, 161–162
Adequate dispersion, 167
Adsorbate, 21–22, 261
Adsorbents, 262
 types and applications of, 254–255
Adsorbers
 types and applications of, 255–259
  fixed bed, 255
  moving bed, 255
  rotating bed, 259
Adsorption, 6, 243, 245–259
 basic principle of, 243–244
 effects of operating parameters, 268–270
 high-gravity adsorption, 259–264
  mechanism of intensification of high-gravity adsorption process, 261
  rotating packed bed for liquid/solid adsorption, 260–261
 kinetics, 266–268
 and mass transfer process, 253–254
 theoretical basis of, 245–253
  equilibrium, 246–251
  kinetics, 251–253
  mechanism, 245
 thermodynamics, 265–266
 toluene-containing exhaust gas, 270–280
 treatment of resorcinol wastewater, 264–270
Adsorption-desorption cycle, 243–244
Adsorption-desorption process, 280
Adsorption equilibrium, 246–251, 261
 gas adsorption equilibrium, 246
 single-component adsorption equilibrium isotherm, 247
Adsorption kinetic model, 251–253, 266–268
 pseudo-first-order kinetic model, 252
 pseudo-second-order kinetic model, 252
 Weber-Morris model, 252
Adsorption rate constant, 274
Adsorption thermodynamics, 265–266
Aerodynamic diameters, 283–284
Air flow rate, 90
Alkaline desulfurizers, 34–39
 for fine removal of H_2S from coal gas, 38–39
 selective removal of H_2S from high-concentration CO_2, 34–38
4-aminoantipyrine spectrophotometry, 186–187, 223
Ammonia, 79–80
AN. *See* Acrylonitrile (AN)
Angular momentum theorem, 120–121
Angular velocity, 2–3, 179
Aniline, 230–231
 high-gravity factor on stability of liquid membrane and removal rate of, 233–234
 initial impact velocity on stability of liquid membrane and removal rate of, 234–235, 234*f*
 liquid membrane stability and removal rate of, 236–237
 surfactant amount on stability of liquid membrane and removal rate of, 232–233
Aqueous solution, 217
Average radial velocity, 179

B

Baffle rotating bed, 19–20, 19*f*
Baffler rotation direction, 111

Beijing University of Chemical Technology, 291–292
BET equation, 248
Biodegradation, 65–66
Bottom liquid distributor, 17–18
Breakthrough curve, 271
Bromine, 91
Brownian diffusion, 294–295
Brownian effect, 294–295
Bubble devolatilization, 87
Buoyancy factor, 211

C

Carbon-based adsorbents, 254–255
Carbon dioxide (CO_2), 33
Cardiovascular disease, 283–284
CAS-RPB. *See* Counter airflow shear rotating packed bed (CAS-RPB)
Centrifugal force, 4, 47–48
Centripetal force, 2–3
CFC. *See* Chlorofluorocarbon (CFC)
Chelated iron desulfurization, 39–40
Chemical absorption process, 24
Chemical industry, 1, 244
Chlorofluorocarbon (CFC), 65
Circular motion, 3
CNPC Auspicious Chemical Group Co, 95–96
Colloid mill, 218–219
Combustion, 284–285
Concurrent flow-RPB, 13–15
Continuous stirred-tank reactor (CSTR), 10
Conventional extractors, 164
Counter airflow shear rotating packed bed (CAS-RPB), 15–16, 46–49, 293
Countercurrent-flow rotating packing bed, 13, 41–46
Coupling of mass transfer and mixing processes, 175–180
  chemical extraction process, 175–176
  mass transfer coefficient of impinging stream-rotating packed bed, 177–180
  physical extraction process, 177
Cross-flow RPB, 13–14
CSTR. *See* Continuous stirred-tank reactor (CSTR)
Cunningham correction coefficient, 289
Cylindrical coordinate system, 3–4

D

Demulsification, 221
Desorption, 75–76, 243–244
  development of rotating packing bed, 81–82
  intensification of heat and mass transfer in thermal desorption by high gravity, 76–80
  matching of desorption and high-gravity technology, 80–81
  temperature, 277
Di-2-ethylhexylphosphoric acid, 198–199
Diffusion coefficient, 209
Diffusion devolatilization, 87
Dimensional analysis, 178
Disk packing, 135–137
Dispersion force, 245
Distillation
  distributed distillation, 151
  of high-viscosity and heat-sensitive materials, 149–150
  liquid distributor, 104–108
    and redistributor for m-stage rotating packed bed, 107–108
    for rotating packed bed, 105–106
  ordinary distillation, 146–148
  overview, 101–103
  packing, 108–111
    fin baffle packing, 110–111, 110f, 111f
    spiral packing, 108–110, 109f
  principle of high-gravity distillation, 103–104
  prospects, 153–154
  recovery of ammonia wastewater, 151–153
  special distillation, 148–149
Distributed distillation, 151
Double-film mass transfer coefficient, 163
Droplets, 166
Drug extraction, 207–208
Dry separation processes, 286–287
Dubinin-Polanyi equation, 247–248
Du Nouy ring method, 207–208
Dust removal efficiency
  effects of various factors on, 301–307
    determination of fractional efficiency, 305–306
    gas flow rate, 301–302
    high-gravity factor, 301
    inlet dust concentration, 305
    liquid flow rate, 302–303
    packing characteristics, 303–305

scanning electron microscopy of inlet and outlet dust, 306–307
separation of different components, 307–308
Dynamic emulsifiers, 218
Dynamic liquid distribution, 107

E

Eddy diffusion, 167
Electrocatalysis, 11–12
Electrostatic bag filters, 286–287
Electrostatic force, 245
Electrostatic interactions, 294–295
Electrostatic precipitators, 286–287
ELM. See Emulsion liquid membrane (ELM)
Emulsion
  particle size and particle size distribution of, 235
  preparation, 222
Emulsion liquid membrane (ELM), 208
  application of, 238
  comparison of impinging stream-rotating packed bed with high speed stirrer, 235–238
    comprehensive comparison, 237–238
    liquid membrane stability and removal rate of aniline, 236–237
    particle size and particle size distribution of emulsion, 235
    stability at different settling times, 236
  demulsification, 221
  in impinging stream-rotating packed bed, 216–221
    composition, 217
    preparation, 216–219
    selection of mobile carrier, 217
  separation process, 219–221, 220f
  structure of, 209f
  treatment of aniline wastewater, 230–238
    high-gravity factor on stability of liquid membrane and removal rate of, 233–234
    initial impact velocity on stability of liquid membrane and removal rate of aniline, 234–235, 234f
    surfactant amount on stability of liquid membrane and removal rate of, 232–233
  treatment of phenol wastewater, 221–230
    effects of material properties on the removal rate of phenol, 226–228
    effects of packing type on the emulsion yield and removal rate of phenol, 223–224

effects of phase ratio R_x on the emulsion yield and removal rate of phenol, 225–226
high-gravity factor β on the emulsion yield and removal rate of phenol, 224–225
Emulsion liquid membrane separation, 212–214
  kinetics of mass transfer, 213–214
  mass transfer mechanism, 212–213
Energy consumption, 244
Environmental pollution, 164
Environmental protection, 10
Equilibrium partition coefficient, 184
Equilibrium separation processes, 2
Exothermic process, 76
Extraction phase, 160

F

Facilitated transport, 209–210
  type I facilitated transport, 210
  type II facilitated transport, 210
Feed flows, 8
Feed loading, 89–90
FGD. See Flue gas desulphurization (FGD)
Filaments, 166
Flash evaporation, 87
Flue gas desulphurization (FGD), 33–34
Fluidized bed, 10–11, 244–245, 257–259
  adsorption of toluene using activated carbon in, 257, 273t
Fluid mechanics, 116–126
Food industry, 244
Fractional distillation, 9
Fractional efficiency, 298
  determination of, 305–306
Freundlich constant, 248–249
Freundlich equation, 248–250

G

Gas adsorption process, 250
Gas diffusion, 2
Gas flow rate, 301–302
Gas-liquid contact, 113–114
Gas-liquid equilibrium, 77–78
Gas-liquid interface, 24, 108
Gas-liquid mass transfer coefficient, 88
Gas-liquid mass transfer equipment, 125
Gas-liquid mass transfer resistance, 24
Gas-liquid-solid separation, 6
Gas phase resistance, 81–82

Gas-phase volumetric mass transfer coefficient, 48, 80
Gas pressure drop, 296–297
Gas separation, 207–208
Gas/solid adsorption process, 243–244
Gas-solid heat transfer, 245
Gas-solid separation, 6, 283
Gas tangential velocity, 119–121, 121f, 122f
Gibbs' adsorption, 227–228
Gibbs-Duhem equation, 161–162
Gibbs equation, 247–248
Gibbs' free energy, 161–162
Grant-Manes equation, 249
Gravitational acceleration, 4
Gravitational wet scrubber, 294–295
Gravity, 3

H

Heat transfer, 76
Henry coefficient, 248
Henry equation, 248
High gravity, 2
  high-gravity factor, 5–6
  high-gravity separation, 2–5
High-gravity absorption, 8–9, 24–30
  gas-liquid mass transfer theory, 24–25
  high-gravity technology for intensification of mass transfer, 25–30
High-gravity adsorption, 10–11, 259–264
  mechanism of intensification of, 261–264
    gas/solid adsorption processes, 263–264
    liquid/solid adsorption processes, 262
  mechanism of intensification of high-gravity adsorption process, 261
  rotating packed bed for liquid/solid adsorption, 260–261
  technology, 245
High-gravity desorption, 9
High-gravity distillation, 9
  characteristics of, 111–146
  equipment, 145–146
  principle of, 103–104
  processes, 9
  in rotating packed bed, 128–144
    fluid mechanics of, 141–144
    mass transfer performance of, 128–141
    structure, 128
  traditional distillation columns, 145–146

High-gravity emulsion liquid membrane, 10
High-gravity extraction, 10
High-gravity factor, 5–6, 89, 223–224, 301
High-gravity field, 4
  acceleration of, 6
  average intensity of, 6
  intensity of, 6
  radial distribution of, 5f
High-gravity gas/solid adsorption processes, 263
High-gravity gas-solid separation, 11, 289–291
  cross-flow and counter current rotating packed bed, 295–299
    gas pressure drop and residence time, 296–297
    separation performance, 297–299
  key problems and theoretical analysis, 289
  principles of high-gravity intensification of, 289–291
  research advances, 291–295
High-gravity separation, 2–5
  classification and characteristics of, 6–7
  equipment for, 12–22
    baffle rotating bed, 19–20, 19f
    rotating packing bed, 12–19
  operating principles of, 7–8
  unit operations of, 8–12
    high-gravity absorption, 8–9
    high-gravity adsorption, 10–11
    high-gravity desorption, 9
    high-gravity distillation, 9
    high-gravity emulsion liquid membrane, 10
    high-gravity extraction, 10
    high-gravity gas-solid separation, 11
    other separation processes, 11–12
High-gravity solvent extraction, 10
High-gravity technology, 65–66, 76, 101–102, 290–291
  application of, 64–65
  for deamination, denitrification, and dehumidification, 63–64
  devolatilization of polymers, 87–91
    operating conditions, 89–90
    removal efficiency, 90
    technological advantages, 90–91
    technological principles, 88
    technological processes, 89
  extraction of bromine from brine, 91–93
    by high-gravity air blowing, 92

technological advantages, 93
recovery of iodine in wet-process phosphoric acid, 93–95
  parameter optimization and removal efficiency, 95
  using high-gravity air blowing, 93–94
removal of nitrobenzene from high-concentration wastewater, 97–98
  parameter optimization and removal efficiency, 98
  technological process, 97–98
stripping of acrylonitrile in wastewater, 95–97
  comparison with other stripping devices, 97
  parameter optimization and removal efficiency, 96–97
  for treatment of acrylonitrile wastewater, 96
stripping of ammonia in denitration of flue gas in power plants, 82–84
  high-gravity stripping of ammonia water, 82–83
  parameter optimization, 83
  stripping efficiency, 83–84
for stripping of ammonia nitrogen wastewater, 84–86
  technological characteristics, 85–86
  technological principles, 84–85
  technological processes, 85
theory of simultaneous heat and mass transfer, 76–79
  characteristics of simultaneous heat and mass transfer, 79–80
  intensification of heat transfer by high-gravity, 78–79
  thermodynamic analysis, 77–78
  water deoxygenation, 98–99
High-gravity wet dust removal technology, 308
  removal and recovery of fine particles in chemical tail gas, 310–311
  removal of dust particles and tar in coal-lock gas, 309
  removal of dust particles and tar in semi-water gas, 308–309
  removal of dust particles in boiler flue gas, 310
High-gravity wet scrubbing, 295
High operating flexibility, 86
High stripping efficiency, 86
Hollow fiber membrane contactor, 165t
Hydrogen sulfide (H_2S), 33

alkali content, 37
CO_2 content, 38
high-gravity factor, 37
high-gravity technology for selective removal of, 35f
liquid-gas ratio, 36–37
PDS concentration, 37
temperature, 37–38
Hydrometallurgy, 10

I

Ibuprofen, 164
Image analytic method, 216
Imperial Chemical Industries (ICI), 101–102
Impinging stream-rotating packed bed, 209–216
  comprehensive comparison of, 238t
  continuous ELM separation process, 220–221
  liquid distribution in packing, 215–216
  methanol-diesel-emulsified fuel, 211–212
  micromixing, 214–215
  preparation process of liquid membrane using, 211–212, 212f
Impinging stream-rotating packed bed (IS-RPB), 10, 20–21, 159, 164
  calculation of extraction operation in, 182–184
    characterization of extraction efficiency, 184
    multi-stage extraction process, 184
    operating line equation, 182–183
  comparison of micromixing efficiency of, 170t
  concentration of acetic acid, 190–193
    effect of β on η, 192
    effect of n on η, 192–193
    effect of u_0 on η, 191–192
  extraction characteristics of, 185–186
    adaptability, 186
    existence form of liquid, 185
    extraction capacity, 186
    residence time, 185
    retention volume of solvent, 185
  extraction of dyes, 196–198
  extraction of indium ions, 198–202
    comparison of different extractors, 201–202
    effect of Q_o on η and $\xi_{In/Fe}$, 199
    effect of R on η and $\xi_{In/Fe}$, 200–201
    effect of β on η and $\xi_{In/Fe}$, 201
  extraction of phenol in wastewater, 186–190
    effect of β on D' and η, 188–189
    effect of R on D', 189–190

Impinging stream-rotating packed bed (IS-RPB)
   effect of solvent compositions on D', 190
   effect of u_0 on D' and η, 187–188
   extraction operation in, 180–186
      multi-stage extraction, 181–182
      single-stage extraction, 180–181
   extraction process in, 166f
   micromixing performance of, 169
   nitrobenzene, 193–196
      comparison experiments, 195–196
      effect of high-gravity factor β on the removal rate and extraction efficiency of, 195
      effect of phase ratio R on the removal rate and extraction efficiency of, 193–194
   other applications, 202–204
      continuous wet extraction of copper, 203
      extraction crystallization of sodium carbonate, 203–204
      extraction of benzoic acid by physical methods, 202
   single-stage extraction process in, 182f
   unique structure and characteristics of, 169
Impinging streams contactor, 165t
Indium (In), 198–199
Indium-tin-oxide (ITO), 198–199
Induction force, 245
Ineffective adsorption layer, 271
Inertial force, 4
Inhalable particulate matter (PM10), 284
Inlet dust concentration, 305
Intermolecular interactions, 243
Internal phase, 210
Intraparticle diffusion rate, 275–277
Iodine, 93
Ionization equilibrium, 77–78
4-isobutyl acetophenone, 164
Isooctanol, 226–227
Isotherm
   C-shaped isotherm, 250
   H-shaped isotherm, 250
   L-shaped isotherm, 250
   S-shaped isotherm, 250
IS-RPB. *See* Impinging stream-rotating packed bed (IS-RPB)

K
Kinetic energy, 166
Knudsen diffusion, 254

L
Langmuir equation, 247–250
Langmuir-Freundlich equation, 248
Linear driving force model, 251–252
Linear velocity, 3
Lipophilic packing, 81–82
Liquid, 262
Liquid adsorption equilibrium, 249, 249f
Liquid distributor, 104–108
   and redistributor for m-stage rotating packed bed, 107–108
   for rotating packed bed, 105–106
Liquid droplets, 113–114, 215–216
Liquid flow direction, 111
Liquid flow rate, 119–120, 302–303
Liquid-liquid extraction, 159
   definition of, 160
   mechanism of process intensification in impinging stream-rotating packed bed, 165–180
      coupling of mass transfer and mixing processes, 175–180
      intensification of mass transfer, 166–169
      intensification of mixing, 169–174
   principle of, 160–164
      equipment for intensification of, 164
      extraction kinetics, 163–164
      phase equilibrium, 161–163
   process, 160f
Liquid-liquid reaction, 20
Liquid-liquid separation, 6
Liquid membrane, 208
   preparation of, 218–219
Liquid membrane separation, 20, 207–208, 238–239
   emulsion liquid membrane separation, 212–214
      kinetics of mass transfer, 213–214
      mass transfer mechanism, 212–213
   by impinging stream-rotating packed bed, 211–212, 214–216
      liquid distribution in packing, 215–216
      micromixing, 214–215
Liquid rotameter, 47–48
Liquid/solid adsorption processes, 262
LMS2, 218
Logarithmic coordinate system, 125
Low power consumption, 85

M

Marine Environmental Protection Committee (MEPC), 46
Mass transfer, 254
  driving force for, 168
  intensification of, 166–169
    increasing interfacial area A, 167
    increasing mass transfer driving force Δc, 168–169
    increasing overall mass transfer coefficient k, 167–168
  kinetics of, 213–214
  mechanism, 212–213
Mass transfer coefficient, 6, 177–178
Mass transfer process, 111
Mass transfer separation, 2
Mathematical models, 294–295
MDEA. See N-methyldiethanolamine (MDEA)
Mechanical separation, 2
Mechanical strength, 217, 244
Media flows, 8
Membrane-internal phase interface, 213
Membrane separation, 2, 10
Methyl tertiary butyl ether (MTBE), 67
Microliquid films, 166
Micromixing, 214–215
Micromorphology, 261
Mixer-settler extractor, 165t
Mixing intensification, 169–174
  comparison of micromixing performance, 174
  effect of fluid viscosity μ, 173
  effect of high-gravity factor β, 171–172
  effect of impact angle α, 170–171
  effect of impact distance L, 172–173
  micromixing time t_m, 173–174
Molecular diffusion, 254
MTBE. See Methyl tertiary butyl ether (MTBE)
Multicomponent adsorption equilibrium isotherm, 249
Multi-layer rotating zigzag bed, 113
Multilevel cross-flow rotating packed bed (MC-RPB), 68, 70
Multi-stage counter-current RPB (MSCC-RPB), 103
Multi-stage cross-flow extraction, 181f
Multi-stage extraction process, 168, 184

N

National Ambient Air Quality Standards, 286
National Coal Chemical Industry Demonstration Project, 311
NB. See Nitrobenzene (NB)
n-butanol, 226–227
n-butyl alcohol, 71
Newton's laws, 4
Nitric oxide (NO_x), 33
  wet absorption techniques of, 55t
Nitrobenzene (NB), 97, 193–196
  comparison experiments, 195–196
  effect of high-gravity factor β on the removal rate and extraction efficiency of, 195
  effect of phase ratio R on the removal rate and extraction efficiency of, 193–194
Nitrogen oxides (NO_x), 82
N-methyldiethanolamine (MDEA), 50
non-Newtonian fluid, 81–82
North University of China, 291–292
NO_x. See Nitric oxide (NO_x)

O

Oil in water in oil (O/W/O) emulsion, 208
Operating line equation, 182–183
Operating parameters, 179, 268–270
Ordinary distillation, 146–148
  of high-viscosity and heat-sensitive materials, 149–150
  special distillation, 148–149
Organic pollutants, 11–12
Orthogonal experimental design, 169–170
Osmotic swelling, 221
Ozonation, 11–12

P

Packed bed scrubber, 287–288
Packed column, 93–94, 165t
Packing, 108–111
  characteristics, 140t, 292–293, 303–305
  fin baffle packing, 110–111, 110f, 111f
  liquid distribution in, 215–216
  spiral packing, 108–110, 109f
Pall ring random packing, 303–304
Parameter optimization, 83
  evaporation of ammonia water, 83
  gas-liquid ratio, 83
  high-gravity factor, 83

Parameter optimization (*Continued*)
  inlet gas temperature, 83
Partial differential equation, 251–252
Particle image velocimetry (PIV), 290–291
Particle size distribution, 296t
Partition coefficient, 10, 161–163, 175–176, 184
Permeability coefficient, 214
  facilitated transport, 209–210
  selective permeation, 209
Petrochemical engineering, 10
Petrochemical industry, 207–208
PetroChina Daqing Chemical Research Center, 53
Pharmaceutical industry, 244
Phase equilibrium, 103–104, 161–163
  in chemical extraction, 162–163
  in physical extraction, 161–162
  principle, 87
Phase ratio, 183
Phenol, 186–187, 221–222
  effects of material properties on removal rate of phenol, 226–228
  effects of packing type on emulsion yield and removal rate of, 223–224
  effects of phase ratio R_x on the emulsion yield and removal rate of, 225–226
  high-gravity factor β on the emulsion yield and removal rate of, 224–225
  material properties on the removal rate of, 226–228
    comparison of impinging stream-rotating packed bed with high-speed stirrer, 229–230, 229f
    effects of additive amounts on emulsion yield, 227–228, 227f
    effects of membrane reinforcing agents on the stability of the liquid membrane, 226–227
    effects of NaOH solution concentration on the removal rate of, 228, 228f
Photocatalysis, 11–12, 65–66
Physical absorption process, 24
Physical adsorption, 245
Physicochemical absorption process, 24
PIV. *See* Particle image velocimetry (PIV)
Plastic wire mesh packing, 223–224
PM2.5, 283–284
  emission standard of, 286
  separation technologies of, 286–289
  sources of, 284–286
    combustion, 284–285
    emissions from industrial processes, 285–286
Polycyclic aromatic hydrocarbons, 283–284
Polymer-based adsorbents, 254–255
Polymer devolatilization, 87
Polymer manufacturing process, 87
Porosity, 244
Preheating temperature, 89
Pseudo-first-order kinetic model, 252
Pseudo-second-order kinetic model, 252

Q
Quasi-structured packing, 108

R
Raffinate phase, 160
Random packing, 108
Rate separation processes, 2
Reactor, 1–2
Redlich-Peterson equation, 250–251
Reinluft process, 255–256
Residence time, 296–297
Resorcinol, 264
  adsorption kinetics of, 266–268
  high-gravity factor on removal rate of, 269
  liquid flow rate on removal rate of, 268
  pH on removal rate of, 270
Resultant force, 4
Reynolds number, 178
Rotating packed bed (RPB), 3, 8, 12–19, 25, 78–79, 101–102, 164, 165t, 245, 289–290
  absorbent and absorption technique, 31
    selectivity and cost, 31
    solubility or absorption capacity, 31
    volatility, 31
  cross-flow and countercurrent, 292
  desulphurization process in, 44f
  development of, 81–82
  distillation process in, 143f
  distribution of liquid droplets in, 290f
  engineering technology of, 32
  expanded view of liquid distributor for, 106f
  and fixed bed, 271–277
  front view of liquid distributor for, 106f
  gas-liquid contact mode in, 292
  gas-liquid mass transfer process in, 26
  for gas/solid adsorption, 261

for intensification of gas-solid mass transfer, 21–22, 21f
for intensification of liquid-liquid mass transfer, 20–21
for liquid/solid adsorption, 260–261, 260f
mass transfer coefficient of, 26
multi-stage cross-flow, 293
operating parameters of, 80–81
parameters for fixed bed and, 279t
production cost of, 32
purification of volatile organic compounds, 65–71
removal of ammonia in production of nitrophosphate fertilizers, 60–65
  case for application of the high-gravity technology, 64–65
  high-gravity technology for deamination, denitrification, and dehumidification, 63–64
  parameter optimization and removal efficiency of the high-gravity technology, 61–62
removal of CO_2 from gas steams, 49–54
  of high-concentration CO_2, 50–53
  of low-concentration CO_2, 54
removal of H_2S, 33–40
  alkaline desulfurizers, 34–39
  chelated iron desulfurization, 39–40
removal of NO_x from flue gas, 54–60
  by high-gravity technology at atmospheric pressure, 56–59
  by high-gravity technology using Fe II U EDTA complex, 59–60
removal of SO_2, 40–49
  counter airflow shear rotating packing bed, 46–49
  countercurrent-flow rotating packing bed, 41–46
with single-block packing, 13–15
with split packing, 15–17
structural optimization of, 32
structure and operating parameters of, 271t
three-dimensional structure of liquid distributor for, 106f
with two-stage or multi-stage packing, 17–19
for waste gas absorption, 33f
Rotating pray column, 165t
Rotating-stream tray scrubber, 287–288

Rotating system, 4
Rotating zigzag bed, 111–128
  application examples, 146–153
  comparison of technological and economic indexes between, 151t
  flooding point curve in, 126f
  flow and contact of gas and liquid phases in, 114f
  fluid mechanics of, 116–126
  gas/liquid flow rate on power consumption of, 127f
  liquid flooding curve for, 125–126
  mass transfer performance, 113–116, 115f
  shaft power of, 126–128
  structure of, 111–113, 112f
Rotation speed, 119–120
Rotor-stator mixer, 174
Routine treatment methods, 264
RPB. See Rotating packed bed (RPB)

S

Scanning electron microscopy (SEM), 306–307
Schmidt number, 178
Second-stage extractor, 181
Segregation index, 169–170, 175–176
Selective non-catalytic reduction (SNCR), 82
Selective permeation, 209
Self-cleaning, 86
Semi-empirical equation, 178
Separation performance, 297–299
  fitting of fractional efficiency, 298–299
  fractional efficiency, 298
  of low-concentration PM2.5 in the gas, 299–307
    effects of various factors on dust removal efficiency, 301–307
    separation process, 300–301
    total dust removal efficiency, 297
Separation processes, 2
Shaft power, 126–128
Sherwood number, 178
Single-layer molecular adsorption, 249
Single-layer rotating zigzag bed, 113
Single-stage extraction, 180–183
Sinopec Anqing Company, 95–96
Sinopec Qilu Petrochemical Company, 95–96
Sinopec Shanghai Petrochemical Co. LTD, 95–96
Sodium dodecylbenzene sulfonate, 294

Solid adsorbent, 251–252
Solid membrane separation, 207–208
Solvent extraction, 186–187
South China University of Technology, 291–292
Span80, 218, 227–228
Spherical membrane, 208
Spray column, 165t
Spray scrubber, 287
   working principle of, 288
Stable emulsion, 218–219
Stirrer, 218–219
Stokes number (St), 289
Structured fin baffle packing, 110–111
Structured packing, 108
Structured spiral packing, 108–110, 109f
Sulfur dioxide (SO_2), 33, 40–41
   from flue gas using Na-Ca alkali solution, 45–46
   from flue gas using sodium citrate, 41–44
   from flue gas using sodium phosphate, 44–45
   from flue gas using sodium sulfite, 45
Swelling, 221
Swelling ratio, 229–230
Swollen emulsion, 223

T

Tangential velocity, 116–119
Taylor vortex extractor, 165t
Temkin equation, 250–251
TETA. See Triethylene tetramine (TETA)
Thermal desorption, 75–76, 81–82
Thermal diffusion, 2
Thermodynamic analysis, 77–78
Thermodynamic equilibrium constant, 162–163
Thermodynamic principle, 161–162
Thermophoresis, 294–295
Toluene, 270–271
Total dust removal efficiency, 297
Total suspended particles (TSP), 284
Traditional wet processes, 291
Tributyl phosphate (TBP), 162–163, 186–187, 226–227
Triethylene tetramine (TETA), 50
T-shaped mixer, 174
TSP. See Total suspended particles (TSP)
Turbulent intensity, 234–235
Type I facilitated transport, 210
Type II facilitated transport, 210

U

ULE. See Ultra-low emission (ULE)
Ultra-low emission (ULE), 40–41
Ultrasonic emulsifier, 218–219
U-shaped liquid distributor, 107

V

van der Waals forces, 245
van der Waals interaction, 243
Van't Hoff equation, 265
Venturi scrubber, 287–289
VOCs. See Volatile organic compounds (VOCs)
Volatile organic compounds (VOCs), 65–71, 66t, 244
   absorption of volatile methanol, 67–70
   mass transfer process of, 71
   recovery of acetic acid in production of energetic compounds, 66–67

W

Wastewater treatment, 65–66, 207–208
Water deoxygenation, 98–99
Water in oil in water (W/O/W), 208
Water vapor, 255
Weber-Morris diffusion model, 275, 276t
Weber-Morris model, 252, 267f
   linear fitting for, 267f
West-to-East Natural Gas Transmission Project of China, 309, 311
Wet gas-solid separation processes, 294
Wet separation processes, 287
WHO. See World Health Organization (WHO)
Wire mesh packing, 135–137
World Health Organization (WHO), 65

X

Xinjiang Uygur Autonomous Region, 309

Y

Yellow dragon, 54–55
Yoon-Nelson mode, 274–275

Z

Zeolite molecular sieves, 244
Zhejiang University, 102
Zigzag channel, 103–104